AF689269

Cancer Biomarkers: Clinical Aspects and Laboratory Determination

Clinical Aspects and Laboratory Determination of Biomarkers Series

Series Editor: Amitava Dasgupta

Volume 1
Alcohol and its Biomarkers: Clinical Aspects and Laboratory Determination

Volume 2
Biomarkers in Inborn Errors of Metabolism: Clinical Aspects and Laboratory Determination

Volume 3
Endocrine Biomarkers: Clinical Aspects and Laboratory Determination

Volume 4
Kidney Biomarkers: Clinical Aspects and Laboratory Determination

Volume 5
Cancer Biomarkers: Clinical Aspects and Laboratory Determination

Cancer Biomarkers: Clinical Aspects and Laboratory Determination

Edited by

Lakshmi V. Ramanathan

Department of Pathology and Laboratory Medicine, Memorial Sloan Kettering Cancer Center, New York, NY, United States

Martin Fleisher

Department of Pathology and Laboratory Medicine, Memorial Sloan Kettering Cancer Center, New York, NY, United States

Michael J. Duffy

UCD School of Medicine, Conway Institute of Biomolecular and Biomedical Research, University College Dublin, Dublin, Ireland; UCD Clinical Research Centre, St. Vincent's University Hospital, Dublin, Ireland

ELSEVIER

Elsevier
Radarweg 29, PO Box 211, 1000 AE Amsterdam, Netherlands
The Boulevard, Langford Lane, Kidlington, Oxford OX5 1GB, United Kingdom
50 Hampshire Street, 5th Floor, Cambridge, MA 02139, United States

Copyright © 2022 Elsevier Inc. All rights reserved.

No part of this publication may be reproduced or transmitted in any form or by any means, electronic or mechanical, including photocopying, recording, or any information storage and retrieval system, without permission in writing from the publisher. Details on how to seek permission, further information about the Publisher's permissions policies and our arrangements with organizations such as the Copyright Clearance Center and the Copyright Licensing Agency, can be found at our website: www.elsevier.com/permissions.

This book and the individual contributions contained in it are protected under copyright by the Publisher (other than as may be noted herein).

Notices
Knowledge and best practice in this field are constantly changing. As new research and experience broaden our understanding, changes in research methods, professional practices, or medical treatment may become necessary.

Practitioners and researchers must always rely on their own experience and knowledge in evaluating and using any information, methods, compounds, or experiments described herein. In using such information or methods they should be mindful of their own safety and the safety of others, including parties for whom they have a professional responsibility.

To the fullest extent of the law, neither the Publisher nor the authors, contributors, or editors, assume any liability for any injury and/or damage to persons or property as a matter of products liability, negligence or otherwise, or from any use or operation of any methods, products, instructions, or ideas contained in the material herein.

ISBN: 978-0-12-824302-2

For Information on all Elsevier publications
visit our website at https://www.elsevier.com/books-and-journals

Publisher: Stacy Masucci
Acquisitions Editor: Patricia M. Osborn
Editorial Project Manager: Matthew Mapes
Production Project Manager: Selvaraj Raviraj
Cover Designer: Bridget Hoette

Typeset by MPS Limited, Chennai, India

Working together to grow libraries in developing countries
www.elsevier.com • www.bookaid.org

Contents

List of contributors

Merav Antman-Passig
Memorial Sloan Kettering Cancer Center, New York, NY, United States

Sigrid V. Carlsson
Departments of Surgery (Urology Service) and Epidemiology and Biostatistics, Memorial Sloan Kettering Cancer Center, New York, NY, United States; Department of Urology, Institute of Clinical Sciences, Sahlgrenska Academy at University of Gothenburg, Gothenburg, Sweden

Daniel C. Danila
Department of Medicine, Memorial Sloan Kettering Cancer Center, New York, NY, United States; Department of Medicine, Weill Cornell Medical College, New York, NY, United States

Maria Diaz
Department of Neurology, Memorial Sloan Kettering Cancer Center, New York, NY, United States

Michael J. Duffy
UCD School of Medicine, Conway Institute of Biomolecular and Biomedical Research, University College Dublin, Dublin, Ireland; UCD Clinical Research Centre, St. Vincent's University Hospital, Dublin, Ireland

Chameera Ekanayake Weeramange
Saliva and Liquid Biopsy Translational Laboratory, Centre for Biomedical Technologies, School of Biomedical Sciences, Faculty of Health, Queensland University of Technology, Brisbane, QLD, Australia; Translational Research Institute, Brisbane, QLD, Australia; Griffith Institute for Drug Discovery (GRIDD), Griffith University, Brisbane, Australia; Department of Medical Laboratory Sciences, Faculty of Health Sciences, The Open University of Sri Lanka, Nugegoda, Sri Lanka

Martin Fleisher
Department of Pathology and Laboratory Medicine, Memorial Sloan Kettering Cancer Center, New York, NY, United States

Emily L. Gill
Department of Pathology and Laboratory Medicine, Children's Hospital of Philadelphia, Philadelphia, PA, United States; Department of Pathology and Laboratory Medicine, University of Pennsylvania, Philadelphia, PA, United States

Daniel A. Heller
Memorial Sloan Kettering Cancer Center, New York, NY, United States; Department of Pharmacology, Weill Cornell Medicine, New York, NY, United States; Department of Physiology and Biophysics, Weill Cornell Medicine, New York, NY, United States

Andrew N. Hoofnagle
Department of Laboratory Medicine and Pathology, University of Washington, Seattle, WA, United States; Department of Medicine, University of Washington, Seattle, WA, United States

Christopher P. Horoszko
Memorial Sloan Kettering Cancer Center, New York, NY, United States; Department of Pharmacology, Weill Cornell Medicine, New York, NY, United States

Shelby M. Hutcherson
Department of Pathology and Laboratory Medicine, Memorial Sloan Kettering Cancer Center, New York, NY, United States

Mijin Kim
Memorial Sloan Kettering Cancer Center, New York, NY, United States

Jieli Li
Department of Pathology, Wexner Medical Center, The Ohio State University, Columbus, OH, United States

Hans Lilja
Department of Pathology and Laboratory Medicine, Memorial Sloan Kettering Cancer Center, New York, NY, United States; Departments of Surgery (Urology Service) and Medicine, Memorial Sloan Kettering Cancer Center, New York, NY, United States; Department of Translational Medicine, Lund University, Malmö, Sweden

Hung S. Luu
Children's Health System of Texas, University of Texas Southwestern Medical Center, Dallas, TX, United States

Samuel I. McCash
Department of Pathology and Laboratory Medicine, Memorial Sloan Kettering Cancer Center, New York, NY, United States

Qing H. Meng
Department of Laboratory Medicine, Division of Pathology and Laboratory Medicine, The University of Texas MD Anderson Cancer Center, Houston, TX, United States

Juliana Müller Bark
Saliva and Liquid Biopsy Translational Laboratory, Centre for Biomedical Technologies, School of Biomedical Sciences, Faculty of Health, Queensland University of Technology, Brisbane, QLD, Australia; Translational Research Institute, Brisbane, QLD, Australia; Griffith Institute for Drug Discovery (GRIDD), Griffith University, Brisbane, Australia

Kazunori Murata
Department of Pathology and Laboratory Medicine, Memorial Sloan Kettering Cancer Center, New York, NY, United States

Freddy T. Nguyen
Institute for Medical Engineering and Science, Massachusetts Institute of Technology, Cambridge, MA, United States; Department of Pathology, Molecular and Cell-Based Medicine, Icahn School of Medicine at Mount Sinai, New York, NY, United States

Khushbu Patel
Department of Pathology and Laboratory Medicine, Children's Hospital of Philadelphia, Philadelphia, PA, United States; Department of Pathology and Laboratory Medicine, University of Pennsylvania, Philadelphia, PA, United States

Elena I. Pentsova
Department of Neurology, Memorial Sloan Kettering Cancer Center, New York, NY, United States

Miguel-Angel Perales
Adult Bone Marrow Transplantation, Department of Medicine, Memorial Sloan Kettering Cancer Center, New York, NY, United States; Department of Medicine, Weil Cornell College of Medicine, New York, NY, United States

Melissa S. Pessin
Department of Pathology and Laboratory Medicine, Memorial Sloan Kettering Cancer Center, New York, NY, United States

William S. Phipps
Department of Laboratory Medicine and Pathology, University of Washington, Seattle, WA, United States

Chamindie Punyadeera
Saliva and Liquid Biopsy Translational Laboratory, Centre for Biomedical Technologies, School of Biomedical Sciences, Faculty of Health, Queensland University of Technology, Brisbane, QLD, Australia; Translational Research Institute, Brisbane, QLD, Australia; Griffith Institute for Drug Discovery (GRIDD), Griffith University, Brisbane, Australia; Menzies Health Institute Queensland (MIHQ), Griffith University, Queensland, Australia

Dinesh Rakheja
Department of Pathology and Laboratory Medicine, Children's Health, Dallas, TX, United States; Department of Pathology, University of Texas Southwestern Medical Center, Dallas, TX, United States

Lakshmi V. Ramanathan
Department of Pathology and Laboratory Medicine, Memorial Sloan Kettering Cancer Center, New York, NY, United States

Mohammad Rasheduzzaman
Saliva and Liquid Biopsy Translational Laboratory, Centre for Biomedical Technologies, School of Biomedical Sciences, Faculty of Health, Queensland University of Technology, Brisbane, QLD, Australia; Translational Research Institute, Brisbane, QLD, Australia; Griffith Institute for Drug Discovery (GRIDD), Griffith University, Brisbane, Australia

Mara Y. Roth
Department of Medicine, University of Washington, Seattle, WA, United States

Christopher M. Shuford
Laboratory Corporation of America Holdings, Burlington, NC, United States

Katie L. Thoren
Department of Pathology and Laboratory Medicine, Memorial Sloan Kettering Cancer Center, New York, NY, United States

Ana Alarcón Tomás
Adult Bone Marrow Transplantation, Department of Medicine, Memorial Sloan Kettering Cancer Center, New York, NY, United States; PhD Program in Signals Integration and Modulation in Biomedicine, Cell Therapy and Translational Medicine, University of Murcia, Murcia, Spain

Lucas Trevisan França de Lima
Saliva and Liquid Biopsy Translational Laboratory, Centre for Biomedical Technologies, School of Biomedical Sciences, Faculty of Health, Queensland University of Technology, Brisbane, QLD, Australia; Translational Research Institute, Brisbane, QLD, Australia; Griffith Institute for Drug Discovery (GRIDD), Griffith University, Brisbane, Australia; Gallipoli Medical Research Institute, Greenslopes Private Hospital, Brisbane, QLD, Australia

Kitsada Wudhikarn
Adult Bone Marrow Transplantation, Department of Medicine, Memorial Sloan Kettering Cancer Center, New York, NY, United States; Division of Hematology and Research Unit in Translational Hematology, Department of Medicine, Chulalongkorn University, Bangkok, Thailand

Zvi Yaari
Memorial Sloan Kettering Cancer Center, New York, NY, United States; School of Pharmacy, Faculty of Medicine, The Hebrew University of Jerusalem, Israel

Preface

Cancer is a leading cause of death worldwide with almost 10 million cases in 2020 per estimates of the WHO. In 2020 the most common causes of cancer deaths were attributed to malignancies of the lung, colon and rectum, liver, stomach, and breast. Approximately, 400,000 children are diagnosed every year, worldwide.

Given the enormous magnitude and complexity of this disease process, there have been remarkable achievements over the years for detecting, monitoring, and treatment of cancer. The earlier the detection, the better are the chances for successful treatment. However, despite decades of basic research involving biomarkers, fewer than 25 have been approved by the US Food and Drug Administration as the journey of a biomarker from the bench to the bedside is complex and challenging. The reason for this is that extensive validation studies are required to be done for any biomarker to be clinically implemented.

In this book, that is part of the series on biomarkers published by Elsevier, the focus is on cancer biomarkers, clinical aspects, and laboratory determination. As the book is targeted for medical students, physicians, oncologists, pathologists, clinical chemists, and medical technologists, we have focused on four areas. The first section, Chapters 1–3, discuss the overview of traditional and nontraditional biomarkers followed by laboratory testing in pediatric as well as adult cancer patients. Emphasis is placed on the unique challenges in the management of these two segments of the cancer patient population. This is followed by Chapters 4–6 that focus on prostate-specific antigen, thyroglobulin, and monoclonal proteins. Chapters 7–9 discuss liquid biopsy, chimeric antigen receptor T-cell therapies, and the validation of circulating tumor cells and cell-free DNA in cerebrospinal fluid. The fourth section that includes Chapters 10–12 assesses emerging technologies as well as different platforms in biomarker testing with summary, conclusions, and further direction. Finally, Chapter 13 broadly summarizes the current status of standard cancer biomarkers and discusses likely future developments with circulating tumor DNA, circulating tumor cells, and extracellular vesicles.

Editors
Lakshmi V. Ramanathan
Martin Fleisher
Michael J. Duffy

Acknowledgments

Our approach to this book dedicated to medical students, oncologists, physicians, clinical chemists, and medical technologists is to give them an idea of challenges faced with adult and pediatric patients in interpreting laboratory test results, steps required for the complete validation of a biomarker before it is introduced in clinical practice, a comprehensive discussion on selected biomarkers and emerging platforms, and testing for biomarkers.

Keeping our overall perspectives of the book in mind, it has been a sheer pleasure for us to work with our authors both within and outside Memorial Sloan Kettering Cancer Center. Their support and enthusiasm for our endeavor is really appreciated as well as their timely submission of chapters and response to our questions. We are grateful to our publishers who have been amazingly patient and helpful throughout this journey.

Not enough can be said regarding our colleagues in clinical chemistry who have worked in our laboratories for many years validating laboratory developed tests, performing required routine tasks, and always giving their 100% to their work. These include our laboratory assistants, bench technologists, supervisors, managers, and fellow clinical chemists. Their hard work and support cannot be measured as well as their dedication to our cancer patients. Our clinical colleagues have been amazing and supportive as is evident from their contributions to our book. We would especially like to acknowledge Drs. Peter Maslak and Mike Tuttle who provided valuable feedback and comments for Chapters 2 and 4. A special vote of thanks to our administrator Evelin Ponce who worked with our schedules and arranged our weekly zoom calls.

Finally, our support from our families over the years cannot be measured and we are truly grateful to them. This was a truly collaborative team effort among the three of us, and it was an honor to be associated with this venture.

Lakshmi V. Ramanathan
Martin Fleisher
Michael J. Duffy

CHAPTER

1 Overview of traditional and nontraditional tumor markers

Jieli Li[1], Qing H. Meng[2] and Lakshmi V. Ramanathan[3]

[1]Department of Pathology, Wexner Medical Center, The Ohio State University, Columbus, OH, United States

[2]Department of Laboratory Medicine, Division of Pathology and Laboratory Medicine, The University of Texas MD Anderson Cancer Center, Houston, TX, United States

[3]Department of Pathology and Laboratory Medicine, Memorial Sloan Kettering Cancer Center, New York, NY, United States

Objectives

- Review laboratory tests used in cancer diagnosis.
- Define characteristics of tumor markers.
- List FDA approved versus non-FDA approved tumor markers.
- Understand overview of different methodologies used for tumor marker measurement.
- Explain role of nontraditional tumor markers and their potential scope.
- Discuss future directions.

1.1 Introduction

About 10 million deaths were reported in 2020 worldwide due to cancer or cancer-related conditions [1]. It has been shown that early diagnosis can dramatically improve 5-year survival from 21% to 89% for all cancers and 5% to 56% for lung cancer [2]. This is mainly due to technological advancements in laboratory medicine

Cancer Biomarkers: Clinical Aspects and Laboratory Determination. DOI: https://doi.org/10.1016/B978-0-12-824302-2.00004-7
© 2022 Elsevier Inc. All rights reserved.

with the measurement of biomarkers, in pathology with examination of tissue and in radiology with imaging studies.

Cancer biomarkers involve measurement of proteins, peptides, biochemicals, DNA, and RNA. They can be further categorized as predictive, diagnostic, or prognostic. These include genetic, transcriptive, epigenetic, and proteomic biomarkers. They can be detected in biological fluids including whole blood, serum, plasma, fluids, urine, stool, sputum, and nipple discharge [3]. Identifying biomarkers can predict response to treatment that is vital for the success of precision and targeted therapy in cancer patients. Currently the US Food and Drug Administration (FDA) has approved 23 protein cancer biomarkers, of which only 16 are detectable in blood [4]. However, these biomarkers do not demonstrate high sensitivity and specificity for clinical diagnosis when employed alone.

Diagnostic testing for tumors is not only limited to clinical chemistry but also includes immunology; hematology; hematopathology; the pathological and histological examination of biopsy, genetic, genomic, and molecular testing (liquid biopsy); imaging tests (X-ray, computerized tomography, ultrasound, magnetic resonance imaging, single-photon emission computerized tomography, and positron emission tomography); and molecular imaging. Blood, tissues, and body fluids are the most common specimens in clinical laboratory testing. Body fluids for tumor marker testing include cerebrospinal fluid (CSF), saliva, effusions, and urine.

1.2 Definition and characteristics of tumor markers

Tumor markers are defined as substances found in tissue, blood, bone marrow, or other body fluids that may be indicative of cancer or certain benign (noncancer) conditions. Many tumor markers are proteins made by both normal cells and cancer cells, but they are made in higher amounts by cancer cells. Genetic changes in tumor tissue, such as gene mutations, patterns of gene expression, and other changes in tumor DNA or RNA, are also being used as tumor markers. A tumor marker may be used with other tests to help diagnose cancer. It may also be used to help plan

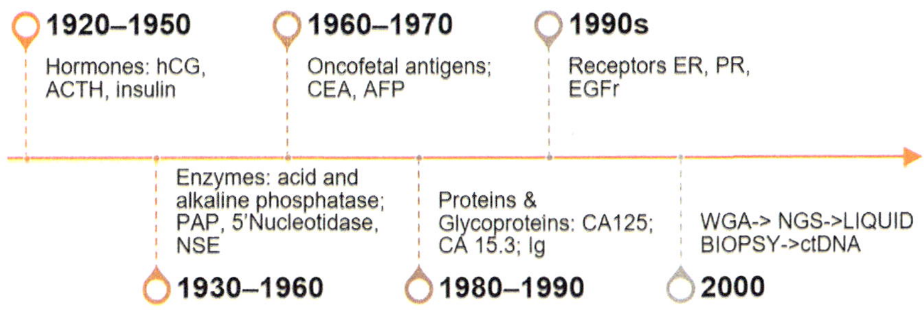

FIGURE 1.1

100 years of tumor marker development. *ACTH*, adrenocorticotropic hormone; *AFP*, alpha fetoprotein; *CA*, carbohydrate antigen; *CEA*, carcinoembryonic antigen; *ct*, circulating tumor; *EGFr*, human epidermal growth factor receptor; *ER*, estrogen receptor; *hCG*, human chorionic gonadotropin; *NGS*, next generation sequencing; *NSE*, neuron specific enolase; *PAP*, prostatic acid phosphatase; *PR*, progesterone receptor; *WGA*, whole genome analysis.

treatment, give a likely prognosis, and monitor the patient's response to therapy.

Traditional tumor marker development over the last few decades has been relatively slow and challenging with the timeline shown in Fig. 1.1.

Tumor markers include:

- Enzymes
- Hormones and hormone metabolites
- Oncofetal antigens
- Carbohydrate markers
- Blood group antigens
- Proteins such as β2-microglobulin and immunoglobulins
- Receptors
- RNAs
- DNA

Table 1.1 presents a summary of tumor markers that have adequate specificity and sensitivity for use in prognosis, detection of recurrence, and monitoring response to treatment.

Table 1.1 The major tumor markers used for prognosis, treatment monitoring, and recurrence detection.

Utility	Tumor markers
Prognosis	β2-Microglobulin
	CA125
	AFP, PSA, CEA
	Lactate dehydrogenase
	HER2/neu
	Estrogen receptor
	Progesterone receptor
Monitoring treatment	CA125
	CA19-9
	CEA
	AFP
	hCG
	PSA
Detection of recurrence	CA15-3
	CA125
	CEA
	AFP
	hCG
	PSA

1.3 Utilization of serum tumor markers in malignant and benign tumors

Despite the challenges and interferences with immunoassay methodologies for tumor markers, their measurements are readily available on automated platforms that are relatively inexpensive and easy to use. Although their potential utility in screening the general population is low, they are widely used for the differential diagnosis of symptomatic patients and monitoring therapy. Overall, they have limited use for early detection, clinical staging of cancer, and estimating tumor volume. Although specific tumor markers are discussed in detail in other chapters in this book, we briefly summarize the utilities of some of the more widely used markers as follows.

1.3.1 Alpha fetoprotein

AFP is used for screening individuals at high risk for hepatocellular carcinoma. It is also used in other malignant conditions including hepatoblastoma, nonseminomatous germ cell tumors of the ovary and testis, and gastrointestinal tract tumors. AFP values are also monitored in other benign conditions that include benign liver disease as well as in newborns and pregnant women and to diagnose neural tube defects.

1.3.2 CEA

This is the only serum marker commonly used for monitoring colorectal cancer. It is used to assess prognosis, postoperative surveillance, and monitor advanced disease. It is also monitored in other malignant conditions that include medullary thyroid carcinoma, breast, liver, ovarian, pancreatic, and prostate cancers. It has been reported to be mildly elevated in benign conditions that include smoking, infections, inflammatory bowel disease, pancreatitis, liver cirrhosis, and some benign tumors of the gastrointestinal tract.

1.3.3 CA19-9

CA19-9 is used in conjunction with imaging studies for the diagnosis of pancreatic cancer. It is also elevated in other malignant conditions that include colon cancer, gastrointestinal malignancies, cholangiocarcinoma, and benign diseases, namely cirrhosis, cholestasis, and pancreatitis. CA 72-4 is a non-FDA approved test that is used as a "lab developed test" for gastric cancer along with CA19-9, CEA, CA125, and CA242. These markers are reported to be elevated in *Helicobacter pylori* infections, gastric ulcer, gastric polyps, atrophic gastritis, benign mucinous cystic neoplasms, and biliary obstruction.

1.3.4 CA15-3/CA27-29

CA15-3/CA27-29 are mostly used to monitor chemotherapy in patients with stage IV breast cancer. In this setting, however, these biomarkers should not be used alone but in combination with imaging [5]. Although sometimes used, the value of these biomarkers in surveillance following curating surgery or following response to adjuvant chemotherapy in stages II or III breast cancer is not clear.

Elevations of CA15-3/CA27-29 may also be observed in other malignancies such as lung, colon, pancreas, liver, ovary, cervix, and endometrium. Like other markers, they may be elevated in benign conditions such as chronic hepatitis, liver cirrhosis, sarcoidosis, tuberculosis, and systemic lupus erythematosus.

1.3.5 CA125, HE4, ROMA score (risk of ovarian malignancy algorithm)

These markers are commonly used in women with transvaginal ultrasound for early detection in women with ovarian cancer and hereditary symptoms. They are also used to monitor treatment, response to chemotherapy, and predict recurrence. Like most other serum markers, they lack tumor and organ specificity. Thus they all can be elevated in cervical, liver, pancreatic, lung, colon, stomach, biliary tract, uterine, fallopian tube, breast, and endometrial carcinomas. Benign conditions where elevated levels may occur include cirrhosis, hepatitis, endometriosis, first trimester of pregnancy, ovarian cysts, and inflammatory disease.

1.3.6 Prostate-specific antigen and prostate-specific antigen derivatives

PSA is currently the most useful marker for prostatic cancer. It is used in screening for early prostate cancer, in patient management, and to monitor posttreatment. An extensive review has been published in JAMA in 2017 [6] summarizing potential benefits and harm of screening for prostate cancer. PSA derivatives that include PSA density, PSA velocity change, percentage-free PSA, prostate health index, the 4 Kallikrein score (4K), and prostate cancer antigen 3 are used extensively as supplementary tests to PSA in aiding prostate cancer diagnosis and assessing prognosis. PSA is reviewed extensively in Chapter 5.

1.4 Understand FDA approved and non-FDA approved tumor markers

Table 1.2 lists the major FDA approved blood cancer biomarkers for clinical use in the United States.

Table 1.2 The major FDA approved blood cancer biomarkers for clinical use in the United States.

Tumor markers	Cancer type
AFP-L3%	Hepatocellular carcinoma
AFP	Testicular cancer
CA27.29	Breast cancer
CA15-3	Breast cancer
Circulating tumor cells (EpCAM, CD45, cytokeratins 8, 18 + , 19 +)	Breast cancer Prostate cancer Colorectal cancer
CA19-9	Pancreatic cancer
CEA	Not specified
EGFR mutation test v2	Nonsmall cell lung cancer
Fibrin/fibrinogen degradation product (Dr-70)	Colorectal cancer
CA125	Ovarian cancer
HE4	Ovarian cancer
OVA1 (multiple proteins)	Ovarian cancer
ROMA (HE4 + CA125)	Ovarian cancer
Thyroglobulin	Thyroid cancer
Total PSA	Prostate cancer
Free PSA (%)	Prostate cancer
Pro2PSA	Prostate cancer

Table 1.3 The major FDA approved urine/tissue cancer biomarkers for clinical use in the United States.

Tumor markers	Cancer type
HER2	Breast Cancer
p63 protein	Prostate cancer
c-Kit	Gastrointestinal stromal tumors
Estrogen receptor	Breast cancer
Progesterone receptor	Breast cancer
EGFR mutation test v2	NSCLC
Nuclear mitotic apparatus protein (NuMA, NMP22)	Bladder cancer
BTA *stat*	Bladder cancer
BTA TRAK	Bladder cancer
ImmunoCyt	Bladder cancer
AFP	Testicular cancer
Human hemoglobin (fecal occult blood)	Colorectal cancer

Table 1.3 lists the major FDA approved urine/tissue cancer biomarkers for clinical use in the United States.

Table 1.4 lists the major non-FDA approved but clinically offered tumor markers [Off-label/laboratory developed test (LDT)].

1.5 Overview of methodologies for measurement of tumor markers

Methods for the quantitation of tumor markers include immunological, high-pressure liquid chromatography and mass spectrometry (HPLC-MS), flow cytometry, cytogenetics, genetic analysis, and proteomics. Immunological methods include the use of monoclonal antibodies that bind to epitopes on tumor marker molecules with either radioactive labels in radioimmunoassay, or enzymes in enzyme-linked immunosorbent assay (ELISA), chemiluminescent labels in enzyme

Table 1.4 The major non-FDA approved but clinically offered tumor markers (off-label/LDT).

Tumor markers	Cancer type
CA242	Pancreatic cancer
CA72-4	Gastric, colorectal cancer
Cytokeratin fragment 19 (CYFRA 21-1)	NSCLC
Squamous cell carcinoma antigen	Lung squamous cell carcinoma, head and neck, cervix carcinoma
Neuron-specific enolase	Small cell lung cancer
Progastrin-releasing peptide (ProGRP)	Small cell lung cancer
Chromogranin A	Carcinoid tumors, neuroendocrine tumors
S-100 proteins	Melanoma, sarcoma, carcinomas
β-hCG	Testicular cancer, trophoblastic tumors, choriocarcinoma
PTHrP	Malignancy-associated hypercalcemia (breast, lung, head, and neck)
β2-Microglobulin	Multiple myeloma, lymphoma (non-Hodgkin)
α-L-fucosidase	Hepatocellular carcinoma
Ferritin	Hematologic malignancies and related anemia
Cerebrospinal fluid circulating tumor cells	Leptomeningeal metastases

immunoassays, and dyes in immunohistochemistry. Table 1.5 summarizes enzyme, protein, and endocrine tumor marker assay methods that rely on the antibody recognition of antigen.

Clinical implementation of biomarker tests necessitates following guidelines specified by Clinical Laboratory Standards Institute (CLSI). These guidelines include evaluation of assay precision, accuracy, linearity/reportable range, analytical specificity, analytical sensitivity, carryover, analyte stability, and reference intervals. As cancer patients are usually monitored over years, it is imperative that these parameters are thoroughly evaluated and understood.

Table 1.5 Overview of methodology for traditional and nontraditional biomarkers measurements.

Tumor marker	Methodology	Specimen	Tumor type/utility
PSA	IA	Serum	Prostate cancer, screening, therapy and recurrence
Lactate dehydrogenase	EA	Serum	Hematologic malignancies, elevation in nonspecific malignancies
Alkaline Phosphatase	EA	Serum	Metastatic carcinoma of bone, hepatocellular carcinoma, osteosarcoma, nonspecific elevation in many bone-related and liver cancers
Neuron specific enolase	RIA/IHC	Serum	Neuroendocrine tumors, monitor disease progression
Serum M-Protein	SPE/IF/ MALDI-TOF	Serum	Plasma cell dyscrasias, monitoring of plasma cell malignancies
Serum-free light chains	IA	Serum	Plasma cell dyscrasias, monitoring of plasma cell malignancies
β2-Microglobulin	IA	Serum	Hematologic malignancies, lymphoproliferative disorders
ACTH	IA	Serum	Pituitary adenoma, ectopic ACTH producing tumor
ADH	IA	Serum	Posterior pituitary tumors, SIADH diagnosis
C-peptide	IA, ELISA	Serum	Insulinoma
Calcitonin	IA	Serum	Medullary thyroid carcinoma and neuroendocrine tumors, therapy response

(*Continued*)

Table 1.5 Overview of methodology for traditional and nontraditional biomarkers measurements. *Continued*

Tumor marker	Methodology	Specimen	Tumor type/utility
CEA	IA	Serum	Colorectal, medullary thyroid carcinoma, breast, liver, ovarian, pancreatic and prostate cancers
Chromogranin A	ELISA, RIA	Serum	Pheochromocytoma, neuroblastoma, carcinoid tumors, small cell lung cancers
Cortisol	IA	Serum, Urine	Cushing's syndrome, adrenal adenoma
Gastrin	IA	Serum	Neuroendocrine tumor, Zollinger-Ellison syndrome, gastrinoma
Growth hormone	IA	Serum	Pituitary adenoma, ectopic GH-secreting tumors, monitoring acromegaly
HVA	HPLC	24 h urine	Neuroblastoma, pheochromocytoma, paraganglioma
5-Hydroxy indole acetic acid	HPLC	24 h urine	Carcinoid tumors
Metanephrines, fractionated	HPLC	24 h urine	Neuroblastoma, pheochromocytoma, paraganglioma
PTH	IA	Serum	Parathyroid adenoma monitoring
Prolactin	IA	Serum	Prolactinoma postsurgical monitoring
VMA	HPLC	24 h urine	Neuroblastoma

ACTH, adrenocorticotropic hormone*; ADH,* antidiuretic hormone*; EA,* enzyme assay*;*
ELISA, enzyme-linked immunosorbent assay*; HPLC,* high-performance liquid chromatography*; HVA,* homo-vanillic acid*; IA,* immunoassay*; IF,* immunofixation*; IHC,* immunohistochemistry*; MALDI-TOF,* matrix-assisted laser desorption/ionization time of flight mass spectrometry*;*
RIA, radioimmunoassay*; SIADH,* syndrome of inappropriate diuretic hormone*;* *SPE,* serum protein electrophoresis*; VMA,* vanylmandelic acid.

1.5.1 Clinical sensitivity and specificity of laboratory tests

Sensitivity refers to the test's ability to correctly detect ill patients who do have the relevant condition. TP indicates true positive; FN, false negative; TN, true negative; and FP, false positive. Mathematically, this can be expressed as:

Diagnostic sensitivity indicates true positive proportion in diseased population, expressed as $= TP/(TP + FN) \times 100\%$.

Diagnostic specificity relates to the test's ability to correctly reject healthy patients without disease, expressed as $= TN/(TN + FP) \times 100\%$.

In the specific example below involving 522 patients, there are 164 people diagnosed with cancer and test results are positive that is TP. There are 86 people with negative results but having disease that is FN. In this case, diagnostic sensitivity $= [164/(164 + 86)] \times 100\% = 65.6\%$. Diagnostic specificity $= [245/(245 + 27)] \times 100\% = 90.1\%$ (Box 1.1).

Box 1.1 Calculation of diagnostic sensitivity and specificity.

	Cancer	No cancer	Total
Test (+)	164 TP	27 FP	191
Test (−)	86 FN	245 TN	331
Total	250 (TP + FN)	272 (TN + FP)	522 grand total

During the evaluation of laboratory tests in cancer, diagnostic efficiency needs to be considered. This includes the concepts of clinical sensitivity, specificity, and predictive values. An ideal biomarker should have high clinical sensitivity (true positive rate—that is to be able to correctly identify individuals with the disease) and specificity (true negative rate—that is the ability to identify individuals without the disease). Other measures of diagnostic efficiency include predictive values. PPV is the likelihood individuals with the disease who are positive. NPV or negative predictive value is the likelihood of test negative individuals who do not have the disease.

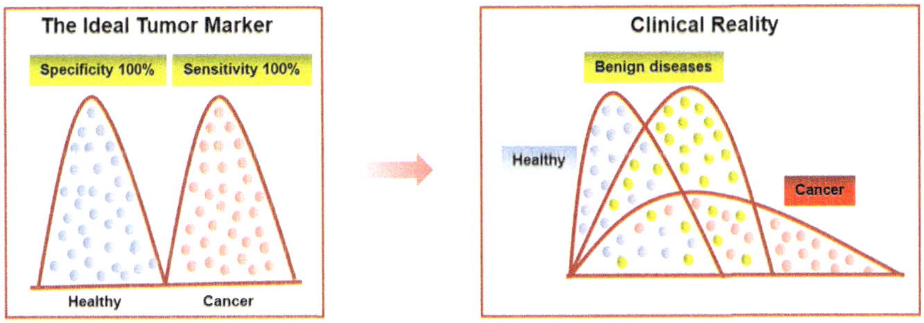

FIGURE 1.2

Traditional cancer biomarkers-perfection versus reality.

Fig. 1.2 summarizes the challenges and utility of traditional cancer biomarkers currently in clinical practice. The reality is that there is no "ideal tumor marker" with 100% sensitivity and 100% specificity, due to the limitation of methodology, inadequate assay standardization, and potential interferences in immunoassays.

Receiver operating curves (ROCs) are widely used for assessing diagnostic tests. The sensitivity and specificity values are plotted for every individual cutoff on a graph with 1 − specificity on the x-axis and sensitivity on the y-axis. The shape of the ROC curve and the area under the curve (AUC) helps determine the discriminatory power of the marker. The closer the curve is located to the upper left-hand corner, the better the marker is in discriminating between disease and nondiseased patients. A perfect biomarker has an AUC of 1.0 whereas a nondiscriminatory biomarker has an AUC of 0.5.

Measures of diagnostic efficiency depend on the distribution of test results in a population with and without the disease and to a large extent on the "cutoff" used to define abnormal values. The cutoff also refers to "medical decision limit," which is defined as the result that separates a "positive" value from a "negative" value. In the case of point of care qualitative devices, the cutoff is assigned by the manufacturer. However, in the case of quantitative results, there is flexibility in changing the cutoff value instead of using the manufacturer's suggestion for a specific patient population. An example of this is beta human chorionic gonadotropin (β-hCG) in cancer patients that will be discussed in Chapter 3.

1.6 Laboratory technologies for tumor marker measurement

These include immunoassays, enzyme assays, and HPLC-MS.

1.6.1 Immunoassay methodologies

An immunoassay is based on the selective binding between an antibody and an antigen. Polyclonal or monoclonal antibodies are used in immunoassays. One of the main reasons' immunoassays enjoy a wide range of popularity is ease of use and relatively low cost.

Polyclonal antibodies have a range of binding strengths and are produced from multiple clones that are multiple cell lines within the body. Monoclonal antibodies are derived from single clones that are identical to each other in sequence and have identical binding strengths. This implies greater precision in immunoassays due to identical binding of antibodies.

Immunoassays are either competitive or noncompetitive. In a competitive immunoassay, a known amount of antigen with a radioactive, enzymatic, fluorescent, or chemiluminescent label competes for a limited number of binding sites on the antibody. The amount of bound labeled antigen is inversely proportional to the amount of antigen present in the sample (Fig. 1.3A). Noncompetitive immunoassays are widely used in cancer biomarker detection. A common example of this technique involves a "sandwich" being created with two different monoclonal antibodies around the antigen (Fig. 1.3B). In this case, the amount of bound labeled antibody is directly proportional to the amount of antigen in the sample.

Separation needs to take place between the bound labeled antigen and the unlabeled antigen, the better the separation, the better is the precision of the assay. Polystyrene and magnetic microparticles are used in immunoassays and they can be easily separated from the solution by using a strong magnet. When the physical separation is required to separate the labeled antibody–antigen complex from the unbound antibody/antigen, the immunoassays are defined as heterogeneous. In contrast, if the physical separation is not required, they are referred to as homogeneous immunoassays. Homogeneous

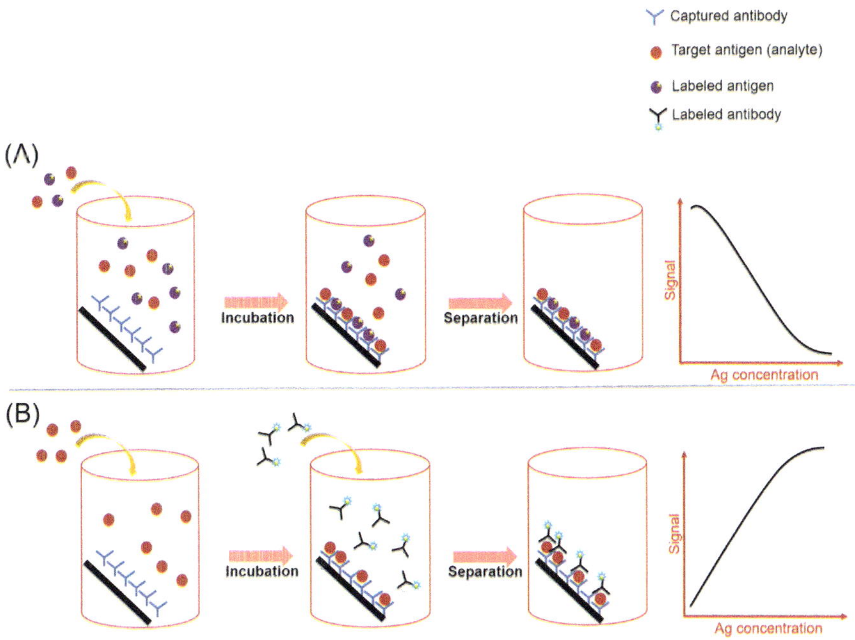

FIGURE 1.3

Schematic illustration of immunoassays. In a competitive immunoassay (A), target antigens (analytes) in specimens compete with labeled antigens to bind with a limited number of specific antibodies. After washing, as more labeled antigens are bound, lesser target antigens remain in the specimen. Therefore the higher the measured signals, lower target antigens are in the specimen, which shows an inversely proportional relationship. In a Sandwich immunoassay (B), target antigens are sandwiched in between two specific antibodies (one is captured antibody and another is labeled antibody) against different epitopes of target antigens. As more target antigens are in the specimen, higher signals are measured, which shows a proportional relationship.

immunoassays are less common and have all components in solution and in these formats, a measurable response is essential.

1.6.2 Other types of immunoassays

Other types of immunoassays include chromatographic immunoassays, lateral flow immunoassays, label-free immunoassays, and electrophoretic immunoassays. Readers are encouraged to consult excellent reviews on these techniques in the literature [7].

Chromatographic immunoassays are a combination of chromatography and immunoassays. The antibody or antibody fragment is immobilized on the stationary phase and the specimen with the compound of interest is injected. The mobile phase carries the sample to the stationary phase. The antibody on the stationary phase selectively binds the compound of interest and the mobile phase carries away unbound components. Once the specimen is applied to the column, the elution buffer disrupts the antibody binding and allows compound to be released from the stationary phase. The compound that is eluted can be detected using ultraviolet, visible absorption, or mass spectrometry. After the compound is quantitated, the mobile phase is run through the column allowing the antibodies to regenerate for additional use. This method is used for the measurement of drugs and other biological compounds [8].

Lateral flow immunoassays or lateral flow immunochromatographic assays are used extensively in point of care testing, home testing, and in the laboratory. These devices are simple to use with results available between 5 and 30 minutes, the most used test being urine pregnancy test. Lateral flow tests operate on the same principles as the ELISA. In essence, these tests run the liquid sample along the surface of a pad with reactive molecules that show a visual positive or negative result. The pads are based on a series of capillary beds, such as pieces of porous paper, microstructured, or sintered polymer [9]. Each of these pads has the capacity to transport fluid (e.g., urine, blood, saliva) spontaneously.

Label-free immunoassay. There are a few immunoassays that do not rely on labels, although the labels are generally employed in immunoassays. Surface plasma resonance (SPR) is a typical label-free immunoassay. Surface plasmon resonance uses refractive index changes on a metal surface without a fluorescent or radioactive label. The analyte bound to the substrate film of the sensor chip is measured by resonant oscillation of conduction electrons stimulated by incident light. No labeled reporter molecule is needed because the binding reaction kinetics is monitored rapidly. It is used extensively for quantitation of proteins, antibodies, DNA, and RNA in research laboratories. This technology is finding applications in the field of cancer biomarkers, cardiac biomarkers, antibodies, hormones, and nucleic acids. Although SPR has been used in the characterization of monoclonal antibodies, it has not been used for circulating antibody concentrations that may assume importance with the explosion of monoclonal antibody therapies and regimens used in the treatment of different cancers.

Electrophoretic immunoassay is a technology which takes advantage of antibody specificity for antigens. One of the most common types of electrophoretic immunoassays is gel electrophoresis for immunofixation. This is an electrophoretic method that applies the sample to a gel support in an electric field, followed by the precipitation by antiserum to the specific protein of interest. Capillary zone electrophoresis/immune substraction detects and immunotypes proteins of interest with specific antibodies (antiserum) in serum or urine samples. Immune substraction can type monoclonal proteins detected by immunofixation.

1.7 Technical limitations of immunoassays

Immunoassays are extensively used in clinical practice as they are rapid, simple, cost effective, and available on most chemistry instrument platforms. Lack of tumor marker standardization makes it challenging to interpret serial patient results on different platforms.

1.7.1 Standardization of tumor marker assays

Despite the ready availability of multiple analytical testing platforms for the commonly available markers, variability between different methodologies is a challenge. The reasons for this can be differences in antibody specificity, analyte heterogeneity, assay design, lack of standard reference materials for calibration, kinetics, and different reference ranges [9]. To address this, international standards have been developed by the World Health Organization (WHO) for several markers including AFP, CEA, hCG, PSA, and free PSA. This information is updated on the WHO website [10]

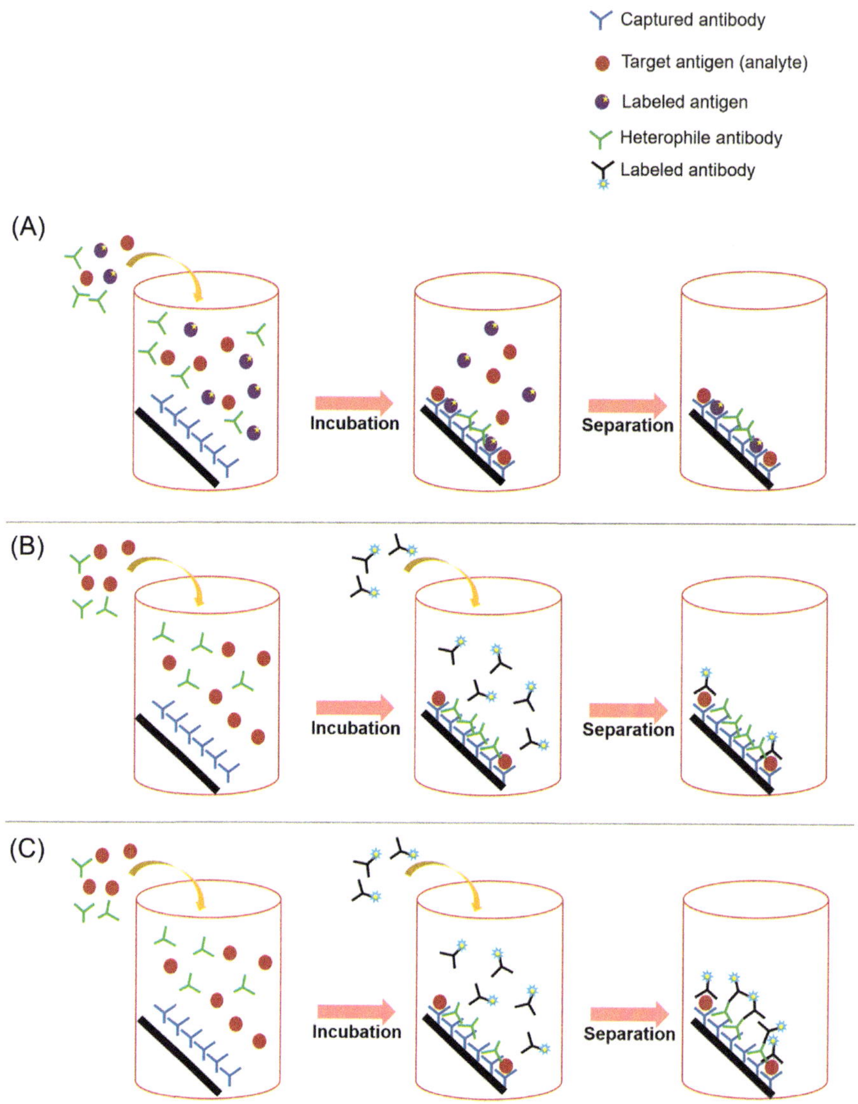

FIGURE 1.4

Mechanism of heterophile antibody interference in both competitive and sandwich immunoassay. In a competitive immunoassay (A), if the heterophile antibodies in the specimen recognize and compete with the labeled antigen to bind with the capture antibodies, less signals are detected, causing falsely high results. In a sandwich immunoassay (B

(*Continued*)

In the case of PSA, Ferraro et al. [11] have recently shown that despite efforts by international agencies, further improvement is still warranted. The authors have shown differences between four commercially available assays for free PSA indicating lack of harmonization. Although the study considered characteristics of antibodies used, it was not considered to be a major factor causing systemic differences between the methods. Despite considerable improvements in methodologies, the authors recommend for surveillance testing, total PSA along with free PSA should be determined using the same methodology.

1.7.2 Interferences in immunoassays

Apart from challenges of standardization, other factors can affect test results. These include sensitivity of assays, quality of antibody, cross reactivity, presence of heterophile antibodies along with hook/prozone and specimen carryover [7]. These issues are addressed below:

1.7.2.1 Heterophile antibodies

Heterophile antibodies are present in patients exposed to animals, patients who have received immunotherapy and have received diagnostic procedures utilizing immunoglobulins or fragments of immunoglobulins. These antibodies can persist for several months. However, interference in competitive assays is usually minimal because the antigens have higher affinity for reagent antibodies than do the heterophile antibody. Information regarding interference by heterophile antibodies is usually included in the package insert of the test method. Fig. 1.4 illustrates how heterophile antibody interference can occur. It is important to bear in mind that interference can be either positive or negative.

◀ and C), if the heterophile antibodies only recognize and bind with the capture antibodies, resulting in blocking the capture antibodies binding sites, less signals are detected, causing falsely low results (B). In contrast, if the heterophile antibodies recognize both capture and labeled antibodies, they work more like the target antigens. Therefore heterophile antibodies are working like the bridge connecting both capture and labeled antibodies, causing falsely high results (C).

Commercial agents and assays are available to remove heterophile antibodies. These can be either mouse serum that can be added to absorb the antibody or commercially available heterophile nonspecific antibody blocking tubes. The procedure for the removal of heterophile antibodies using heterophile blocking tubes (HBTs) involves, incubating 500 μL of serum that is added to an HBT and incubated for 1 hour at room temperature. The treated sample is then analyzed, and the results are compared with the sample with no pretreatment to obtain a result that is free from heterophile antibody interference [7].

1.7.2.2 Hook/prozone effects

Hook/prozone effects occur when the concentration of analyte/antibody is too high that both the capture and labeled antibodies are saturated. Therefore the assay response will drop off after the maximum response, with the typical "hook" shape in the curve (Fig. 1.5). Hook effects usually occur when the analyte concentration is far above the analytical measurement range, which can lead to falsely low results. To overcome the problem, sample dilution and retesting is recommended.

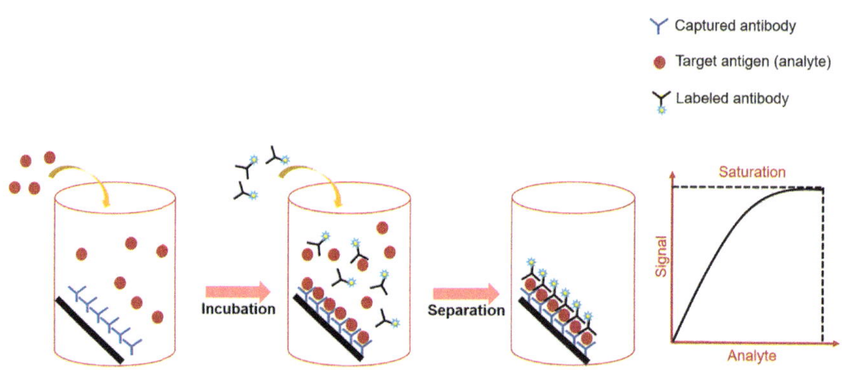

FIGURE 1.5

Schematic illustration of hook effect. As the concentration of target antigens continue to increase, when the antigens exceed the amount of the captured antibodies and all the captured antibodies are bound, a saturated curved will be observed, causing a falsely low result.

1.7.2.3 Biotin interference

The strong affinity of biotin to streptavidin allows the rapid separation of the analyte and is used in some immunoassay formats. The biotinylated antibodies/analytes are captured on a streptavidin microparticle or another solid phase. When biotin–streptavidin binding is used in competitive immunoassays, excess biotin in the specimen can block the binding of biotinylated analyte on the streptavidin-coated solid phase, which causes falsely high results (Fig. 1.6A). In contrast, when biotin–streptavidin binding is used in "sandwich" immunoassays, excess biotin in the specimen can block the binding of biotinylated antibodies on the streptavidin-coated solid phase, which causes falsely low results (Fig. 1.6B). It should be noted that biotin interference depends on the methodology of immunoassays and not all immunoassays are affected by excess biotin.

1.8 High-pressure liquid chromatography mass spectrometry

This technique is commonly used for the detection of small molecules such as catecholamines to diagnose carcinoid tumors, pheochromocytoma, and neuroblastoma.

The combination of liquid chromatography and mass spectrometry is a powerful platform for protein identification and quantitation, particularly in cancer biomarker discovery. High resolution liquid chromatography separations with fast and sensitive mass spectrometry detection is being used to monitor hundreds of proteins simultaneously in a targeted manner. This technique will be covered in Chapter 4.

There are several reports in the literature that mass spectrometric profiles of plasma or serum can be used to distinguish between patients who have cancer and those who do not have cancer. These methods include surface-enhanced laser desorption/ionization time of flight mass spectrometry and matrix-assisted laser desorption/ionization time of flight mass spectrometry (MALDI-TOF-MS). Further analytical and clinical validation of these techniques are

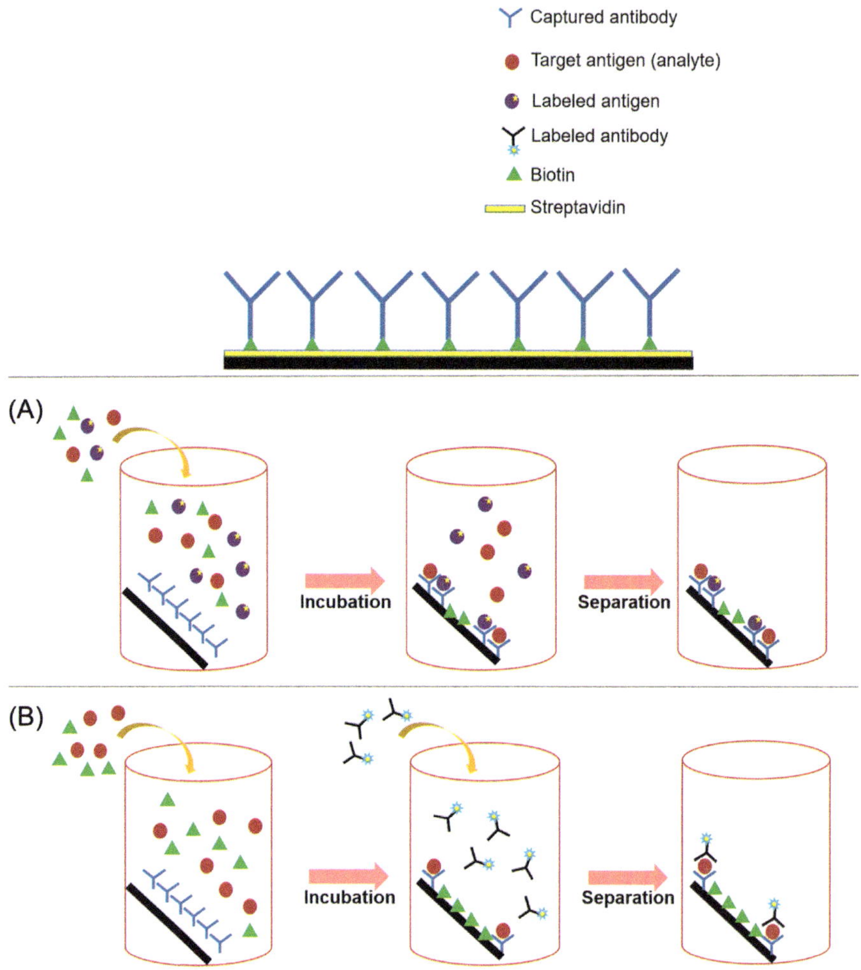

FIGURE 1.6

Schematic illustration of biotin interference in immunoassays. In a competitive immunoassay (A), endogenous and labeled antigens compete for single biotinylated-captured antibody binding sites. The biotinylated-captured antibodies are then bound to the streptavidin-coated solid phase. In the presence of high biotin concentrations, excess biotin binds to the solid phase and prevents the binding of antibody with endogenous and labeled antigens, causing a falsely decreased signal and thus falsely high results. In a sandwich immunoassay (B), the target antigens are bound by the labeled and biotinylated captured antibodies.

(*Continued*)

warranted. Other mass spectrometric techniques that will be discussed in detail in Chapter 11 are used primarily for biomarker discovery.

1.9 Methods for detecting nontraditional and emerging tumor markers

While traditional tumor markers have been largely used in monitoring treatment response and surveillance for recurrence or remission in cancer patient care, little progress has been made in discovery and clinical application of traditional tumor markers during the past 30 years. After entering the 21st century, especially the era of precision medicine or personalized medicine, new technologies including proteomics and molecular diagnostics in body fluids became available that resulted in the validation of several new markers, especially for guiding cancer therapy. The new technologies represent genetic, genomic, proteomic, metabolomic, microarray, mass spectrometry, nanotechnology, and liquid biopsy. In addition, molecular diagnostics play a key role in precision medicine or personalized cancer medicine. The breakthrough of major technologies and key representative biomarkers are highlighted as follows.

1.9.1 Metabolomics and proteomics

Proteomic technology platforms have developed rapidly during the recent years with the understanding of cancer metabolomics. Mass spectrometry is used for protein-based tumor biomarker discovery focusing on the metabolic pathways such as pyruvate kinase, L-lactate dehydrogenase, Warburg effect, glutamine metabolism, and oxidative stress [12,13]. Mass spectrometry–based proteomics is one

◀ The biotinylated-captured antibodies link the sandwich complex to a streptavidin-coated solid phase. In the presence of high biotin concentrations, excess biotin competes with biotinylated-captured antibodies for the streptavidin binding sites, preventing the link of sandwich complex to solid phase, causing falsely low results.

of the most promising new approaches for cancer biomarker discovery.

MALDI laser-based imaging mass spectrometry (IMS) technology has been developed for the rapid analysis of tissues or for imaging tissue specimens of a mass spectrum with specific *m/z* value. This technology of IMS can now be translated to the anatomic pathology laboratory [14]. Recently MasSpec Pen has been developed and is used in ex vivo and in vivo cancer diagnosis. The MasSpec Pen is a hand-held device that can rapidly identify the molecular profile (metabolites, lipids, and proteins) of tumors from healthy tissue using a small volume water droplet and MS analysis. This technology has been validated for thyroid, breast, lung, and ovarian cancer, and it has been demonstrated to have high sensitivity, specificity, and overall accuracy [15–16]. Although these techniques and numerous biomarkers being discovered are promising, both the method and biomarkers need to be externally and thoroughly translated and validated in clinical and laboratory settings before clinical implementation [17].

1.9.2 Molecular and companion diagnostics

Genetic testing and (whole) gene sequencing provide greater insight into the link between DNA sequence and disease. The sequencing of the human genome and the detection of genetic mutation or a panel of mutations, especially those actionable mutations, are being used for targeted therapy in precision cancer medicine such as HER2, EGFR, BRAC, BRAF, PD-L1, KRAS, and ALK. A companion diagnostic (CDx) is a medical device, often an in vitro device, which provides information that is essential for the safe and effective use of a corresponding drug or biological product. The test helps a health care professional determine whether a particular therapeutic product's benefits to patients will outweigh any potential serious side effects or risks [18]. Currently there are over 42 CDx tests are FDA approved. For example, there are 10 approved HER2/neu tests are used for guiding Herceptin therapy. BRCA1 and BRCA2 mutations are used in guiding therapy for breast and ovarian cancer. EGFR mutations (insertion/deletion) are used for guiding selection of tyrosine kinase inhibitors for nonsmall cell lung cancer

(NSCLC). PD-L1 is for guiding antibody therapy for NSCLC and other types of carcinoma. There are several other FDA-approved CDx tests such as BRAF, KIT, KRAS, and ALK. A list of FDA-cleared or approved CDx tests can be found on the FDA website [19]. With the further understanding of tumor biology, the ongoing evolution of CDx and more approvals of CDx tests by FDA, CDx will shed light in oncology practice.

1.9.3 Liquid biopsy

Liquid biopsy is an approach to identify or isolate liquid biomarker from body fluids such as blood, plasma/serum, urine, CSF, saliva, ascites, or pleural effusion. Refer Chapter 7 for a detailed discussion on this subject. It is used along with tissue biopsy. Initially liquid biopsy referred to analysis of circulating tumor cells (CTCs) and circulating tumor DNA (ctDNA) or circulating free DNA (cfDNA) in peripheral blood for diagnosis, prognosis, real-time monitoring of therapy, and guiding personalized therapy in cancer patients. Now this has been advanced to other nucleic acids, mRNA, miRNA, lncRNA, and most recently extracellular vesicles (EVs) including exosomes (40–120 nm), microvesicles (150–1000 nm), and metabolites [20–21].

Detection of CTCs in peripheral blood of patients with metastatic cancer has been used for prognosis and monitoring therapeutic response of patients with metastatic cancer. Detection of CTCs has only been approved by FDA for metastatic breast cancer, colorectal cancer, and prostate cancer. Moreover, the development of epithelial mesenchymal transition of CTCs during the disease limits the application of CTCs [22]. The development of more advanced technology and expanded application of CTCs are needed. Single cell molecular analysis of CTC may shed light in advancing its utility.

Analysis of circulating tumor DNA (ctDNA) may offer more advantages over CTC. The detection of ctDNA in early-stage tumors may be useful for screening and management of patients with cancer [23]. Liquid biopsies have become increasingly used clinically in recent years in risk assessment, cancer screening, diagnosis, prognosis, prediction, monitoring therapeutic response, and guiding targeted therapy [21]. The FDA has already approved several single-gene assays and, more recently, several multigene assays to detect genetic alterations in plasma circulating-free DNA (cfDNA) for use as CDx matched to specific molecularly targeted therapies for cancer such as EGFR gene mutations for targeted therapy of NSCLC [24,25].

There is no doubt that these new and emerging tumor markers are playing and will continue to play significant roles in cancer care, particularly in risk assessment, early diagnosis, prognosis, prediction, and most importantly in guiding personalized cancer therapy and targeted therapy. Nevertheless, one must recognize the challenges and limitations of these tumor markers. Despite the enormous potential in the field of oncology for the use of liquid biopsy of CTCs and ctDNA even with comprehensive tumor genome analysis, liquid biopsy cannot yet replace gold standard diagnostic techniques in clinical application. With exception of FDA-approved CDx and EGFR-based liquid biopsy, the majority of these tumor markers are not FDA-approved and are considered as LDTs. Advanced technologies are needed to improve the analytical sensitivity and specificity. Novel approaches to enable cellular and molecular characterization of CTCs, ctDNA, RNA, miRNA, lncRNA, peptides, and EVs need to be developed. Further clinical and laboratory validations are needed to validate the clinical utilities.

Details of these new and emerging technologies and tumor markers are described in Chapters 10 and 11.

1.10 Summary Points

1. Virtually no tumor marker has been shown to be specific or sensitive enough to use in the detection of cancer in the general population. We believe that both traditional and nontraditional tumor markers in cancer need to be considered together for more effective screening, diagnosis, prognosis, and response to treatment.
2. Pearls for interpreting and ordering tumor marker tests include not relying on the results of a single test, not changing methodology and trying to stay consistently ordering the test from the same lab.

3. If immunoassay methodology is used, attention needs to be paid to interferences from heterophile antibodies as well as other substances along with prozone and the hook effect.
4. Laboratory-developed tests that are not uncommon need to be carefully validated before being used in clinical situations.

1.11 Future direction

Is there a future diagnostic test for cancer, is the question? While traditional tumor markers will continue to play significant roles in cancer patient care, new tumor markers are becoming available with recent biological and technical advances, including genomic, epigenomic, transcriptomic, proteomic, metabolomic, array analysis (DNA, proteins, tissues), mass spectrometry, automated DNA sequencing/polymerase chain reaction/NGS and liquid biopsies (CTCs, micro RNS, circulating nucleic acids (ctDNA), extracellular vesicles for personalized cancer medicine. A combined panel of traditional and nontraditional biomarkers should be developed for optimal patient care.

References

[1] World Health Organization. Cancer, <https://www.who.int/newsroom/fact-sheets/detail/cancer>; 2022 [accessed date February 3, 2022].

[2] National Cancer Institute, Bethesda, MD, <http://seer.cancer.gov/csr/1975_2015/>, based on November 2017 SEER data submission, posted to the SEER website April 2018.

[3] Kulasingam V, Diamandis E. Strategies for discovering novel cancer biomarkers through utilization of emerging technologies. Nat Rev Clin Oncol 2008;5:588–99.

[4] Mordente A, Meucci E, Martorana GE, Silvestrini A. Cancer biomarkers discovery and validation: state of the art, problems and future perspectives. Adv Exp Med Biol 2015;867:9–26.

[5] Duffy MJ, McDermott EW, Crown J. Blood-based biomarkers in breast cancer: from proteins to circulating tumor cells to circulating tumor DNA. Tumour Biol 2018;40(5) 1010428318776169.

[6] Fenton JJ, Weyrich MS, Durbin S, Liu Y, Bang H, Melnikow J. Prostate-specific antigen-based screening for prostate cancer: evidence report and systematic review for the US preventive services task force. JAMA. 2018;319(18):1914–31.

[7] Ward G, Simpson A, Boscato L, Hickman PE. The investigation of interferences in immunoassay. Clin Biochem 2017;50(18):1306–11.

[8] Matsuda R, Rodriguez E, Suresh D, Hage DS. Chromatographic immunoassays: strategies and recent developments in the analysis of drugs and biological agents. Bioanalysis 2015;7(22):2947–66.

[9] Moser AC, Carlson LTL. General principles of immunoassays. Future Med 2014;. Available from: https://doi.org/10.2217/ebo.13.617.

[10] WHO international biological reference preparations. <https://www.ifcc.org/media/478523/alph2018-who_reference_materials.pdf>; 2018 [accessed date March 30, 2018].

[11] Ferraro S, Bussetti M, Rizzardi S, Braga F, Panteghini M. Verification of harmonization of serum total and free prostate-specific antigen (PSA) measurements and implications for medical decisions. Clin Chem 2021;67(3):543–53.

[12] Zhou W, Liotta LA, Petricoin EF. Cancer metabolism: what we can learn from proteomic analysis by mass spectrometry. Cancer Genomics Proteom 2012;9(6):373–81.

[13] Hristova VA, Chan DW. Cancer biomarker discovery and translation: proteomics and beyond. Expert Rev Proteom 2019;16 (2):93–103.

[14] Aichler M, Walch A. MALDI Imaging mass spectrometry: current frontiers and perspectives in pathology research and practice. Lab Invest 2015;95(4):422–31.

[15] Zhang J, Rector J, Lin JQ, Young JH, Sans M, Katta N, et al. Nondestructive tissue analysis for ex vivo and in vivo cancer diagnosis using a handheld mass spectrometry system. Sci Transl Med 2017;9(406):eaan3968.

[16] Sans M, Zhang J, Lin JQ, Feider CL, Giese N, Breen MT, et al. Performance of the MasSpec Pen for rapid diagnosis of ovarian cancer. Clin Chem 2019;65(5):674–83.

[17] Diamandis EP. Mass spectrometry as a diagnostic and a cancer biomarker discovery tool: opportunities and potential limitations. Mol Cell Proteom 2004;3(4):367–78.

[18] US Food and Drug Administration. Companion diagnostics, <https://www.fda.gov/medical-devices/in-vitro-diagnostics/companion-diagnostics>; 2018 [accessed date December 7, 2018].

[19] US Food and Drug Administration. List of cleared or approved companion diagnostic devices (in vitro and imaging tools), <https://www.fda.gov/medical-devices/vitro-diagnostics/list-cleared-or-approved-companion-diagnostic-devices-vitro-and-imaging-tools>; 2021 [accessed date March 17, 2022].

[20] Diaz Jr LA, Bardelli A. Liquid biopsies: genotyping circulating tumor DNA. J Clin Oncol 2014;32(6):579–86.

[21] Alix-Panabières C, Pantel K. Clinical applications of circulating tumor cells and circulating tumor DNA as liquid biopsy. Cancer Discov 2016;6(5):479–91.

[22] Satelli A, Brownlee Z, Mitra A, Meng QH, Li S. Circulating tumor cell enumeration with a combination of epithelial cell adhesion molecule- and cell-surface vimentin-based methods for monitoring breast cancer therapeutic response. Clin Chem 2015;61(1):259–66.

[23] Phallen J, Sausen M, Adleff V, Leal A, Hruban C, White J, et al. Direct detection of early-stage cancers using circulating tumor DNA. Sci Transl Med 2017;9(403):eaan2415.

[24] Ignatiadis M, Sledge GW, Jeffrey SS. Liquid biopsy enters the clinic – implementation issues and future challenges. Nat Rev Clin Oncol 2021;18(5):297–312.

[25] Testing.com. FDA approves first liquid biopsies that test for mutations in multiple genes, <https://labtestsonline.org/news/fda-approves-first-liquid-biopsies-test-mutations-multiple-genes>; 2021 [accessed date Jan 14, 2021].

CHAPTER

Laboratory testing in pediatric cancer patients

2

Khushbu Patel[1,2], William S. Phipps[3], Hung S. Luu[4] and Dinesh Rakheja[5,6]

[1]*Department of Pathology and Laboratory Medicine, Children's Hospital of Philadelphia, Philadelphia, PA, United States*

[2]*Department of Pathology and Laboratory Medicine, University of Pennsylvania, Philadelphia, PA, United States*

[3]*Department of Laboratory Medicine and Pathology, University of Washington, Seattle, WA, United States*

[4]*Children's Health System of Texas, University of Texas Southwestern Medical Center, Dallas, TX, United States*

[5]*Department of Pathology and Laboratory Medicine, Children's Health, Dallas, TX, United States*

[6]*Department of Pathology, University of Texas Southwestern Medical Center, Dallas, TX, United States*

Objectives

- List common types of childhood malignancies and their clinical features.
- Identify the key diagnostic features of the common childhood malignancies.
- Discuss ancillary laboratory testing often used for management and differential diagnosis of pediatric cancers.
- Describe the methodology and limitations of biochemical tumor markers routinely used in the management of pediatric cancers.

2.1 Introduction

Cancer is a leading cause of death in children and adolescents in the United States and worldwide [1,2]. In the United States, an

Cancer Biomarkers: Clinical Aspects and Laboratory Determination. DOI: https://doi.org/10.1016/B978-0-12-824302-2.00010-2
© 2022 Elsevier Inc. All rights reserved.

estimated 10,500 individuals under 14 years and 5090 between ages 15 and 19 years were diagnosed with cancer in 2021 [3]. Most children and adolescents survive (84%–85% at 5 years), reflecting the continual decline ($>60\%$) in death rates over the past half century with only a smaller recorded increase in cancer incidence over the same time period [4]. Improved mortality reflects the continuous optimization of standard therapies through clinical trials, in addition to the development of novel targeted therapies, as our understanding of underlying molecular features has evolved [4].

As with adult cancers, laboratory testing plays a conspicuous role in the management of pediatric cancers. Cancer biomarkers are biological molecules (e.g., proteins, nucleic acids) that find utility in cancer screening, risk assessment, therapy selection, prognostic determination, and monitoring [5]. Biomarkers in pediatric cancers overlap to some extent with those of relevance in adult cancers. However, differences exist because of, at least in part, the differing types and proportions of cancers that occur in adults and children. Hematologic malignancies, for example, constitute a greater share of cancer in children compared to adults. Acute leukemia is the most common type of cancer affecting children 0–14 years of age, accounting for a quarter to a third of all malignancies. Neoplasms of the central nervous system (CNS) are next most common ($\sim 27\%$), followed by lymphoma, which, when including both non-Hodgkin and Hodgkin subtypes, accounts for more than 10% of all cases. Meanwhile, some extra-CNS solid tumors, although less common in children than hematologic malignancies and CNS tumors, are almost exclusively cancers of pediatrics (e.g., neuroblastoma and Wilms tumor). In adolescents 15–19 years old, the distributions of cancers overlap with the distributions in younger patients, but the predominance of leukemias diminishes to $<15\%$, with an increasing share of some solid tumors usually associated with adults (e.g., thyroid carcinoma) [4]. In addition to differing types of cancer, the mechanisms (e.g., genetic alterations) underlying given cancer types may vary between children and adults. In this chapter, we review the key tumor classifications in pediatric cancers, focusing on relevant cancer biomarkers, and highlighting key differences with adult cancers.

2.2 Childhood leukemia

Acute lymphoblastic leukemia/lymphoma (ALL, 3000–6000 cases every year in the United States), with a peak incidence between ages 2 and 4 years, is the most common cancer diagnosed in children and represents approximately 25% of cancer diagnoses among patients 15–19 years of age. As noted in the most recent WHO guidelines, ALL is considered a disease of children, with the vast majority (75%) occurring in children under the age of 6 years [6]. Between 1975 and 2010, the 5-year survival rate for pediatric ALL increased from 60% to approximately 90% for children younger than 15 years and from 28% to more than 75% for adolescents aged 15–19 years [7]. Acute myeloid leukemia (AML) also occurs in children and adolescents, but, unlike in adults, it is substantially less common compared to ALL (<1000 AML cases per year in the United States in patients under the age 20 years). In contrast to ALL, the overall 5-year survival of pediatric AML patients has increased from approximately 30% to more than 65%. About half of pediatric AML patients will relapse and die from the disease, and the main treatment issues stem from the high frequency of treatment-related deaths and the long-term side-effects of high treatment doses [8].

ALL and AML originate from immature cells (blasts) in the bone marrow. ALL is the proliferation of blasts arising from precursors to B- or T-lymphocytes. These cancers are predominantly B-cell (85%) lineage, followed by T-cell lineage (~15%), with natural killer (NK) lineage being extremely rare [9]. AML is postulated to arise from a hematopoietic stem cell and can display granulocytic, monocytic, megakaryocytic, and erythroid differentiation. Rapid expansion of disease from within the bone marrow stifles production of other marrow elements (e.g., red bloods, platelets) leading to cytopenias and associated findings such as pallor (from anemia), easy bleeding/bruising (from thrombocytopenia), and infection (diminished normal white blood cells). Blasts can vary from small blasts, with scant cytoplasm and condensed chromatin, to larger cells, with moderately abundant cytoplasm and open chromatin and prominent nucleoli. Cytoplasmic granules are more commonly seen in myeloid blasts; however, they can also be seen in 2%–7% of pediatric B-ALL cases but are exceedingly rare in adult cases [10]. In clinical practices, ALL occurs along a spectrum with what has historically been referred to as lymphoblastic lymphoma (LBL).

LBL is a type of non-Hodgkin lymphoma (NHL) characterized by rapid expansion of lymphoblasts as a mass, often mediastinal, with associated lymph node involvement. The term lymphoma has been used traditionally for nodal disease or extranodal tissue involvement with <25% bone marrow lymphoblasts [11]. The responsible cell population is morphologically indistinguishable from that of ALL, and thus ALL and LBL are typically considered a single disorder (ALL/LBL). The "leukemic" presentation is typically of the B-cell type, with the "lymphoma" presentation more commonly seen in T-cell forms of ALL/LBL. AML arises from a myeloid precursor cell in the bone marrow and can present similarly to ALL/LBL. The presence of ≥20% myeloid blasts in either the peripheral blood or bone marrow is usually required for the diagnosis of AML. An absolute cutoff for percentage of blasts has not been established in the diagnosis of ALL/LBL (unlike AML), although a value of >25% marrow blasts is used to define ALL in many treatment protocols, and the WHO guidelines suggest avoiding the diagnosis when there are fewer than 20% marrow lymphoblasts [6].

2.2.1 Diagnosis

Formal diagnosis of ALL/LBL and AML is based on morphologic and immunophenotypic examination of the blood and bone marrow, with examination of other tissues (e.g., lymph nodes) performed where indicated. Evaluation of the cerebral spinal fluid is also typically evaluated using similar techniques to assess for involvement of the CNS. The biomarkers in hematologic malignancies comprise the surface protein molecules, or cluster of differentiation (CD) markers, which enable classification of WBCs based on lineage and provide a signature, or "immunophenotype," of the cancer cell population moving forward (Table 2.1).

Immunophenotyping is performed by flow cytometry on peripheral blood and tissue aspirates. This technique provides quick,

Table 2.1 Flow cytometry antigen panel.

Lineage	Antigen markers
Hematopoietic precursors	CD34, HLA-Dr, TdT, CD45
Myeloid	CD13, CD33, CD15, MPO, CD117
– Monocytic	CD14, CD64, CD11c, CD36
– Megakaryoblastic	CD41, CD61
– Erythroid	CD71, Glycophorin A
B-lineage	CD19, CD20, CD79a, CD10, Cyto CD22, kappa, and lambda
T-lineage	CD2, CD3, CD5, CD7, Cyto CD3, CD1a, CD4, CD8

multi-parametric analyses of cells in a solution [12]. Cells are first treated with fluorescent probe–linked antibodies against surface markers of interest and run through an analyzer that uses lasers to generate fluorescence and light scatter readouts for individually passing cells, one by one. In turn, this generates detailed phenotypic characterizations of normal and abnormal cell populations, along with their proportions. The limitations of flow cytometry include the need for fresh tissue, limited morphologic correlation, inability to assess nonviable cells, and loss of viable tissue with transport [13].

2.2.2 Cell surface markers

A primary application of immunophenotyping acute leukemia is distinguishing between B- and T-cell forms of ALL/LBL, as well as differentiating ALL/LBL from AML. T-cell ALL/LBL is derived from T-cell progenitors and usually expresses TdT with variable expression of CD1a, CD2, CD3, CD4, CD5, CD7, and CD8. Markers in AML reflect that the blasts can show myeloid, monocytic, megakaryoblastic, and erythroid differentiation (Table 2.1).

Many of these markers reflect surface proteins that are critical to the function of the normal immune cell. CD19, for example, which is expressed by B-cells and plasma cells (most), forms a complex with other surface membrane proteins and plays roles with respect to B-cell development and activation. In addition, CD19 and CD22, another member of the B-cell antigen family, make attractive targets for immunotherapy therapy such as chimeric antigen receptor (CAR) T-cell because normal tissue expression of these antigens is restricted only to the B-cell lineage, and clinical tolerance for B-cell ablation with supportive therapy is high [14]. CAR T-cell therapy, in which the patient's own mature T-lymphocytes are engineered to recognize major histocompatibility complex nonrestricted tumor antigens by transduction of CARs, has been used as a promising treatment of pediatric B-ALL. Initial response rates in patients treated with CAR T-cells have been impressive compared with historical outcomes. However, more recent studies have shown that 50%–60% of patients will not achieve clinical remission or will relapse [15]. Two mechanisms for CAR T-cell treatment failure have been identified and include the loss of target antigen and lack of persistence of CD19 CAR T-cells in patients. Loss of the CD19 and CD22 target antigens have been proposed to be due to possible mechanisms including the development of mutations, alternative splicing of the antigen, antigenic masking, and lineage switching [16]. Further discussion of CAR-T-cell therapy is available in Chapter 8.

2.2.3 Other biomarkers

Additional biomarkers in pediatric acute leukemia include genetic features, including detectable cytogenetic alterations, which constitute elements in formal tumor classification by the WHO. These are assessed using standard cytogenetic techniques, that is, karyotyping, and florescence in situ hybridization (FISH), as well as multicolor FISH. Standard cytogenetic analysis (karyotyping) is able to detect all microscopically detectable chromosomal aberrations occurring simultaneously in leukemic cells, regardless of whether these aberrations are numerical or structural, or, in the case of the latter, balanced or unbalanced. FISH employs cDNA or genomic DNA fragments (probes) complementary to specific sequences in the human genome that are labeled with fluorochromes and hybridized to fixed metaphase chromosomes and/or interphase nuclei and can detect recurrent structural abnormalities such as translocations, inversions, or deletions. Meaningful results of standard cytogenetic

Table 2.2 Major chromosome aberrations in pediatric B cell acute lymphoblastic leukemia.

Chromosomal change	Frequency	Risk category
Normal karyotype	31%–41%	Relatively favorable
t(12;21)(p13;q22); *ETV6-RUNX1*	25%	Favorable
Trisomy 4 and 10	25%	Favorable
t(1;19)(q23;p13); *TCF3-PBX1*	6%	Adverse
t(v;11q23)/*KMT2A*	80% of infantile ALL	Adverse
t(9;22)(q34;q11); *BCR-ABL1*	2%–4%	Adverse
Hypodiplody (<45 chromosomes)	5%	Adverse
t(5;14)(q31.1;q32.1); *IGH-IL3*	< 1%	Standard
iAMP21	2%	Adverse
BCR-ABL1-like	10%–25%	Adverse

analysis were obtained in 83% and 91% cases of childhood in two large studies compared to between 70% and 75% of samples analyzed in adult ALL studies. Among the successfully analyzed cases, one or more clonal aberration was detected in 60%–79% of adult cases, and 57%–82% of pediatric ALL cases [17].

The Third International Workshop on Chromosomes in Leukemia was the first major study to demonstrate the independent prognostic significance of cytogenetic findings in diagnosing B-ALL. Subsequent studies have confirmed these findings and have added data on clinical relevance of additional recurrent aberrations as well as elucidated the molecular basis and biological consequences of many of these aberrations [17]. Identified alterations have significant prognostic and treatment implications and often implicate specific disturbances in normal cell biology with respect to proliferation/survival (mutations of specific genes, e.g., FLT3) or impairment of differentiation (translocations) [18].

Table 2.2 summarizes the incidence and associated prognosis of the major cytogenetic findings in childhood B-ALL. Table 2.3 enumerates the genes that are frequently mutated in specific subgroups

Table 2.3 Genes commonly mutated in acute lymphoblastic leukemia.

Disease	Gene	Function
B-ALL or T-ALL	*CDKN2A/2B*	Cell cycle modulator
B-ALL or T-ALL	*PAX5*	B-cell differentiation
B-ALL or T-ALL	*EBF*	B-cell differentiation
B-ALL or T-ALL	*RB1*	Cell cycle modulator
T-ALL	*NOTCH1*	T-cell differentiation
B-ALL	*ETV6*	Transcription factor
B-ALL	*NRAS*	Ras pathway
B-ALL	*KRAS*	Ras pathway
B-ALL	*IKZF1*	B-cell differentiation
B-ALL	*PTPN11*	Ras pathway
B-ALL	*E2-2*	B-cell differentiation
B-ALL	*FLT3*	Ras pathway

of ALL. As noted earlier, prognostic factors and risk classification differ in T-ALL compared with B-ALL, and blast cytogenetics features are not presently used for risk stratification. The frequency and prognosis of common cytogenetic abnormalities in pediatric AML are summarized in Table 2.4. Notably, genetic changes within leukemic populations can be heterogeneous, and some genetic alterations are shared between ALL and AML.

During treatment for leukemia, patients undergo re-evaluation of disease at specified time intervals with respect to the start of therapy. Reassessment by flow cytometry is required to ascertain the presence of minimal residual disease (MRD). The key prognostic determinant in B-ALL and T-ALL is MRD response at the end of induction and at a second point at the end of the consolidation treatment phase [19]. MRD is also a key prognostic indicator for AML. Long-term monitoring also requires reassessment for the previous cancer population, looking for cells with previously identified phenotype. Additional laboratory testing is performed before and during the treatment of leukemia. Both leukemia and its treatment may incur complications related to renal and hepatic function, as well as bone marrow, and thus ongoing testing is required with respect to

Table 2.4 Frequency and prognosis of common cytogenetic abnormalities in pediatric acute myeloid leukemia (AML).

Chromosomal change	Frequency	Risk category
Normal karyotype	20%–26%	According to cryptic cytogenetic abnormalities or mutations
t(1;22)(p13;q13); *RBM15-MLK1*	0.3%	Intermediate
inv(3)(q21q26) / t(3;3)(q21;q26); *GATA2; MECOM (EVI1)*	2%	Adverse
-5 / del(5q)	1.2%	Adverse
t(6;9)(p23;q34); *DEK-NUP214*	1%–2%	Adverse
−7	3%	Adverse
del(7q)	3%	Intermediate
t(7;12)(q36;p13); *ETV6; MNX1*	1%	Adverse
+8	10%–14%	Adverse
t(8;16)(p11;p13); *KAT6A-CREBBP*	<1%	Intermediate
t(8;21)(q22;q22); *RUNX1-RUNX1T1*	12%–15%	Favorable
del(9q)	0%–4.6%	Adverse
t(9;22)(q34;q11.2); *BCR-ABL1*	0.6%	Adverse
t(11q23)/ *KMT2A*	16%–21%	Adverse or intermediate
t(9;11)(p22;q23); *KMT2A-MLLT3*	6%–9%	Intermediate
t(15;17)(q22;q21); *PML-RARA*	6%–10%	Favorable
inv(16)(p13q22) / t(16;16)(p13;q22); *CBFB-MYH11*	5%–9%	Favorable
t(16;21)(p11;q22); *FUS-ERG*	0.4%	Adverse

electrolytes, creatinine, liver enzymes, cell counts, and coagulation parameters.

Several non-specific biomarkers provide measures of tumor burden in leukemias and other cancers. Lactate dehydrogenase (LDH) is a potential biomarker with this utility described in a number of contexts, in and out of pediatrics [20]. It has broad tissue expression because of its involvement with glucose metabolism. Pui et al. performed a study of 293 children with ALL (so-called "standard risk" disease), with LDH performed at the time of diagnosis. LDH level was associated with some indicators of disease burden (e.g., leukocyte count, severity of thrombocytopenia) but not others. However, patients with the highest LDH levels such as >1000 U/L, were most likely to fail treatment [21].

2.2.4 Treatment

Treatment of ALL/LBL, as with other cancers, consists of combination chemotherapy separated into stages (induction, consolidation, and maintenance). Monitoring is critical to ensure adequate dosage but also prevent toxicities. Methotrexate (MTX) is a folic acid antagonist that appears to act through multiple mechanisms in different therapeutic contexts but is often given in the treatment of leukemias. Each potential adverse effect requires separate monitoring, often facilitated by laboratory testing. When treating with high-dose MTX, monitoring of the drug itself is needed due to narrow therapeutic window and broad range of potentially fatal toxic effects. Specifically, blood levels can be measured to evaluate for potential rescue by leucovorin. Several assays are available to measure MTX. Immunoassays are fast and cheap but subject to interference from metabolites (deoxyaminopteroic acid) and other compounds, including from the administration of rescue agents (glucarpidsae) [22]. These interferences differ from assay to assay.

The use of liquid chromatography-tandem mass spectrometry (LC–MS/MS) overcomes these limitations but is subject to other challenges, particularly relating to the speed of assay. First, samples must undergo preparation constituting protein precipitation and/or solid phase extraction. Second, quantitation requires calibration, typically performed using external standards. Because these standards must be prepared alongside a batch, throughput is hampered. A potential workaround is utilization of multipoint internal standardization [22]. This is a modification of the isotope-dilution approach using multiple, differentially labeled internal standards. The standards have different isotopic labeling corresponding to different masses.

2.3 Childhood lymphoma

Lymphoma is the third most common childhood malignancy (prevalence rate 10%–12%) followed by acute leukemia and CNS tumors [23]. There are two types of clinicopathologic lymphoma: Hodgkin lymphoma (HL) and NHL, each with distinct clinical subtypes. HL accounts for approximately 38% of lymphoma cases in children (ages 0–14 years) and approximately 65% of cases in adolescents (ages 15–19 years), while NHL accounts for 62% of lymphoma cases in children and approximately 35% of lymphoma cases in adolescents [24].

2.3.1 Hodgkin lymphoma

HL can be divided into two biologically distinct variants: classic HL (cHL) and nodular lymphocyte-predominant HL (nLPHL). There are four subtypes of cHL: nodular sclerosis (most frequent in older patients [> 10 years of age]), mixed cellularity (more frequent in younger children [< 10 years of age]), lymphocyte rich (rare in pediatric patients), and lymphocyte depleted (also rare in pediatric patients) [25]. cHL is characterized by large binucleate (Reed–Sternberg) or multinucleated neoplastic cells or mononuclear variants (collectively termed Hodgkin Reed–Sternberg [HRS] cells) with large macronucleoli. Despite being of B-cell lineage, the HRS cells are only weakly positive for some B-cell cell markers such as PAX-5 and predominantly negative for others such as CD20 and CD79a. The malignant HRS cells are positive for CD30 and CD15 and constitute <1% of the tumor bulk. Most of the tumor mass is composed of a background of non-neoplastic T-cells, B-cells, eosinophils, histiocytes, and plasma cells.

The two subtypes of cHL most likely to present in pediatric patients are the nodular sclerosis and mixed cellularity subtypes. In nodular sclerosis cHL, the lymph node shows thickened capsule with a nodular infiltrate composed of mixed inflammatory cells admixed with the large HRS cells with sclerotic collagen bands separating the nodules. This subtype affects 70% of children with cHL. Mixed-cellularity cHL presents with the inflammatory cell background and scattered admixed HRS cells without the sclerotic bands. This subtype is most commonly diagnosed in children younger than 10 years of age. The histologic subtypes have some prognostic implications with the mixed-cellularity subtype appearing to have a better outcome.

NLPHL accounts for only 5%–10% of childhood HL and is recognized as a distinct entity by the WHO [26]. The large, atypical cells with lobed nuclei in NLPHL, referred to as lymphocyte-predominant (LP) cells, are typically sparse and smaller than HRS cells and lack the macronucleolus. They express B-cell markers such as CD20 + , CD79a, Pax5, Oct-2, and Bob-1; cHL markers CD30 and CD15 are negative. Lymphoma typically shows a nodular growth pattern with macrofollicles expanded by follicular dendritic cells. The LP cells are rimmed by T helper cells that express CD3, CD4, and CD57.

2.3.2 Non-Hodgkin lymphoma

NHL is a heterogeneous group of neoplasms that accounts for approximately 7% of all pediatric malignancies. The neoplasms originate from either immature (lymphoblastic) or mature B, T, or NK cells. The features of LBL have previously been discussed in the prior section. Burkitt lymphoma (BL), diffuse large B-cell lymphoma (DLBCL), and anaplastic large-cell lymphoma (ALCL) are the most common NHL types in the pediatric population (Table 2.5). BL and DLBCL constitute the majority of non-lymphoblastic B-cell lymphomas in the pediatric population.

2.3.3 Markers

For evaluation and initial workup of pediatric lymphomas, it is recommended that an excisional biopsy be performed. Although

Table 2.5 Pediatric lymphoma incidence (per 100,000 person-years).

	<5 years	5–9 years	10–14 years	15–19 years
Hodgkin lymphoma	<1	2–6	9–12	28–31
Non-Hodgkin lymphoma				
Burkitt lymphoma	0.8–3.2	1.1–6	0.8–6.1	1.2–2.8
Lymphoblastic	0.9–1.6	1–2.2	0.7–2.8	0.9–2.2
Diffuse large B-cell lymphoma	0.6	< 1–1.2	1.4–2.5	4.9–6.1
Anaplastic large cell lymphoma and other	1.5–2.3	1.6–3.3	2.8–4.3	3.4–7.8

lymphoma is traditionally regarded as a malignancy that predominantly involves the lymph nodes, it may also involve primary lymphoid tissues (bone marrow and thymus) or various secondary lymphoid tissues other than lymph nodes (spleen, mucosa-associated lymphoid tissue) or nonlymphoid organs (skin, bone, brain, lungs, liver, salivary glands, etc.). Lymphoma cannot be distinguished from benign conditions solely on the basis of clinical presentation. The immunohistochemical markers used in the evaluation of tissue samples are summarized in Table 2.6.

The workup should include a thorough history and physical examination, including determination of one or more B symptoms (unexplained recurrent fevers of at least 100.4 F within the last month, drenching night sweats within the last month, weight loss of 10% of body weight within 6 months of diagnosis), and examination of lymphoid regions and spleen [26]. Other essential components of the workup include standard laboratory tests such as CBC, serum LDH, alkaline phosphatase, uric acid (UA), C-reactive protein levels, and erythrocyte sedimentation rate. Pediatric patients newly diagnosed with lymphoma may present with life-threatening conditions resulting from either the mass effect of the tumor or metabolic derangements due to rapid tumor cell breakdown in the settings of compromised renal function. The latter is also known as tumor lysis

Table 2.6 Immunophenotypic markers in common pediatric lymphomas.

Marker	cHL	NLPHL	DLBCL	BL	ALCL-ALK +	ALCL-ALK−
CD20	−/+	+	+	+	-	-
CD79a	−/+	+/−	+/−	+	-	-
PAX-5	+ (weak)	+	+	+	-	-
MUM-1	+	−/+ (weak)	Variable	−/+	-	-
BLC-2	Variable	−/+	+/−	-	-	-
BCL-6	+/− (weaker, fewer cells +)	+	+/−	+	-	-
CD10	-	-	+/−	+	-	-
CD15	+	-	-	-	-	-
CD30	+	−/+ (usually <10%)	-	-	+	+
EMA	−/+	+/−	−/+	-	+/−	+/−
ALK-1	-	-	-	-	+	-
CD3	-	-	-	-	−/+	−/+
CD2	-	-	-	-	−/+	−/+
Perforin/ granzyme B	-	-	-	-	+	+

syndrome (TLS). TLS is a life-threatening condition that can be seen in patients diagnosed with neoplasms with a high proliferation rate and large tumor burden such as BL, DLBCL, or B-cell LBL.

While these components are normally excreted in the urine, the excretory capacity of the kidneys may be overwhelmed in patients with bulky lymphomas. Laboratory TLS should be differentiated from clinical TLS. Laboratory TLS is defined as asymptomatic metabolic disarray and includes two or more of the following laboratory abnormalities: hyperuricemia [>8.0 mg/dL (>475.88 mmol/L)]; hyperkalemia (>6.0 mmol/L); hyperphosphatemia (>4.5 mg/L); and/or hypocalcemia [corrected calcium <7.0 mg/dL (<1.75 mmol/

L), ionized calcium <1.12 mg/dL (<0.28 mmol/L)]. These must occur simultaneously within 3 days prior to and up to 7 days after therapy onset. The hypocalcemia results from an excess of serum UA and phosphorus precipitating calcium into calcium urate and calcium phosphate crystals in the kidneys. Clinical TLS is defined as the laboratory findings along with clinical evidence of end-organ damage.

These can include neuromuscular symptoms secondary to hypocalcemia, uremia, increased creatinine level, oliguria, hypertension, cardiac dysrhythmia, pulmonary edema, altered mental status, seizures, and death. Serum LDH levels can be used as the surrogate marker for tumor bulk. Indeed, patients with LDH levels more than twice the upper limit of normal are at a high risk for TLS [25].

2.3.4 Treatment

In addition to conventional chemotherapy and radiation therapy, biological therapy has advanced lymphoma treatment. Lymphoma cell antigen–specific monoclonal antibodies and cytotoxic T-cells, as well as small molecules that target intracellular signaling pathways, have been developed in recent decades and are in use. These include monoclonal antibodies that target CD19 [denintuzumab mafodotin (CD19a antibody-cytotoxic drug conjugate); blinatumomab (CD19/CD3 bispecific T-cell engager)], anti-CD20 monoclonal antibodies (rituximab, ofatumumab, obinutuzumab), and anti-CD30 monoclonal antibody (brentuximab vedotin antibody-cytotoxic drug conjugate) are used as either single agents or in combination with conventional chemotherapy [27]. CAR cytotoxic T-cells, which are capable of recognizing and selectively killing CD19-, CD20-, and CD30-expressing lymphoma cells, were found to be most effective in the setting of MRD and are used for treatment of relapsed lymphomas that are resistant to conventional therapies.

2.4 Brain and central nervous system tumors (gliomas)

The most common pediatric brain tumors are pilocytic astrocytoma and medulloblastoma. Pilocytic astrocytoma is a low grade (WHO grade 1) glial neoplasms that make up almost a fifth of all pediatric

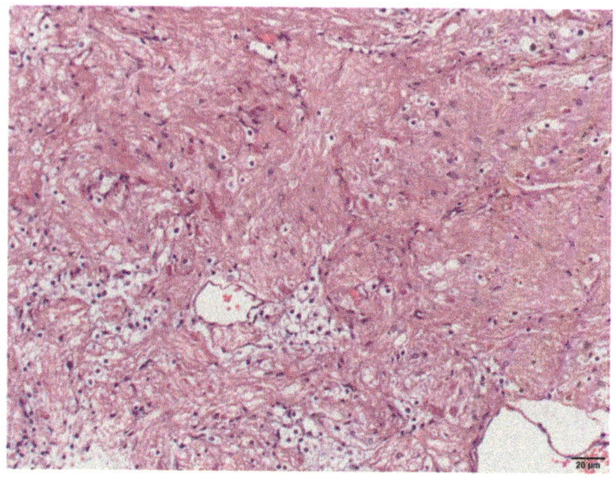

FIGURE 2.1

A pilocytic astrocytoma with bland neoplastic astrocytes and scattered Rosenthal fibers.

primary brain tumors [28,29]. It is a slow-growing tumor that occurs infratentorially in children. Depending on the precise location, it presents with macrocephaly, headache, increased intracranial pressure, endocrinopathy, visual deficits, and proptosis. Histologically, pilocytic astrocytoma is sparsely to modestly cellular and often biphasic with neoplastic glial cells embedded in loose areas with microcysts and compact areas with Rosenthal fibers (Fig. 2.1). Approximately 85% of pilocytic astrocytomas area now known to harbor genetic changes in *BRAF*, most commonly BRAF-KIAA1549 gene fusion, which can be detected by fluorescence in situ hybridization [30].

Medulloblastma is the most common embryonal tumor of the CNS and the most common pediatric malignant brain tumor (WHO grade 4). It is characterized by sheets of small, round, mitotically active cells that show no obvious differentiation on hematoxylin and eosin stain. Histomorphologically, medulloblastomas may be large cell/anaplastic (Fig. 2.2), nodular/desmoplastic, extensively nodular, or classic. The histologic features have prognostic significance with large/cell anaplastic medulloblastomas being aggressive tumors and desmoplastc/nodular medulloblastomas and medulloblastomas with extensive nodularity having an excellent outcome [31].

More recently, transcriptome and methylation profiling has resulted in incorporation of molecular signatures into the

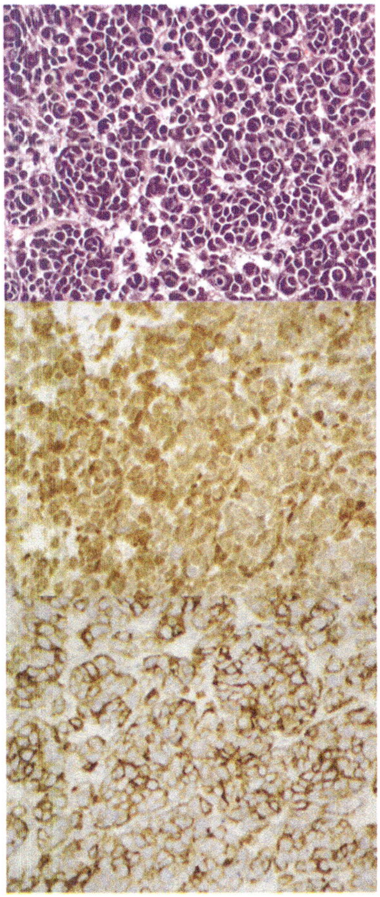

FIGURE 2.2

Large cell/anaplastic medulloblastoma with SHH activation indicated by nuclear staining for YAP1 and cytoplasmic staining for GAB1.

medulloblastomas classification [32]. For routine clinical practice, immunohistochemical biomarkers serve as surrogates for the molecular classes [33]. Thus medulloblastomas with immunostaining profile YAP1 +, GAB1 +, beta-catenin non-nuclear are sonic hedgehog (SHH) pathway–activated, while medulloblastomas with nuclear beta-catenin staining are Wingless-related integration site (Wnt) pathway activated. The remaining medulloblastomas are classified as non-Wnt/non-SHH. Abnormal immunostaining patterns for P53 (nuclear overexpression or complete absence of staining) characterize medulloblastomas with *TP53* mutations. A combination of molecular and histomorphologic features are used for prognostic subgrouping and guiding therapy [34].

2.5 Neuroblastoma

Neuroblastomas account for approximately 8% of all childhood cancers and is the most common malignancy diagnosed in the first year of life, with >80% of cases occurring in early childhood before the age of 4. They fall within the category of monoamine-producing tumors. Neuroblastoma tumors develop from neural crest cells originating from embryonal neuroblasts of the sympathetic peripheral nervous system. The tumors can occur in the adrenal medulla or anywhere along the sympathetic chain [35]. Presenting symptoms vary depending on the sight and aggressiveness of the primary tumor. Patients can present with pain from abdominal distension, hepatomegaly, neck mass, limping or abnormal gait due to compression of the spinal cord, weight loss and/or anemia. The hepatomegaly can cause urinary or bowel obstructions and respiratory distress in infants. Common sites for metastasis include liver, bone marrow, and bone. Metastatic disease can cause proptosis, periorbital ecchymosis, bone pain, or skin lesions. Neuroblastoma has also been associated with opsoclonus myoclonus ataxia syndrome (OMS), a paraneoplastic inflammatory disorder that presents with ataxia, opsoclonus, myoclonic jerks, irritability, and behavioral changes [36].

The MYCN oncogene is amplified in approximately 20% of neuroblastomas [37,38]. MYCN amplification is associated with rapid tumor progression and poor prognosis. MYCN is a transcription factor that modulates other downstream genetic targets and pathways with proliferative functions. Notably, these targets include *Mouse Double Minute 2 Homolog* (*MDM2*), ornithine decarboxylase, and mTOR, all of which can be targeted therapeutically.

Rare cases of familial form for neuroblastoma have been described and account for <2% of all cases. Two tumor susceptibility genes paired-like *homeobox 2b (PHOX2B)* and *anaplastic lymphoma kinase (ALK)* have been liked to familial forms of neuroblastomas, with the latter representing a majority of the familial neuroblastoma cases

[39,40]. *ALK* mutations are therapeutically actionable and are found in 15% of sporadic neuroblastomas. Activating mutations in the *ALK* gene also cause other malignancies such as large cell lymphoma, non-small cell lung cancer, and squamous cell carcinoma. Mutations in the *ATRX* gene, which encodes for a helicase, are most common in children and young adults affecting 44% of patients over 12 years, who have a poor prognosis [41].

Diagnosis of neuroblastoma is confirmed by histologic determination of the primary or metastatic tumor. Additionally, radiographic and functional imagining can assist in detecting the tumors. Neuroblastic tumors are histologically subclassified on the basis of the presence of neuropil, the percentage of ganglionic differentiation, and the percentage of Schwannian stroma into undifferentiated neuroblastoma, poorly differentiated neuroblastoma, differentiating neuroblastoma, intermixed ganglioneuroblastoma, and ganglioneuroma. In addition, nodular ganglioneuroblastoma is a neuroblastic tumor composed of multiple tumor clones that may have different histologic appearances (Fig. 2.3).

The tumors are further stratified histologically into favorable and unfavorable histopathology according to the above diagnostic categories, patient's age, and mitosis-karyorrhexis index [42,43]. Immunohistochemical stains are typically not needed, except for undifferentiated neuroblastomas where positivity for synaptophysin, neuron-specific enolase (NSE), or tyrosine hydroxylase is helpful in

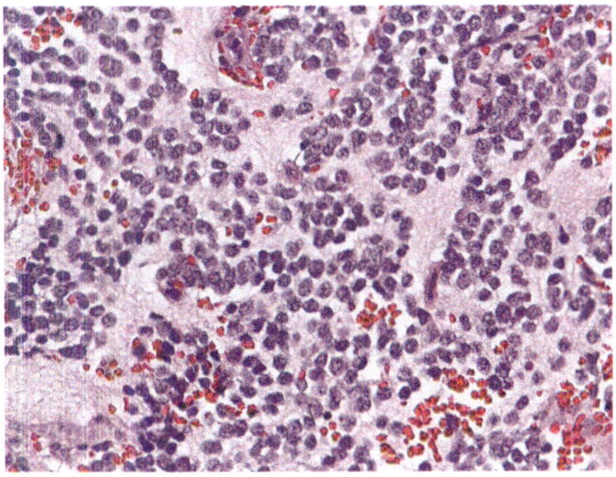

FIGURE 2.3

Poorly differentiated neuroblastoma.

establishing neuroblastic differentiation. Urinary vanillylmandelic acid (VMA) and homovanillic acid (HVA) are the primary biochemical markers used for diagnosis. However, these markers can be low in early stages when the primary tumor is small. HVA and VMA are metabolites of norepinephrine and dopamine and are elevated in more than 90% of neuroblastoma cases. Several studies have investigated the utility of screening programs for infants to detect undiagnosed neuroblastomas and have found that screening does not result in decreased mortality. Many of the tumors identified in these studies were early-stage tumors that spontaneously regressed [44]. Patients who present with diarrhea may have elevated vasoactive intestinal peptide (VIP) secreted by the tumor [45].

HVA and VMA are excreted in high concentrations in most patients with neuroblastoma and can be measured using either gas chromatography, high-pressure liquid chromatography, or mass spectrometry (either GC–MS or LC–MS/MS) [46,47]. HVA and VMA concentrations in random urine collections correlate well with 24-hour collections, making urine collections easier in young children and infants. Typically, HVA and VMA concentrations are normalized to creatinine concentrations to allow for ease of interpretation and sample collection. Diet and certain medications such as Bactrim (sulfamethoxazole and trimethoprim) can result in elevations of HVA. Serum HVA/VMA measurements have been shown to be less sensitive than urine [48].

Reference intervals for HVA and VMA can vary depending on the method of analysis. Both HVA and VMA increase with age throughout childhood; however, values corrected with creatinine decline with age due to increased excretion of creatinine during growth [49]. A high ratio of HVA to VMA results from reduction in tumor cell conversion of dopamine to norepinephrine [50]. This metabolic pattern has been associated with aggressive tumors. However, the use of HVA and VMA measurement for treatment monitoring is limited and follow-up of patients after initial treatment requires tumor imaging and clinical assessment [51].

Ferritin, an acute phase reactant, is secreted by neuroblastoma cells, and increase in ferritin is also associated with other cancers suck as leukemia and breast cancers. Interestingly, ferritin secreted by tumor cells differs biochemically and has altered glycosylation and electrophoretic properties [52]. Though elevated levels of ferritin can be found in other disease states and becomes elevated with blood cell transfusions and chemotherapy, it is indicative of more severe disease. LDH also becomes elevated in patients with neuroblastoma and can be useful for monitoring disease; however, it is also not tumor-specific. Elevated serum NSE is also seen in neuroblastomas as well as in other cancers such as ALL and NHL.

Owing to the limitations of both urinary and serum biochemical markers, diagnosis and monitoring of neuroblastoma rely heavily on imaging from ultrasound and CT to MRI and PET, which are useful for confirming the presence of disease as well as monitoring cancer progression and treatment. ^{131}I-Meta-iodobenzylguanidine (MIBG) is a radioactive tracer for the cell membrane of catecholamine transporter and vesicular storage pathways and is useful for visualizing metastatic spread. From a biosafety perspective, it is important that laboratories have a risk assessment and appropriate protocols for handling samples drawn from patients who have received radioactive ^{131}I MIBG.

2.6 Wilms tumor

Wilms tumor is the most common childhood genitourinary tract cancer and the third most common solid tumor of childhood, comprising 95% of all kidney cancers and 6% of all cancers diagnosed in children younger than 15 years and presenting as an abdominal mass [53,54]. It is characterized histologically by triphasic histology comprising blastemal (undifferentiated mesenchyme), epithelial (tubular and glomerular structures), and stromal (fibroblastic, rhabdomyoblastic) elements (Fig. 2.4). Immunohistochemistry plays no role in the histopathologic diagnosis of classical appearing Wilms tumors.

Wilms tumors are classified histologically into favorable and unfavorable histology on the basis of the absence or presence, respectively, of anaplasia. The latter predicts poor response to chemotherapy, is histologically defined by the presence of large hyperchromatic nuclei and atypical multipolar mitotic figures, and correlates with TP53 mutations (Fig. 2.5) [55,56].

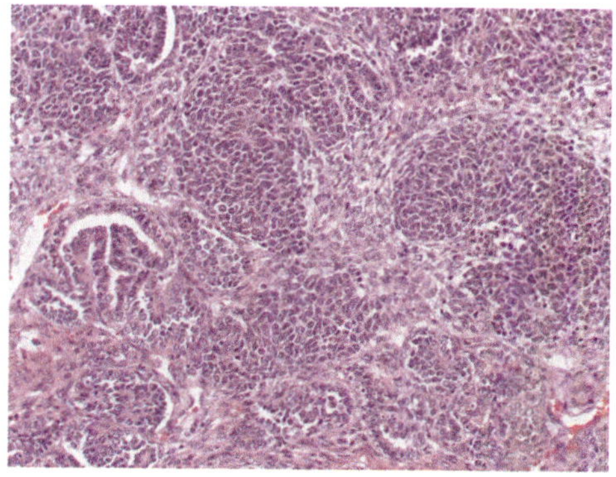

FIGURE 2.4

Wilms tumor with triphasic histology.

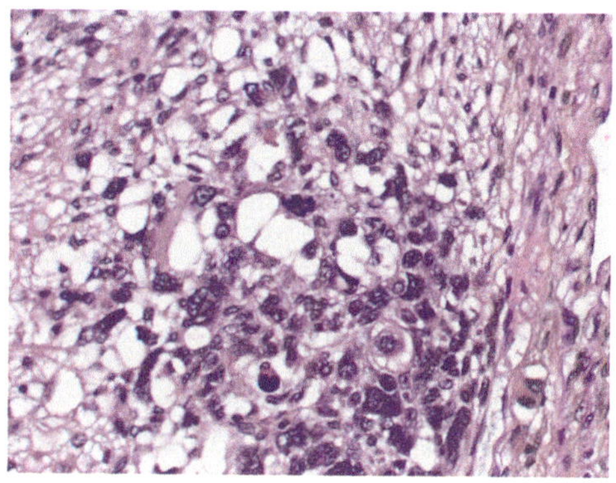

FIGURE 2.5

Wilms tumor with anaplasia.

Loss of heterozygosity at 16q and 1p are associated with poor outcomes in favorable histology Wilms tumors [57,58]. Gain of chromosome 1q is a stronger predictor of poor outcomes in patients with favorable histology [57,59]. Recent studies have expanded the genetic landscape of Wilms tumors beyond mutations in CTNNB1, WT1, and AMER1 (WTX) to include mutations in components of microRNA biogenesis such as DROSHA and DICER1, and mutations in genes involved in kidney development such as SIX1/2 [60–63]. A small subset of Wilms tumors occurs in the context of a genetic syndrome such as Beckwith–Weidemann, Denys–Drash,

Simpson–Golabi–Behmel, and Perlman. In addition, familial Wilms tumors are associated with germline mutations in WT1, CTR9, REST, TRIM28, FBXW7, NYNRIN, and KDM3B [64]. At present, the known genomic drivers of Wilms tumors play no role in treatment, and there are no circulating or urinary biomarkers that are used clinically for diagnosis or follow-up of patients with Wilms tumors.

2.7 Hepatoblastoma

Hepatoblastoma is a rare type of childhood cancer accounting for about 1% of all pediatric cancers. However, it is the most common liver malignancy in children [65]. Hepatoblastoma affects infants to children <5 years and usually present with an abdominal mass. Patients are usually otherwise asymptomatic at diagnosis. Other symptoms may include weight loss, decreased appetite, vomiting, abdominal pain, jaundice, fever, anemia, and back pain due to compression of the tumor. The tumor originates from immature liver precursor cells or hepatoblasts. They are categorized on the basis of their level of cell differentiation by histological subtyping as epithelial type and mixed epithelial–mesenchymal elements including bone, cartilage, or skeletal muscle tissue. The epithelial components resemble fetal liver cells, mature liver cells or bile duct cells (Fig. 2.6). Histologic classification is important for predicting prognosis [66]. The most mature, well-differentiated fetal histology with the lowest mitotic index has the best prognosis with local disease that can be easily resected surgically. The pure fetal subtype has close to 100% survival rate. Overall, an 80% survival rate for hepatoblastoma can be achieved if the tumor is completely resected. Adjuvant chemotherapy is also administered to patients. Subtypes with embryonal histology have poor prognosis with greater risk for invasion into multiple liver segments, lymphatics and greater risk of metastasis [67]. Common sites of metastasis include the lung, abdomen and abdominal structures, but rarely to the bone, bone marrow, and CNS.

Only a limited number of cancer-causing genetic aberrations have been identified for hepatoblastoma [68]. A WES study revealed 1–7 mutations/case, many of which affect the Wnt signaling pathway. Almost 90% of cases contain mutations or deletions in

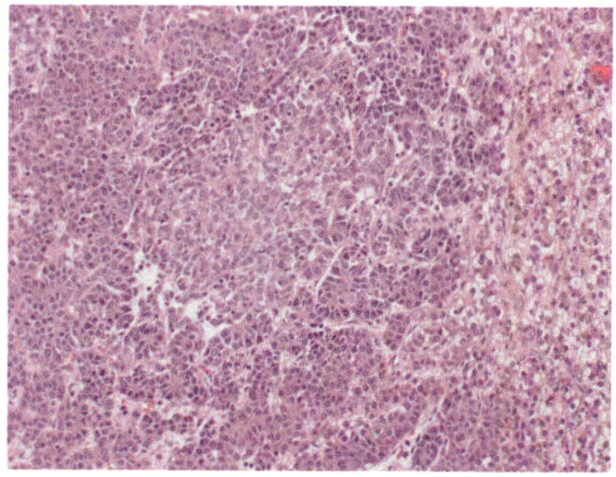

FIGURE 2.6

Hepatoblastoma with embryonal and fetal histology.

CTNNB1, *APC*, or *AXIN1*. Mutations in *NFE2L2*, which encodes for Nrf2, have been found in a small group of hepatoblastoma cases. Nrf2 is a transcription factor with target genes involving the antioxidant response. Loss of heterozygosity on chromosome 11p is also found commonly in 30% of patients with hepatoblastoma and is associated with Beckwith–Wiedemann syndrome.

α-Fetoprotein (AFP) is elevated in over 80% of hepatoblastoma cases, and AFP-negative tumors are associated with a poor prognosis [69]. AFP is also elevated in yolk sac tumors, teratomas, and some embryonal carcinomas. Being an oncofetal antigen, healthy neonates and premature infants show large elevations in AFP compared to adults. AFP decreases to baseline during the first 12–24 months of life and is re-expressed at high levels in malignant conditions. Thus, in preterm infants and newborns with AFP-secreting malignancy, interpretation of total AFP concentrations becomes difficult as it is often unclear whether decline in total AFP concentrations are due to increasing age or response to treatment [70,71].

AFP is a 68-kDa glycoprotein synthesized in the fetal yolk sac, liver, and intestine and processed into a mature 591 amino acid protein. AFP contains a single biantennary polysaccharide chain N-linked to asparagine-232 of the protein. Glycosylation variants characteristic of malignant and benign tissue have been described [72].

A subset of hepatoblastoma cases (18%) secrete hCG and most of these patients develop precocious puberty [73,74]. In these cases, virilization occurs because of hCG-induced testosterone production, and consequently, serum testosterone levels are also elevated. After treatment, hCG and testosterone levels normalize.

2.8 Germ cell tumors

Germ cell tumors (GCTs) are rare, accounting for 3%–4% of pediatric malignancies but with an increasing percentage with age (15%–16% of adolescent malignancies) [75]. Overall incidence is about 900 new cases per year in the United States. GCTs may occur in gonadal and extragonadal sites including retroperitoenum, mediastinum, and CNS.

GCTs encompass heterogenous benign and malignant neoplasms with five primary histologic subtypes—teratoma, yolk sac tumor, germinoma (seminoma in males, dysgerminoma in females), embryonal carcinoma, and choriocarcinoma (Fig. 2.7). Of note, tumors can contain mixed histology, with corresponding impact on measured tumor markers. Peak incidences vary with the underlying subtype. Teratomas, despite being rare overall, are the most common neoplasm of the newborn. Before puberty, yolk sac tumors (malignant, by definition) and teratoma predominate in males with peak age of onset at 0–4 years. In females, benign teratomas predominate. Clinical presentation depends on the site of origin, such as painless swelling of the testicle, or pelvic mass in females. The production of beta-HCG can produce phenotypic effects. Because extragonadal tumors (mainly localized to sacrococcygeal or cervical areas) are common, associated effects may be observed. The underlying molecular drivers, predisposing genetic syndromes, and environmental factors behind GCTs are varied and complex [76].

Potential serum markers of GCT include hCG, AFP, and LDH [77]. These markers have potential utility with respect to diagnosis, risk assessment, prognostication, and monitoring response to therapy. hCG and AFP are tumor markers of most nongerminoma GCTs. hCG belongs to a family of glycoprotein hormones consisting also LH and FSH, all of which share the same alpha subunit. The beta subunits determine the biological activity and are unique to each hormone. hCG may also be elevated in the setting of

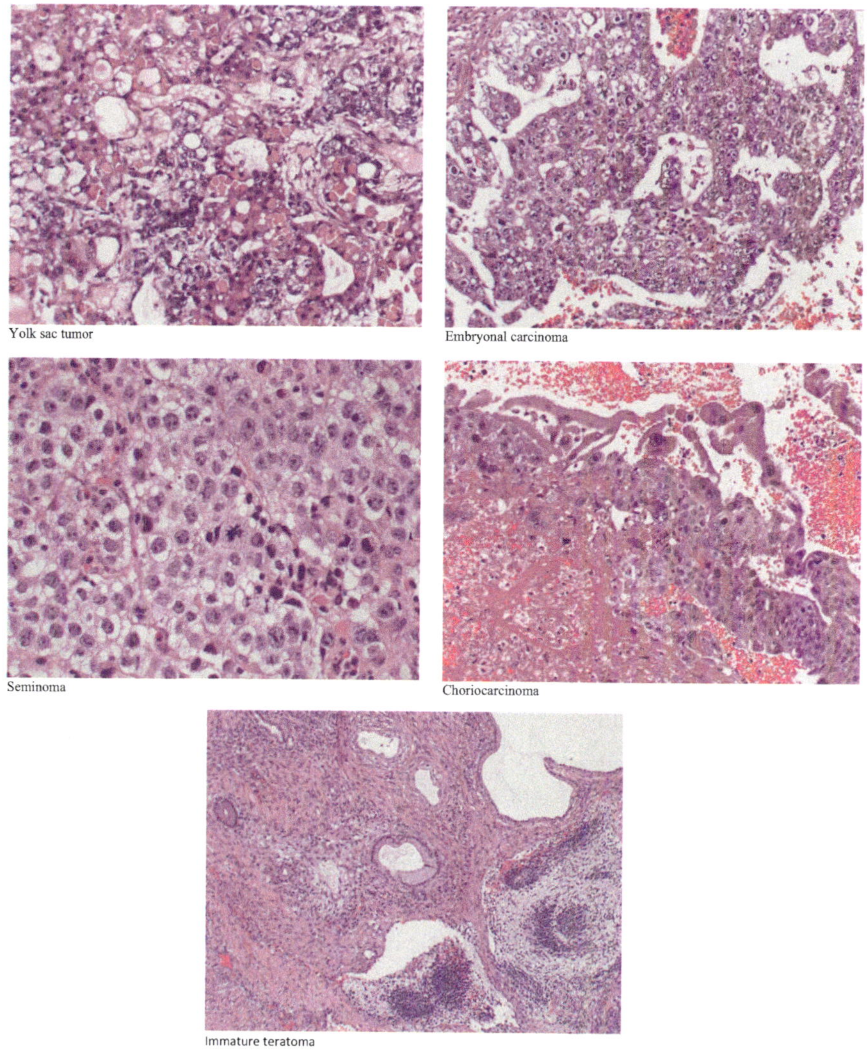

FIGURE 2.7

Germ cell tumors: yolk sac tumor, seminoma, embryonal carcinoma, choriocarcinoma, and immature teratoma.

nongerminomas but not pure germinomas. Assays for the measurement of hCG measure either intact hCG, or total hCG (total beta hCG), which measure intact hCG and free hCGbeta. Most clinical laboratories use the latter type of hCG assay and intact hCG assays should not be used outside of pregnancy detection as some cancers only secrete hCGbeta [78]. hCG has a half-life of about 1.5 days and prolonged half-life >3.5 days during treatment is associated with recurrence and adverse prognosis. hCGbeta has a longer half-life than intact hCG. Changes in hCG concentrations following treatment also depend on stage of disease. Elevations of hCG are also possible because of non-neoplastic conditions including hypogonadal states, while artifactual elevations may occur in the presence of heterophile antibodies. hCG is expressed in a variety of GCTs which includes 20% of seminomas, 50% of nonseminomatous GCTs, dysgerminomas, in which hCG expression may result in precocious puberty in young girls.

AFP (normal half-life ~ a week) is normally produced by the yolk sac. Extreme elevations (> 10,000 ng/mL) are relatively specific for underlying malignancy. Elevation AFP in the context of GCT is consistent with yolk sac tumor or embryonal carcinomas. Yolk sac tumors rarely express hCG. AFP levels correlate with stage.

LDH, as in leukemia and other disorders, can reflect tumor burden in GCT, as evidenced by multiple studies, and provide a prognostic indicator for GCT. Specific isoenzymes show greater specificity, in particular LDH-1, which may be relatively increased in the setting of seminoma or ovarian dysgerminoma [77].

2.9 Conclusions and future directions

Pediatric malignancies individually are rare, and often the types of cancers have little overlap with those seen in adults. Leukemias, lymphomas, gliomas/medulloblastoma, neuroblastoma and Wilms tumors are the most common types of cancer in children. In the United States, more than 80% of children with malignancies are cured. Cancer survival rates are much higher in children compared to adults. Early diagnosis and prompt treatment is vital for reducing cancer burden and improving outcomes. Currently, childhood cancers cannot be detected or prevented through screening or surveillance programs. Early diagnosis is primary based on awareness of symptoms by family and primary care providers, which is crucial for timely diagnosis, staging and treatment. Histologic, cytometric and molecular evaluation of cancer tissue aids in the diagnosis and staging of pediatric cancer. Serological and urinary tumor markers primarily play a supportive role in diagnosis; however, are often essential tools for cost-effective and noninvasive monitoring. Laboratory techniques such as sequencing/genomic profiling, FISH, and cytogenetics are becoming a standard of care for diagnosis and management. Most childhood cancers can be cured or treated with surgery and/or radiotherapy. Immunotherapy is a newer class of treatment that is now available for certain types of cancer. Advances in omics technologies and data analytics are anticipated to not only foster the development of newer, more sensitive and specific cancer biomarkers that will aid in early detection and diagnosis but also treatment of cancers.

2.10 Summary points

- Currently, no biomarkers have demonstrated success in screening for pediatric malignancies.
- Initial workup of pediatric malignancies is primarily based on symptomatic evaluation and supporting laboratory findings.
- Histological, cytometric, and molecular characterization of cancer tissue and cells is crucial for diagnosis, staging and/or prognostication of pediatric malignancies.
- Serological and urinary tumor markers largely have a role in prognostication and monitoring of pediatric malignancies. These biomarkers also play a supportive role in diagnosis in conjunction with clinical findings, histological, radiographic, and molecular evidence.

References

[1] Kyu HH, Stein CE, Boschi Pinto C, Rakovac I, Weber MW, Dannemann Purnat T, et al. Causes of death among children aged 5–14 years in the WHO European Region: a systematic analysis for the Global Burden of Disease Study 2016. Lancet Child Adolesc Health 2018;2:321–37.

[2] Ward ZJ, Yeh JM, Bhakta N, Frazier AL, Atun R. Estimating the total incidence of global childhood cancer: a simulation-based analysis. Lancet Oncol 2019;20:483–93.

[3] Siegel RL, Miller KD, Fuchs HE, Jemal A. Cancer statistics, 2021. CA Cancer J Clin 2021;71:7–33.

[4] Siegel RL, Miller KD, Jemal A. Cancer statistics, 2020. CA Cancer J Clin 2020;70:7–30.

[5] Henry NL, Hayes DF. Cancer biomarkers. Mol Oncol 2012;6:140–6.

[6] Arber DA, Orazi A, Hasserjian R, Thiele J, Borowitz MJ, Le Beau MM, et al. The 2016 revision to the World Health Organization classification of myeloid neoplasms and acute leukemia. Blood. 2016;127:2391–405.

[7] PDQ Pediatric Treatment Editorial Board. Childhood acute lymphoblastic leukemia treatment (PDQ®): health professional version. PDQ Cancer Information Summaries [Internet]. Bethesda, MD: National Cancer Institute; 2002[cited 2021 Dec 18]. Available from. Available from: http://www.ncbi.nlm.nih.gov/books/NBK65763/.

[8] Manola KN. Cytogenetics of pediatric acute myeloid leukemia. Eur J Haematol 2009;83:391–405.

[9] El Hussein S, Medeiros LJ, Khoury JD, Aggressive NK. Cell leukemia: current state of the art. Cancers (Basel) 2020;12:E2900.

[10] Song JY, Khojeini EV, Dwyre DM, Jonas BA. B lymphoblastic leukemia with granules mimicking acute myeloid leukemia. Int J Hematol 2015;102:251–2.

[11] Cortelazzo S, Ferreri A, Hoelzer D, Ponzoni M. Lymphoblastic lymphoma. Crit Rev Oncol Hematol 2017;113:304–17.

[12] McKinnon KM. Flow cytometry: an overview. Curr Protoc Immunol 2018;120:5.1.1–5.1.11.

[13] Olsen RJ, Chang C-C, Herrick JL, Zu Y, Ehsan A. Acute leukemia immunohistochemistry: a systematic diagnostic approach. Arch Pathol Lab Med 2008;132:462–75.

[14] Haso W, Lee DW, Shah NN, Stetler-Stevenson M, Yuan CM, Pastan IH, et al. Anti-CD22-chimeric antigen receptors targeting B-cell precursor acute lymphoblastic leukemia. Blood. 2013;121:1165–74.

[15] Dufner V, Sayehli CM, Chatterjee M, Hummel HD, Gelbrich G, Bargou RC, et al. Long-term outcome of patients with relapsed/refractory B-cell non-Hodgkin lymphoma treated with blinatumomab. Blood Adv 2019;3:2491–8.

[16] Shah NN, Maatman T, Hari P, Johnson B. Multi targeted CAR-T cell therapies for B-cell malignancies. Front Oncol 2019;9:146.

[17] Mrózek K, Harper DP, Aplan PD. Cytogenetics and molecular genetics of acute lymphoblastic leukemia. Hematol Oncol Clin North Am 2009;23:991–1010 v.

[18] de Rooij JDE, Zwaan CM, van den Heuvel-Eibrink M. Pediatric AML: from biology to clinical management. J Clin Med 2015;4:127–49.

[19] Raetz EA, Teachey DT. T-cell acute lymphoblastic leukemia. Hematol Am Soc Hematol Educ Program 2016;2016:580–8.

[20] Kornberg A, Polliack A. Serum lactic dehydrogenase (LDH) levels in acute leukemia: marked elevations in lymphoblastic leukemia. Blood. 1980;56:351–5.

[21] Pui CH, Dodge RK, Dahl GV, Rivera G, Look AT, Kalwinsky D, et al. Serum lactic dehydrogenase level has prognostic value in childhood acute lymphoblastic leukemia. Blood. 1985;66:778–82.

[22] Hoffman MA, Schmeling M, Dahlin JL, Bevins NJ, Cooper DP, Jarolim P, et al. Calibrating from within: multipoint internal calibration of a quantitative mass spectrometric assay of serum methotrexate. Clin Chem 2020;66:474–82.

[23] SEER Cancer Statistics Review 1975–2006. Previous version – SEER Cancer Statistics [Internet]. SEER. [cited 2021 Dec 18]. Available from: <https://seer.cancer.gov/archive/csr/1975_2006/index.html>.

[24] Ward E, DeSantis C, Robbins A, Kohler B, Jemal A. Childhood and adolescent cancer statistics, 2014. CA Cancer J Clin 2014;64:83–103.

[25] Buhtoiarov IN. Pediatric lymphoma. Pediatr Rev 2017;38:410–23.

[26] Flerlage JE, Hiniker SM, Armenian S, Benya EC, Bobbey AJ, Chang V, et al. Pediatric Hodgkin lymphoma, Version 3.2021. J Natl Compr Canc Netw 2021;19:733–54.

[27] Mauz-Körholz C, Metzger ML, Kelly KM, Schwartz CL, Castellanos ME, Dieckmann K, et al. Pediatric Hodgkin lymphoma. J Clin Oncol 2015;33:2975–85.

[28] Ostrom QT, Gittleman H, Liao P, Rouse C, Chen Y, Dowling J, et al. CBTRUS statistical report: primary brain and central nervous system tumors diagnosed in the United States in 2007–2011. Neuro Oncol 2014;16(Suppl 4):iv1–63.

[29] Rosemberg S, Fujiwara D. Epidemiology of pediatric tumors of the nervous system according to the WHO 2000 classification: a report of 1,195 cases from a single institution. Childs Nerv Syst 2005;21:940–4.

[30] Collins VP, Jones DTW, Giannini C. Pilocytic astrocytoma: pathology, molecular mechanisms and markers. Acta Neuropathol 2015;129:775–88.

[31] Rutkowski S, von Hoff K, Emser A, Zwiener I, Pietsch T, Figarella-Branger D, et al. Survival and prognostic factors of early childhood medulloblastoma: an international meta-analysis. J Clin Oncol 2010;28:4961–8.
[32] Taylor MD, Northcott PA, Korshunov A, Remke M, Cho Y-J, Clifford SC, et al. Molecular subgroups of medulloblastoma: the current consensus. Acta Neuropathol 2012;123:465–72.
[33] Shuangshoti S, Tadadontip P, Techavichit P, Thorner PS, Shuangshoti S, Teerapakpinyo C. Simplified molecular subtyping of medulloblastoma for reduced cost and improved turnaround time. Appl Immunohistochem Mol Morphol 2020;28:538–43.
[34] Louis DN, Perry A, Burger P, Ellison DW, Reifenberger G, von Deimling A, et al. International society of neuropathology—Haarlem consensus guidelines for nervous system tumor classification and grading. Brain Pathol 2014;24:429–35.
[35] Newman EA, Nuchtern JG. Recent biologic and genetic advances in neuroblastoma: Implications for diagnostic, risk stratification, and treatment strategies. SemPediatric Surg 2016;25:257–64.
[36] Galstyan A, Wilbur C, Selby K, Hukin J. Opsoclonus-myoclonus syndrome: a new era of improved prognosis? Pediatr Neurol 2017;72:65–9.
[37] Brodeur GM. Neuroblastoma: biological insights into a clinical enigma. Nat Rev Cancer 2003;3:203–16.
[38] Louis CU, Shohet JM. Neuroblastoma: molecular pathogenesis and therapy. Annu Rev Med 2015;66:49–63.
[39] Mosse YP, Laudenslager M, Khazi D, Carlisle AJ, Winter CL, Rappaport E, et al. Germline PHOX2B mutation in hereditary neuroblastoma. Am J Hum Genet 2004;75:727–30.
[40] Mossé YP, Laudenslager M, Longo L, Cole KA, Wood A, Attiyeh EF, et al. Identification of ALK as a major familial neuroblastoma predisposition gene. Nature. 2008;455:930–5.
[41] Molenaar JJ, Koster J, Zwijnenburg DA, van Sluis P, Valentijn LJ, van der Ploeg I, et al. Sequencing of neuroblastoma identifies chromothripsis and defects in neuritogenesis genes. Nature. 2012;483:589–93.
[42] Sano H, Bonadio J, Gerbing RB, London WB, Matthay KK, Lukens JN, et al. International neuroblastoma pathology classification adds independent prognostic information beyond the prognostic contribution of age. Eur J Cancer 2006;42:1113–19.

[43] Shimada H, Ambros IM, Dehner LP, Hata J, Joshi VV, Roald B. Terminology and morphologic criteria of neuroblastic tumors: recommendations by the International Neuroblastoma Pathology Committee. Cancer. 1999;86:349–63.

[44] Katanoda K. Neuroblastoma mass screening—what can we learn from it? J Epidemiol 2016;26:163–5.

[45] Bourdeaut F, de Carli E, Timsit S, Coze C, Chastagner P, Sarnacki S, et al. VIP hypersecretion as primary or secondary syndrome in neuroblastoma: a retrospective study by the Société Française des Cancers de l'Enfant (SFCE). Pediatr Blood Cancer 2009;52:585–90.

[46] Fauler G, Leis HJ, Huber E, Schellauf C, Kerbl R, Urban C, et al. Determination of homovanillic acid and vanillylmandelic acid in neuroblastoma screening by stable isotope dilution GC-MS. J Mass Spectrom 1997;32:507–14.

[47] Lionetto L, Lostia AM, Stigliano A, Cardelli P, Simmaco M. HPLC-mass spectrometry method for quantitative detection of neuroendocrine tumor markers: vanillylmandelic acid, homovanillic acid and 5-hydroxyindoleacetic acid. Clin Chim Acta 2008;398:53–6.

[48] Berthold F, Hunneman DH, Harms D, Käser H, Zieschang J. Serum vanillylmandelic acid/homovanillic acid contributes to prognosis estimation in patients with localised but not with metastatic neuroblastoma. Eur J Cancer 1992;28A:1950–4.

[49] Pussard E, Neveux M, Guigueno N. Reference intervals for urinary catecholamines and metabolites from birth to adulthood. Clin Biochem 2009;42:536–9.

[50] Nakagawara A, Ikeda K, Higashi K, Sasazuki T. Inverse correlation between N-myc amplification and catecholamine metabolism in children with advanced neuroblastoma. Surgery. 1990;107:43–9.

[51] Simon T, Hero B, Hunneman DH, Berthold F. Tumour markers are poor predictors for relapse or progression in neuroblastoma. Eur J Cancer 2003;39:1899–903.

[52] Hann HW, Stahlhut MW, Evans AE. Source of increased ferritin in neuroblastoma: studies with concanavalin A-sepharose binding. J Natl Cancer Inst 1986;76:1031–3.

[53] Chu A, Heck JE, Ribeiro KB, Brennan P, Boffetta P, Buffler P, et al. Wilms' tumour: a systematic review of risk factors and meta-analysis. Paediatr Perinat Epidemiol 2010;24:449–69.

[54] Rivera MN, Haber DA. Wilms' tumour: connecting tumorigenesis and organ development in the kidney. Nat Rev Cancer 2005;5:699–712.

[55] Maschietto M, Williams RD, Chagtai T, Popov SD, Sebire NJ, Vujanic G, et al. TP53 mutational status is a potential marker for risk stratification in Wilms tumour with diffuse anaplasia. PLoS One 2014;9:e109924.

[56] Dome JS, Cotton CA, Perlman EJ, Breslow NE, Kalapurakal JA, Ritchey ML, et al. Treatment of anaplastic histology Wilms' tumor: results from the fifth National Wilms' Tumor Study. J Clin Oncol 2006;24:2352−8.

[57] Gratias EJ, Dome JS, Jennings LJ, Chi Y-Y, Tian J, Anderson J, et al. Association of chromosome 1q gain with inferior survival in favorable-histology Wilms tumor: a report from the Children's Oncology Group. J Clin Oncol 2016;34:3189−94.

[58] Grundy PE, Breslow NE, Li S, Perlman E, Beckwith JB, Ritchey ML, et al. Loss of heterozygosity for chromosomes 1p and 16q is an adverse prognostic factor in favorable-histology Wilms tumor: a report from the National Wilms Tumor Study Group. J Clin Oncol 2005;23:7312−21.

[59] Segers H, van den Heuvel-Eibrink MM, Williams RD, van Tinteren H, Vujanic G, Pieters R, et al. Gain of 1q is a marker of poor prognosis in Wilms' tumors. Genes Chromosomes Cancer 2013;52:1065−74.

[60] Wegert J, Ishaque N, Vardapour R, Geörg C, Gu Z, Bieg M, et al. Mutations in the SIX1/2 pathway and the DROSHA/DGCR8 miRNA microprocessor complex underlie high-risk blastemal type Wilms tumors. Cancer Cell 2015;27:298−311.

[61] Walz AL, Ooms A, Gadd S, Gerhard DS, Smith MA, Guidry Auvil JM, et al. Recurrent DGCR8, DROSHA, and SIX homeodomain mutations in favorable histology Wilms tumors. Cancer Cell 2015;27:286−97.

[62] Rakheja D, Chen KS, Liu Y, Shukla AA, Schmid V, Chang T-C, et al. Somatic mutations in DROSHA and DICER1 impair microRNA biogenesis through distinct mechanisms in Wilms tumours. Nat Commun 2014;2:4802.

[63] Torrezan GT, Ferreira EN, Nakahata AM, Barros BDF, Castro MTM, Correa BR, et al. Recurrent somatic mutation in DROSHA induces microRNA profile changes in Wilms tumour. Nat Commun 2014;5:4039.

[64] Mahamdallie S, Yost S, Poyastro-Pearson E, Holt E, Zachariou A, Seal S, et al. Identification of new Wilms tumour predisposition genes: an exome sequencing study. Lancet Child Adolesc Health 2019;3:322−31.

[65] Musick SR, Smith M, Rouster AS, Babiker HM. Hepatoblastoma. StatPearls [Internet]. Treasure Island, FL: StatPearls Publishing; 2021 [cited 2021 Sep 13]. Available from: <http://www.ncbi.nlm.nih.gov/books/NBK534795/>.

[66] López-Terrada D, Alaggio R, de Dávila MT, Czauderna P, Hiyama E, Katzenstein H, et al. Towards an international pediatric liver tumor consensus classification: proceedings of the Los Angeles COG liver tumors symposium. Mod Pathol 2014;27:472–91.

[67] Haas JE, Muczynski KA, Krailo M, Ablin A, Land V, Vietti TJ, et al. Histopathology and prognosis in childhood hepatoblastoma and hepatocarcinoma. Cancer. 1989;64:1082–95.

[68] Bell D, Ranganathan S, Tao J, Monga SP. Novel advances in understanding of molecular pathogenesis of hepatoblastoma: a Wnt/β-catenin perspective. Gene Expr 2017;17:141–54.

[69] De Ioris M, Brugieres L, Zimmermann A, Keeling J, Brock P, Maibach R, et al. Hepatoblastoma with a low serum alpha-fetoprotein level at diagnosis: the SIOPEL group experience. Eur J Cancer 2008;44:545–50.

[70] Maruyama K. Serum α-fetoprotein concentration in extremely low-birthweight infants. Pediatr Int 2017;59:159–62.

[71] Wu JT, Book L, Sudar K. Serum alpha fetoprotein (AFP) levels in normal infants. Pediatr Res 1981;15:50–2.

[72] Taketa K, Hirai H. Lectin affinity electrophoresis of alpha-fetoprotein in cancer diagnosis. Electrophoresis. 1989;10:562–7.

[73] Nakagawara A, Ikeda K, Tsuneyoshi M, Daimaru Y, Enjoji M, Watanabe I, et al. Hepatoblastoma producing both alpha-fetoprotein and human chorionic gonadotropin. Clinicopathologic analysis of four cases and a review of the literature. Cancer 1985;56:1636–42.

[74] Braunstein GD, Bridson WE, Glass A, Hull EW, McIntire KR. In vivo and in vitro production of human chorionic gonadotropin and alpha-feteoprotein by a virilizing hepatoblastoma. J Clin Endocrinol Metab 1972;35:857–62.

[75] Pierce JL, Frazier AL, Amatruda JF. Pediatric germ cell tumors: a developmental perspective. Adv Urol 2018;2018:9059382.

[76] Nathan and Oski's Hematology and Oncology of Infancy and Childhood. Clinical key [Internet]. [cited 2021 Dec 19]. Available from: <https://www-clinicalkey-com.proxy.library.upenn.edu/#!/browse/book/3-s2.0-C20120011015>.

[77] von Eyben FE. Laboratory markers and germ cell tumors. Crit Rev Clin Lab Sci 2003;40:377–427.
[78] Gronowski AM. Clinical assays for human chorionic gonadotropin: what should we measure and how? Clin Chem 2009;55:1900–4.

CHAPTER

3 Challenges with common laboratory tests in the adult cancer patient

Samuel I. McCash and Melissa S. Pessin

Department of Pathology and Laboratory Medicine, Memorial Sloan Kettering Cancer Center, New York, NY, United States

Objectives

- Review changes to normal/expected values for laboratory tests in cancer patients due to inflammation and how to interpret them.
- Understand interferences to lab assays that are produced by cancers and their treatments.
- Explain physiologic changes that occur in cancer patients due to treatment and how they affect lab results.
- Discuss preanalytical issues more common in cancer patients and how to mitigate them.
- Describe analytical challenges in cancer patients and discuss how they can be addressed.
- Consider the challenges of critical values in the cancer patient population.

3.1 Introduction

Laboratory testing in cancer patients poses some unique challenges due to both the patient's cancer itself as well as the treatments for the patient's cancer. Cancers hijack the patient's normal physiological processes such as cell growth, cell death, vessel growth, inflammation, and coagulation in ways that are generally abnormal; thus the laboratory measurements of such

Cancer Biomarkers: Clinical Aspects and Laboratory Determination. DOI: https://doi.org/10.1016/B978-0-12-824302-2.00002-3
© 2022 Elsevier Inc. All rights reserved.

processes reflect these abnormalities. In addition, various treatments of cancer, while killing off the cancer cells, also have effects on the patient's normal tissues which can be reflected in expected and unexpected abnormal laboratory measurements. Furthermore, some treatments directly interfere with the testing methods. In this chapter, we describe a wide range of challenges to the performance and interpretation of common laboratory tests in the adult cancer patient.

3.2 The effect of inflammation on laboratory results

The body's natural response to cancer is to illicit an immune response. In healthy individuals, white blood cells (WBCs) activate to destroy abnormal cells that are or could become cancerous. They do this, in part, by releasing cytokines, which activate other cellular defenses to combat this abnormality or neoplasia. These cytokines induce inflammatory responses against the abnormal cells as well as other changes in the body including altering the levels of specific proteins in the blood. This set of proteins is called acute phase reactant proteins (i.e., acute phase reactants, acute phase proteins) which are defined as proteins whose plasma or serum concentration increases or decreases by at least 25% during inflammatory processes [1] (Table 3.1). Most acute phase reactant proteins (APRPs) increase with inflammation and are referred to as positive APRPs.

Table 3.1 Acute phase proteins, positive and negative.

Positive acute phase reactant proteins	Negative acute phase reactant proteins
Complement	Albumin
Fibrinogen	Transferrin
Haptoglobin	Transthyretin
Procalcitonin	
C-reactive protein	
Ferritin	

Conversely, APRPs that decrease during inflammation are called negative APRPs. These changes are in part due to the increased or decreased production of APRPs by the liver, but the complete mechanism of these changes is not fully understood. The following are some of the more important markers to note that are affected by inflammation in the cancer patient.

3.2.1 Fibrinogen

Fibrinogen is often elevated, sometimes quite markedly in cancer patients. Though notable on test results to be elevated, there are a few other items that health practitioners must be aware of.

Elevated fibrinogen can cause problems with serum samples. Typically, fibrinogen would not be present in serum samples as it is usually removed by the clotting and centrifugation processes. However, if the sample is not given enough time to fully clot, residual fibrinogen may remain in the supernatant. This is further compounded when fibrinogen levels are particularly high and thus the fibrinogen is not entirely consumed within the standard 30-min clotting time prior to centrifugation. After centrifugation any residual fibrinogen in the serum portion can form micro-clots which adversely affect the analyzer by clogging the aspiration and tubing systems and interfering in test signal as aggregates in transmittance-based assays.

Residual fibrinogen may also be seen on serum protein electrophoresis (SPEP) studies, appearing as a band on gel electrophoresis in the anodal gamma region adjacent to the beta region, or as a peak between the beta 2 and gamma regions on capillary electrophoresis (Figs. 3.1 and 3.2) [2]. It is important to note that fibrinogen can mimic a monoclonal spike; however, this misinterpretation can easily be identified using a follow-up immunofixation study. It is also possible to prove the presence of fibrinogen using an antifibrinogen reagent on immunofixation; however, this is not a standard reagent within most clinical labs. The presence of fibrinogen on SPEP studies should not be reported as they are preanalytical issues and not indicative of an abnormal finding in the patient.

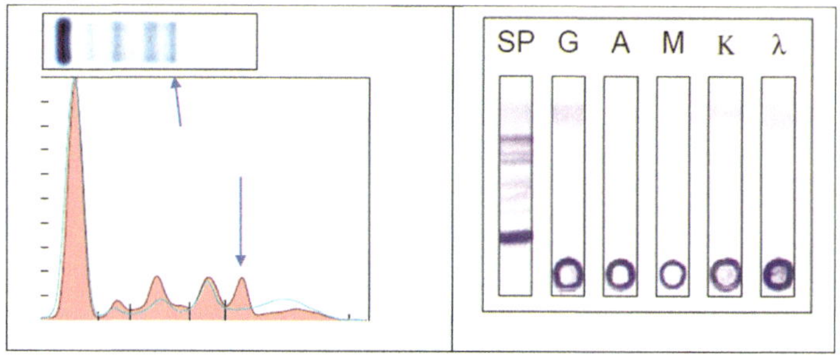

FIGURE 3.1

Gel protein electrophoresis shows a band (arrows) in the gamma region adjacent to the beta region; however, the negative immunofixation electrophoresis shows that it is not a monoclonal antibody. The band is fibrinogen.

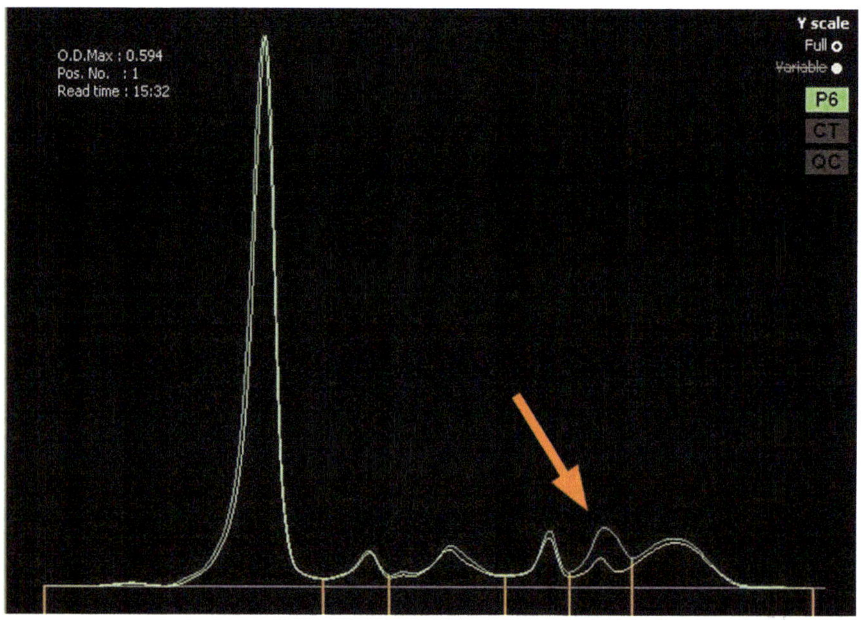

FIGURE 3.2

Image showing an overlay of a plasma tracing over a serum tracing. The taller peak of the plasma is due to fibrinogen (arrow) and is not present in the serum (shorter peak). Fibrinogen may interfere with proper M-spike quantitation of IgA monoclonal proteins, as they can migrate similarly.

Image and samples courtesy of Karen Anderson from Sebia.

3.2.2 D-Dimer

The D-dimer assay can be useful in ruling out deep vein thrombosis (DVT) and pulmonary embolism (PE) in individuals with a low pre-test probability (PTP) where the negative predictive value (NPV) is 96%–100% (median = 99%). PTP can be assessed using the Wells clinical score [3]. Patients with active cancer have a higher PTP than the general population as they are often in a hypercoagulable state due to expression of procoagulants, tissue factor exposure with invasive disease, and increased inflammatory cytokines. Therefore micro-clot formation followed by fibrinolysis can be constantly active in cancer patients making this assay less useful for ruling out DVT or PE. D-dimers, as well as other fibrin split products, will likely be elevated in a cancer patient population. This issue is accounted for in the Wells clinical score which increases by +1 for both DVT and PE in patients with active cancer. A higher score correlates with a higher PTP which decreases the NPV. Cancer patients automatically start at a DVT score of 1 which corresponds to at least a moderate PTP, thus increasing the risk of a false negative. It is also important to note that the type of assay being used to measure D-dimers will affect the NPV. It has been shown that a high sensitivity assay can safely rule out DVT in low to moderate PTP conditions; however, one must be more cautious interpreting results if a moderately sensitive assay is being used.

3.2.3 Ferritin

Ferritin has been touted as an important indicator of iron status due to its affordable cost and high availability on automated chemistry analyzers. Low levels are indicative of iron deficiency even if the patient has concurrent anemia of chronic disease. In cancer patients, however, ferritin is often elevated as it is a positive acute phase reactant making it almost useless in differentiating between iron deficiency and anemia of chronic disease in this patient population. Iron levels and iron binding–capacity assays provide more useful information. Newer tests such as reticulocyte-hemoglobin (Ret-He) are also more informative. Since the Ret-He assay measures the hemoglobin content of reticulocytes, it is a good indicator of red blood cell (RBC) production as this requires adequate stores of iron within the bone marrow. A low Ret-He suggests iron-deficient hematopoiesis. Studies at Memorial Sloan Kettering (MSK) have determined that a Ret-He above 32 pg rules out iron deficiency with an NPV of 98.5% [4]. Ret-He is also very convenient as it can be added to a complete blood count (CBC) order as part of the reticulocyte count.

3.2.4 Haptoglobin

Haptoglobin is also increased in cancer patients. This is not usually an issue since elevated levels are not indicative of other disease conditions. However, at increased levels, haptoglobin can cause changes to SPEP studies mimicking monoclonal proteins. Haptoglobin runs in the alpha 2 region with alpha 2 macroglobulin. This is further complicated in that haptoglobin can migrate anodally or cathodally to alpha 2 macroglobulin depending on its phenotype (Fig. 3.3). In such cases a monoclonal protein can be ruled out by a negative immunofixation study.

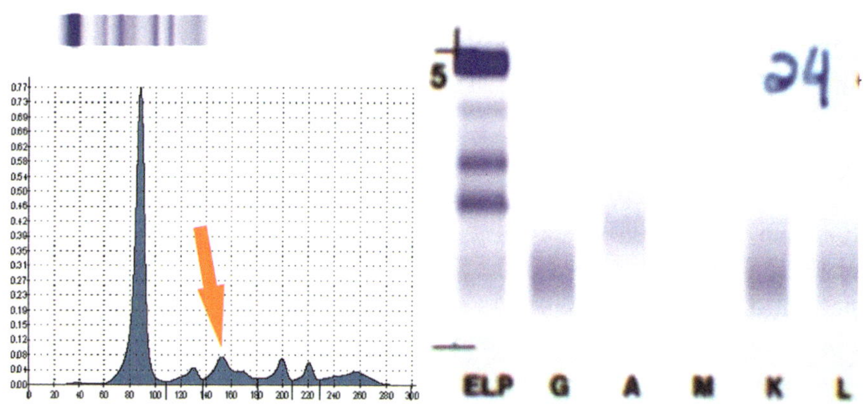

FIGURE 3.3

Elevated haptoglobin may appear as a peak just anodal to alpha-2-macroglobulin in the alpha 2 region mimicking a monoclonal antibody. The negative immunofixation shows that it is not a monoclonal antibody.

3.2.5 Procalcitonin

As an indicator of bacterial infection, procalcitonin has been found to have several uses including early detection of sepsis. Though the cutoff values, sensitivity, and specificity have been defined in the general population, evidence is sparse and sometimes controversial when it comes to specific patient populations including those with neutropenia or cancer. Pastores and Blouin showed that procalcitonin is a good predictor of culture confirmed bacteremia early in critically ill cancer patients when using the cutoff of 2.0 ng/mL; however, its usefulness in implementing and ceasing antibiotic therapy is yet to be determined [5]. Overall the usefulness of procalcitonin in cancer patients is controversial as it can be elevated in patients with metastatic disease or with a cancer-induced inflammatory state [6].

3.2.6 Albumin and its effect on total calcium

While the majority of APRPs are increased with inflammation, there are a handful that are decreased. These are called negative APRPs, the most notable being albumin. Albumin production by the liver is decreased during inflammation. Not only are results low, but test methods affected by albumin levels may be altered as well. For instance, SPEP studies are routinely normalized using the albumin level. In the presence of a low albumin, the other fractions are stretched out vertically (appear taller) on the tracing making them appear deceivingly large (Fig. 3.4).

The decreased amount of albumin also influences the measured levels of other analytes that normally bind to albumin in the blood, with calcium being the most common and strongly affected. In many instances, calcium is measured as total calcium because of the ease of doing so with automated methods. Total calcium is comprised primarily of that bound to albumin, thus in patients with low albumin, the total calcium is proportionately low. This does not, however, indicate clinical hypocalcemia as this bound calcium is not active. Only ionized calcium is active; therefore ionized calcium levels are appropriately indicative of proper functioning within the body. In these patients with hypoalbuminemia,

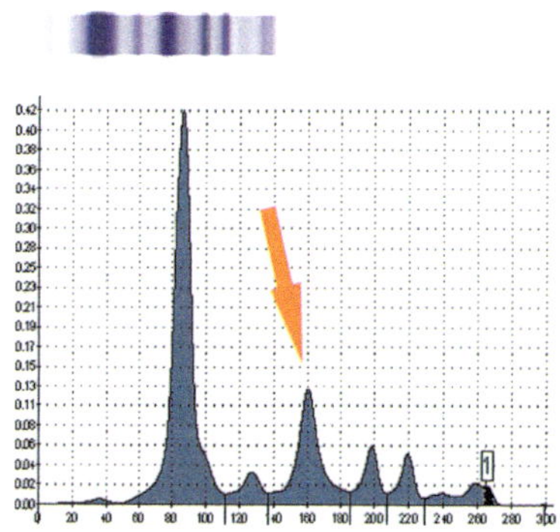

Fractions	%	g/dl		Ref. g/dl
Albumin	56.8	2.78	<	3.70 - 4.90
Alpha 1	5.0	0.25		0.20 - 0.40
Alpha 2	20.5	1.00		0.50 - 1.00
Beta 1	6.6	0.32		0.30 - 0.60
Beta 2	5.5	0.27		0.20 - 0.50
Gamma	5.6	0.27	<	0.60 - 1.50

Bands	%	g/dl
1	1.0	0.05

FIGURE 3.4

Alpha 2 peak (arrow) appears deceptively large because albumin is very low. When albumin is low any of the remaining peaks may be mistaken for paraprotein. Also seen is a very small peak of paraprotein and a second peak anodal to it which is daratumumab.

ionized calcium is often at normal levels and its measurement would be the preferred method if readily available. However, methods for measuring ionized calcium are not part of the automated chemistry analyzer compendium and thus not included in routine chemistry panels such as the basic or comprehensive metabolic profiles. An alternative solution is to report out a corrected calcium which is calculated using the Payne equation: corrected calcium (mg/dL) = measured total Ca (mg/dL) +0.8 [4.0 − serum albumin (g/dL)], where 4.0 represents the average albumin level. The laboratory information system can perform this calculation automatically to make this value readily available for the clinical team and help them to identify the true presence of either hyper or hypocalcemia in this patient population [7].

3.3 Interferences produced by cancer

3.3.1 Monoclonal proteins

Cancer can produce different substances that interfere with lab assays. One such substance is monoclonal antibodies produced by neoplastic B-cells/plasma cells. At high concentrations, these paraproteins may interfere in several different methods. There are five basic mechanisms of interference: (1) protein-induced precipitation, (2) binding to reaction components, (3) water volume displacement, (4) increasing sample viscosity, and (5) RBC agglutination (Tietz, Chapter 5) [8]. Turbidimetric methods can be affected by unexpected precipitants that scatter the measurable light signal. Colorimetric methods, such as that used for bilirubin measurement, can also be affected. The precipitation of these proteins is influenced by several properties including substance concentration, pH, sample additives, and the chemical characteristics of the paraprotein. Due to these many variables, it is difficult to predict when a paraprotein will precipitate and how much is acceptable as not to interfere with an assay. Therefore laboratories must do their own interference testing to determine what levels of paraprotein are acceptable in their own test assays. Monoclonal antibodies from cancer can also bind nonspecifically to the analyte or to other reagents interfering with the test chemical reactions and causing inaccurate results. Both mechanisms may be overcome by performing dilutional studies or running the samples on different methods.

Volume displacement is another issue that can occur in samples from patients with high concentrations of monoclonal protein. The best-known example of this concept is the effect on electrolytes and other analytes when using ion-selective electrodes for indirect measurements. Normally, blood consists of both a liquid portion and a solid portion. For indirect methods, the ratio of the liquid to solid portions is changed because the sample is diluted. This is corrected for by a calculation in most analyzer software, based on the average water portion of a sample. When paraprotein is present in substantial amounts, it decreases the actual water volume. Thus the volume of the water in the sample is smaller than what is assumed by the analyzer. This effectively decreases the result for the analyte concentration, such as sodium, and can make it appear artificially low when it is in fact normal in the body. When this happens with sodium, the phenomenon is called pseudohyponatremia. This same displacement can happen when a large amount of lipid is present in the sample such as seen in patients with hypercholesterolemia or hypertriglyceridemia. The solution of such cases is to run the sample using a direct method, typically found on blood gas analyzers or point of care devices.

Viscosity of blood increases as the concentration of paraprotein increases, both in the patient as well as in the patient's blood sample. Viscosity affects the actual volume of sample pipetted, as it takes longer and more pressure to aspirate a given volume as the viscosity increases. Samples with unacceptable high viscosity need to be first diluted to ensure the proper volume for measurement. Other solutions include removing monoclonal antibody using antibody neutralizing containers or filtering the specimen. Such processes need to be validated by the lab to ensure that these additional steps do not interfere with test results.

Finally, some antibodies, most commonly IgM, generated by malignant antibody-producing cells bind to the RBCs in a sample and crosslink them, causing hemagglutination in the tube. These antibodies are generally more reactive at lower-than-body-temperatures and are called cold agglutinins. Some cold agglutinins are reactive at room temperature, causing the RBCs in the tube to clump. Sometimes the clumps of cells will be counted by an analyzer as single large cells which drastically increases the MCV and decreases the RBC count falsely. In more extreme cases, the sample will not be able to be aspirated for measurement. In both situations, prewarming the tube immediately prior to running the sample often corrects this problem, but in cases with high thermal amplitude antibodies, a manual hematocrit is all that can be performed.

3.3.2 Choriogonadotropin

Cancers can also produce proteins that are normally part of different processes in the body. Human choriogonadotropin (hCG) is known to be produced by a number of different cancer types such as that of the lung, pancreas, breast, and colon (Table 3.2), as well as those more expected to produce this protein such as germ cell tumors and choriocarcinoma. In pregnancy, hCG is produced by the placenta and is a useful marker to detect the onset and progression of pregnancy. Pregnancy tests are often done in female patients of childbearing age prior to a procedure to ensure that there is not a fetus also present that could be potentially harmed by that procedure. This includes all patients undergoing imaging studies, chemotherapy, and surgical procedures, procedures which are common for the cancer patient

Table 3.2 List of cancers producing human choriogonadotropin (hCG).

Cancer type	Total hCG (mIU/mL)	Intact hCG (mIU/mL)
Anaplastic astrocytoma	17.7	<0.5
Breast mixed ductal/lobular carcinoma	13.5	0.7
Cervical adenocarcinoma	11.1	<0.5
Cervical squamous cell carcinoma	40.3	<0.5
Cholangiocarcinoma	12.5	<0.5
Cholangiocarcinoma	11.4	<0.5
Colonic adenocarcinoma	25.4	1.2
Colonic adenocarcinoma	757.5	2.7
Colonic adenocarcinoma	102.8	<0.5
Gastric adenocarcinoma	37.3	5.5
Lung squamous cell carcinoma	7.6	<0.5
Pancreatic adenocarcinoma	77.4	0.6
Pancreatic adenocarcinoma	35.4	2.7
Pancreatic adenocarcinoma	77.3	<0.5
Pancreatic adenocarcinoma	16.2	<0.5
Sacral giant cell tumor	48.5	<0.5
Thyroid medullary carcinoma	124.9	<0.5

Data originally published in McCash SI, Goldfrank DJ, Pessin MS, Ramanathan LV. Reducing false-positive pregnancy test results in patients with cancer. Obstet Gynecol 2017;130:825–829.

population. At our institution, there have been many cases where a falsely positive pregnancy test has delayed a patient's treatment unnecessarily, causing unneeded anxiety and stress for the patient and their family. It is important to note that the hCG molecule produced by cancer is most often only the beta subunit, beta-hCG. This subunit is typically the target of detection in most pregnancy tests in the United States. At our institution, we began to use a pregnancy test specific for intact hCG, to help decrease the number of these false-positive pregnancy tests. Intact hCG consists of an alpha and a beta subunit and can be detected when the capture antibody is against the alpha subunit. Using the intact hCG assay reduced our false positive rate for this test by 90% [9]. However, not all false positive cases are caught this way, as there are a few cancers, such as choriocarcinoma, that produce the intact hCG. Despite this, the implementation of the intact hCG assay for the majority of our cancer patient population has been appreciated by our patients and clinicians.

3.4 Physiologic changes due to cancer treatment

3.4.1 Menopause

Another cause of false-positive pregnancy tests is hCG production by the pituitary gland due to menopause. Menopause can occur early in cancer patients if their ovaries are damaged significantly by chemotherapy. The ovaries are responsible for the production of estrogen and progesterone which, among many other things, provides a negative feedback loop to the pituitary gland decreasing its output of luteinizing hormone (LH) and follicle stimulating hormone (FSH). hCG is closely related to LH and FSH and can also be produced by the pituitary. When the negative feedback is lost due to decreased levels of estrogen and progesterone, the pituitary gland produces hormones unabated, including intact hCG. This hCG is detected both by qualitative and quantitative pregnancy tests. Sometimes it is helpful to get an accompanied FSH level to better determine that the elevated hCG levels are most likely due to chemotherapy-induced menopause [10].

Another way to address this elevated hCG issue is to adjust the normal range of hCG. Typically, hCG is lower than 5 mIU/mL in patients who are not pregnant [8]. In an unpublished study by Fleischer et al. that looked at a few hundred pregnancy tests, they found a significant number of cancer patients with results between 5 and 14 mIU/mL that were found not to be pregnant. These low elevations of hCG were most likely due to pituitary production in the setting of ovarian dysfunction/dysregulation. For this reason, at MSK, the upper end of the reference range (not pregnant) for hCG was adjusted up to 14 mIU/mL, resulting in a decrease of the number of false positive tests and lessening the interruptions to patient care without missing any truly pregnant patients.

3.4.2 Stem cell transplant

Hematopoietic cancers are often treated with bone marrow ablation followed by stem cell transplant. After the stem cells are introduced, the bone marrow enters a period of recovering which may cause some changes in laboratory testing. For instance, SPEP studies will often demonstrate small bands/peaks in the beta to gamma regions. This is because a few lymphocyte clones may predominate in the early recovery and their antibodies are seen as oligoclonal bands on protein electrophoresis studies. Therefore it is critical to know the typing and migration pattern of the patient's original clone when evaluating myeloma patient SPEP studies post stem cell transplantation. One must not mistake an oligoclonal band as the patient's original disease, as that may affect the disease response category or may categorize the patient as having recurring disease. Oligoclonal bands are considered as a good prognostic factor in multiple myeloma patients' status post stem cell transplant.

3.4.3 Bone marrow recovery

Bone marrow recovery also influences the types of circulating hemoglobin seen on hemoglobin electrophoresis. Typically, in adults, most of the hemoglobin present is Hgb A at around 97% with a small portion being Hgb A2 around 2%–3.5%, and little to no Hgb F, 0%–2%. Hgb F is the predominant hemoglobin in the fetus and newborn infant [11]. Functionally, Hgb F is a survival requirement for the fetus as it has two delta chains instead of two beta chains in Hgb A. The delta chain provides an increased oxygen affinity of fetal blood to aid the fetus in oxygen exchange from the pregnant mother's blood. As the infant ages to adulthood, Hgb F is replaced by Hgb A. After bone marrow suppression during bone marrow recovery, pathways for the production of Hgb F can be reactivated and can increase the amount of Hgb F up to 5%–10% [11]. This reactivation can also occur in patients with cancers involving the bone marrow such as leukemia, as well as with solid organ cancers metastasizing to the bone marrow.

3.4.4 Filgrastim

Cancer patients may often become neutropenic due to bone marrow depletion from chemotherapy. This puts these patients at an elevated risk for infection. Filgrastim, a granulocyte colony stimulating factor analog can be given to such patients to increase their WBC count and absolute neutrophil count. It does this by stimulating the production, maturation, and deployment of granulocytes into the peripheral blood. This activation of granulocytes is accompanied by a few changes in the CBC.

One notable change is the presence of immature granulocytic series including blasts seen on manual differentials. Blasts are generally never a normal finding in patients and require verification by a hematologist or pathologist. Some healthcare facilities treat the presence of blasts as a critical value; however, this is inappropriate in cases where the patient is on filgrastim. For patients on filgrastim, it is quite common to see 1%–5% blasts circulating in the peripheral blood. Rarely do blasts reach greater 20% of the WBCs on the manual differential if due to filgrastim, but when they do, this is more likely to occur in younger pediatric patients. Blasts appear in the peripheral blood not only because of the increased production of neutrophil precursors, but also from the enhancement of cell mobilization from the marrow.

With the increase of proliferation and mobilization of WBCs, it is common to also see a left shift of the myeloid series. Analyzers with an automated differential may count and report out an immature granulocyte fraction in these cases. Promyelocytes, myelocytes, metamyelocytes, and bands will be seen on the manual differential. Neutrophils are often hypergranulated (toxic granulation), contain blue Dohle bodies, and occasionally are hypersegmented [12]. Another feature noticed is an increased incidence of monocytosis, which has also been reported in the literature [13].

3.4.4.1 Nucleated red blood cells

Filgrastim appears to have the ability to mobilize most hematopoietic cell types in the bone marrow, and that includes nucleated RBC precursors. Typically, these precursors are more mature, having a small round nucleus containing highly condensed chromatin and slightly light blue cytoplasm, similar to that of a reticulocyte. Anecdotally, it has been observed that some of these nucleated red blood cells have irregular nuclear contours or nuclear folding mimicking dysplasia, but this is not typically a predominant feature.

3.4.4.2 Malignant plasma cells

In patients with severe resistant multiple myeloma, it is currently standard of care to perform ablative therapy followed by autologous hematopoietic stem cell transplant. Most commonly, the stem cells are acquired by mobilizing the bone marrow with filgrastim so that stem cells can be harvested from the peripheral blood. In these patients, it is not uncommon to find circulating plasmacytoid cells from the patient's multiple myeloma as they are mobilized as well. This may be concerning for the presence of a plasma cell leukemia, but these plasma cells disappear from the peripheral blood a few days after filgrastim was administered.

3.5 Preanalytical issues

3.5.1 Pseudohyperkalemia

Due to the different mechanisms and causes of hyperkalemia in samples there are different definitions of pseudohyperkalemia. For our purposes in this chapter, we will define pseudohyperkalemia as an elevation of potassium in the sample that is not representative of the actual state of the patient. Patients with high WBC counts due to hematopoietic malignancies have the highest risk of this issue, and it occurs more often in plasma specimens. Leukemia and lymphoma cells tend to be more fragile than normal WBCs, particularly when being treated with chemotherapeutic agents, and have a high propensity to lyse. This lysis releases intracellular potassium into the serum/plasma in vitro which increases over time until centrifugation. The likelihood and effect of leukemia and lymphoma cells lysing increase at higher WBC. Conventionally, pseudohyperkalemia has been defined as high potassium in serum, but normal in plasma [14]. This definition was based on the experience with patients who had extremely high platelet counts in which potassium was falsely elevated due to platelet release during the clotting process. The solution for this issue is to run a potassium test on a plasma sample where the clotting process does not take place and the platelets remain intact. For leukemia patients, the issue is just the opposite. In these patients, the plasma potassium is elevated but potassium run on a serum sample is not, hence the term reverse pseudohyperkalemia which is sometimes refered to in the literature.

Several hypotheses have been made as to what causes pseudohyperkalemia in plasma specimens with high WBC. One concept is that heparin damages the cell membrane and allows faster leakage of potassium into the plasma [15]. Another is that leukemia cells, such as those in chronic lymphocytic leukemia, are more fragile and likely to break which is seen as smudge cells on a peripheral blood smear. These issues may be compounded when combined with mechanical forces such as improper phlebotomy or transit via pneumatic tube. Another possibility is that in plasma specimens with separator gel, the leukemia cells break down rapidly when passing through the gel. In the case of serum, these fragile cells are trapped in the clotted matrix in a serum tube and thus are protected during centrifugation. Further studies will be necessary to identify more precisely the true cause(s) and to prevent the occurrence in these patients, which currently leads to unnecessary worry and repeat testing in this patient group.

3.5.2 Lactate dehydrogenase

Similar to pseudohyperkalemia described earlier, lactate dehydrogenase (LDH) can be falsely elevated in cases of in vitro hemolysis, WBC lysis, or platelet activation which are more likely to occur in the cancer patient population. Since LDH is predominantly found intracellularly, it will be falsely elevated with in vitro destruction of cells. This may erroneously indicate the presence of intravascular hemolysis or end organ damage. In addition, due to cancer cells being abnormally fragile as well as multiplying at an elevated rate, LDH levels are used for certain cancers as a tumor marker to follow disease progression. Falsely elevated LDH levels can lead to unnecessary worry and increased testing for ruling out such progression. Thus testing for LDH in serum only has been shown to improve the reliability of its measurement (MSK unpublished data).

3.5.3 Hemolysis

Hemolysis during phlebotomy also appears to be more common in the cancer patient population. This is partly because the chemotherapeutic agents and other medications the cancer patients are given make the RBC membrane more fragile. Another reason is that cancer patients are more likely to have intravascular accesses for intravenous (IV) fluid or chemotherapy administration, and thus are more likely to have blood drawn through those IV lines or ports. The problem with this use is that many IV access devices are designed to push fluid into the patient and are not made for fluid removal. This design typically requires a greater force to retrieve blood through it creating shear forces strong enough to hemolyze cells. Because of this issue and the possibility of line contamination from the medications and fluids administered in these lines (see below), direct venipuncture is often a better choice for accurate blood testing results.

3.5.4 Contamination

Contamination by other substances is another common cause of preanalytical issues. Though not limited to just cancer patients, it tends to happen more frequently since they often have implanted ports or central lines allowing regular doses of IV chemotherapy. Any substance that has been injected into the port or line has a risk of contaminating blood samples taken from that port or line. One common contaminate is IV fluid used to flush the ports. Residual fluid in the port will dilute out analytes affecting those with highest concentrations the most. Of the chemistry tests, sodium and chloride have the highest propensity to appear low, although potassium can appear low as well. The hematocrit and hemoglobin may also appear falsely low. If the port is not flushed out appropriately after chemotherapy infusion, some agent can still be in the port. This is particularly problematic when drawing a trough level for the drug as it could appear quite high. Another issue is if

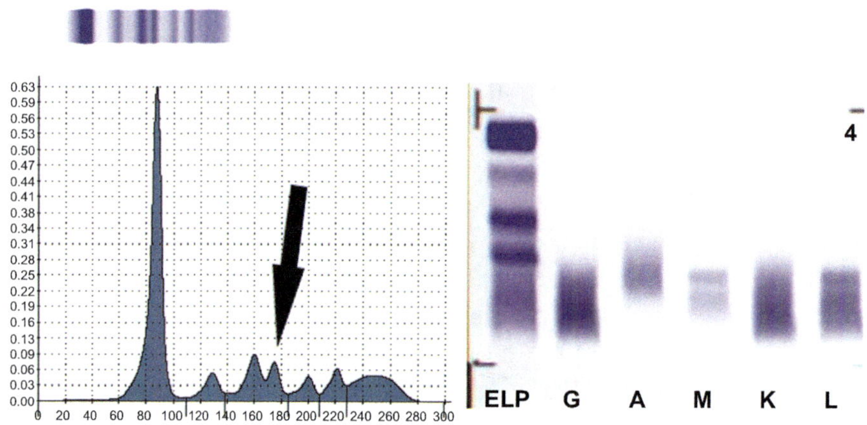

FIGURE 3.5

Patient who underwent computed tomography scan with contrast. The contrast appears as a peak cathodal to the alpha 2 peak mimicking a monoclonal antibody on capillary electrophoresis studies. The contrast does not show up on gel electrophoresis since it does not stain like other proteins, nor does it show up on immunofixation. The patient does, however, have a small IgM lambda band on immunofixation.

heparin flushes were used, particularly when drawing samples for coagulation studies. Residual heparin could falsely lengthen the prothrombin time and partial thromboplastin time. Radiographic contrast may also be a contaminant if blood is drawn from a line from which contrast was delivered through. In high concentrations, contrast interferes with gel separator tubes in that it increases the density of plasma or serum causing it to separate to below the gel. At normal doses, contrast can also be seen as a peak cathodal to the alpha 2 region on capillary electrophoreses mimicking a monoclonal antibody (Fig. 3.5) [16].

3.6 Analytical issues

3.6.1 Glucose-6-phosphate dehydrogenase

Glucose-6-phosphate dehydrogenase (G6PD) is a cytoplasmic enzyme that is essential for an erythrocyte's capacity to withstand oxidant stress. G6PD activity is significantly decreased in the genetically inherited disorder, G6PD deficiency. It is typically measured in RBCs during workup for an unknown cause of hemolytic anemia or neonatal jaundice. In the cancer population, it is used to screen for G6PD deficiency in patients who may receive oxidative chemotherapies or are at a higher risk of tumor lysis syndrome (TLS) which might necessitate treatment with rasburicase. Rasburicase is an oxidative agent used to metabolize uric acid which at elevated levels can cause acute kidney injury. Sometimes, the onset of TLS is unanticipated, and thus a rapid G6PD assay is needed. Accurate G6PD measurements are not performed in many laboratories and may require being sent out to a reference laboratory. Due to the possible unanticipated need for a rapid G6PD level in the cancer patient population, institutions that treat many leukemia patients may need to have this test readily available. Qualitative tests are challenging in this population as they are affected by low hematocrits. Quantitative tests can also be challenging in this patient population as high WBCs falsely elevate G6PD levels, and thus the WBCs need to be removed prior to determining the G6PD levels in the RBCs. G6PD activity is calculated using both a measurement of G6PD and a measurement of the amount of hemoglobin, each of which requires its own method, and which may be performed in two separate sections of the laboratory, an additional challenge to coordinate.

3.6.2 Therapeutic antibodies

Both intravenous immunoglobulins (IVIG) as well as monoclonal antibodies used in the treatment of cancer patients can cause interferences/false results in a number of tests. It is well known that IVIG preparations can contain antibodies to a number of specific antigens including blood group antigens and pathogens [17]. This is specifically called out in IVIG package inserts. "After infusion of IgG, the transitory rise of the various passively transferred antibodies in the patient's blood may yield false-positive serological testing results, with the potential for misleading interpretation. Passive transmission of antibodies to erythrocyte antigens (e.g., A, B, and D) may cause a positive direct or indirect antiglobulin (Coombs') test" [18]. Similarly, monoclonal antibodies, such as anti-CD38 (daratumumab) for the treatment of multiple myeloma can cause false-positive indirect antiglobulin tests that resemble autoantibodies and might result in a false diagnosis of autoimmune hemolytic anemia if administration of this treatment is not identified by the blood bank [19].

Additionally, daratumumab (as well as other monoclonal therapeutic antibodies) is detected on protein electrophoresis and immunofixation studies. Its half-life is about 18 days and it can be detected for several months depending on the dosage and number of administrations. In such cases, it is important to know what the patient's original clone typing and migration are, and if the patient has recently received daratumumab. With this information, one can often interpret results appropriately except in instances where the patient's original clone is IgG kappa and migrates at a location similar to that of daratumumab on protein electrophoresis [20]. In these instances, an assay that modifies the migration of daratumumab can be used to remove its interference enabling accurate interpretation of the assay [21]. Please refer to Chapter 6 for an excellent discussion on this topic.

Therapeutic monoclonal antibodies may also decrease analytes that it is against, in vitro. For instance, in patients with cytokine release syndrome being treated with siltuximab, an anti-IL6 antibody, IL6 levels could decrease in a specimen over time. Any delay in testing could cause falsely low IL6 level results and may negatively affect proper clinical decision making.

3.6.3 Analytic sensitivity lower range

In the cancer patient population as part of the disease or its treatment, many analytes fall markedly outside the reference ranges—and often outside of the normal reportable range of many methods. Assays for cancer markers like Prostate Specific Antigen (PSA), which had their start as cancer screening tests, were developed to detect elevations of the protein, and thus did not require levels of detection that were below what were found in a normal person. However, in patients who have undergone prostatectomy, even a minute amount of the cancer marker might signify recurrence. For PSA, some assays go down to below 0.1 or even 0.01 ng/mL, and thus methods will need to be used that have an analytical sensitivity that is acceptable to your oncologists. For cancers that are stimulated by hormones, such as testosterone or estradiol, there are patients who are on medications that suppress these hormones. For these patients, detection of any amount of those hormones might mean the suppressive medication is not working sufficiently. Again, traditional assays for many of these hormones were not developed to measure accurately to such low levels, and therefore specialized assays, such as are often performed by mass spectroscopy, will be needed. For further discussion, please refer to Chapters 1 and 5.

3.6.4 Analytic sensitivity higher range

Tumor markers can also reach levels well above the reportable range for many assays. Tests are developed to report up to a certain level, a level that can be easily and routinely validated. Generally, this is done using materials with a known high concentration to show that measurements are accurate at that high level. If a cancer patient has concentrations of a tumor marker above this reportable range, which is not uncommon, the result is often reported as greater than the highest reportable value. For instance, if the highest reportable value is 1000 mg/dL, then any value above that concentration would be reported as >1000 mg/dL, even if the concentrations were 10,000 mg/dL or even 1,000,000 mg/dL. This poses a problem for clinicians who want to see if a treatment is working for a given patient. If the patient starts with an extremely high concentration of tumor marker such as 10× greater the high reportable range, even a reduction of 75% of the tumor marker would not be detected with this standard reporting scheme. The solution to this endeavor is to dilute the sample until it comes within the reportable range, and then multiply by the dilution factor. The challenge with this process is that the laboratory must have an extremely accurate and precise method for making the dilutions to ensure results are accurate. Laboratories should have a rigorous validation of their dilution processes before reporting out results above their normal reportable range.

Another issue with extremely high concentrations of a tumor marker is the hook effect in noncompetitive sandwich immunoassays. The hook effect (also discussed in Chapter 1) occurs when the analyte concentration is so elevated that both the capture and signal antibody become saturated, inhibiting the bridging between the two antibodies which is essential for measurement [22]. In effect, the measured level is significantly lower that the true level of that analyte. This is often detected when a patient with a relatively high concentration of tumor marker suddenly shows a decrease of that marker even though the patient is clinically worsening. The actual situation is that the tumor marker has increased to such a high level as to cause a hook effect in the assay. Potentially, progression could be missed if not suspected clinically. It is also possible that if the patient had started a new therapy, this might be interpreted that the therapy was working when in reality it was not.

3.7 Postanalytical issues

3.7.1 Critical values

Critical values, as defined by Lundberg, are life-threatening results that require immediate notification to the patient's healthcare provider [23,24]. To simplify the identification of these critical values, the levels are often set at a single level without considering the condition of an individual patient type. Cancer centers have additional challenges with this concept as many of their patients have laboratory test values that would typically be critical in the general population. However, in the cancer population, these same findings are often expected and do not warrant immediate attention. One such example is the CBC parameters of platelet count and WBC count. Often platelets <20 K/mcL or WBC <1.0 K/mcL are considered critical values. However, in patients undergoing bone marrow ablation, this is an expected finding. Using these conventional cutoffs would create an enormous number of critical value notifications which are not clinically necessary and would create an unneeded amount of burden on laboratory staff and clinical care providers. In this situation, it is important to create more complex rules that prevent unneeded calls for these expected low values. This can be done by creating a rule set that looks at the delta rather than the absolute result and checks to see if the patient is inpatient or outpatient. Therefore platelet counts <20 K/mcL are only critical if they occur in the outpatient setting, or in an inpatient setting where within the last 7 days there was either a platelet count of >30 K/mcL or no available platelet count. This rule can dramatically decrease the number of clinically unimportant critical results and alleviate the workload for both lab staff and clinicians.

Another critical value that should be evaluated is total calcium. As we described before, albumin tends to run low in cancer patients due to chronic inflammation caused by cancer. A low albumin will reduce the total calcium even to levels that would trigger a critical value. One way to address this is to report out the corrected calcium and have critical results rules apply to only this corrected result. This was instituted at MSK and resulted in a marked reduction in the number of unnecessary critical calls [21]. Another consideration would be to only use ionized calcium for this patient population; however, the lack of automated methods makes this exceedingly difficult in a high-volume setting.

Another consideration is to look at how often to call a critical value of the same type on the same patient. Again, many of the patients will have values that would otherwise be called "critical" due to chemotherapy or bone marrow ablation. Calling these values repeatedly does not offer any value to the patient or the clinician, so it is best to have rules to avoid these situations. One effective way of dealing with this is to only call a critical value of a particular analyte once in a certain time period. An easy method is to only call the first of consecutive critical values for a respective analyte each 24 h. For example, if a patient's hemoglobin level were critically low and remained critically low in repeated measurements, it would not be called multiple times. Only if the level returned to a noncritical level or a designated period had passed, aka 24 h, would another critical value be called to the ordering provider.

3.8 Conclusion and future directions

In conclusion, clinicians and laboratorians must be aware of the nuanced challenges that come with lab testing in cancer patients be it due to physiologic changes, cancer produced interference, or therapy-related changes/interferences. One must always be on the look out for such issues as new tests, therapeutics, and cancers that are continuously being discovered which might affect laboratory results. Test vendors are not required to check for every possible interfering substance out there, so labs have to run their own inference studies. This is even more important in cancer centers where clinical trials using experimental drugs are taking place on a regular basis.

3.9 Summary points

- Cancer patients are usually in an inflammatory state increasing or decreasing acute phase proteins beyond what is expected in the general population.
- Albumin, a negative acute phase protein, is often low in cancer patients which affects other analytes including total calcium and could contribute to false low critical values.
- A variety of cancers can produce beta-hCG which may cause falsely positive pregnancy tests used to clear cancer patients for treatments and procedures.
- General population analyte cutoffs for sepsis screening are not applicable to the cancer patient population
- Laboratory results may be affected by different therapies given to cancer patients.
- Clinicians must be aware that several types of preanalytical errors are more prone in cancer patients including pseudohyperkalemia, hemolysis, and contamination/dilution by venous line contents.

References

[1] Morley JJ, Kushner I. Serum C-reactive protein levels in disease. Ann N Y Acad Sci 1982;389:406–18.

[2] Keren DF. Protein electrophoresis in clinical diagnosis. Chicago, IL: American Society for Clinical Pathology Press; 2012.

[3] Kottke-Marchant K. An algorithmic approach to hemostasis testing. 2nd ed. Northfield, IL: CAP Press; 2016.

[4] Peerschke EI, Pessin MS, Maslak P. Using the hemoglobin content of reticulocytes (RET-He) to evaluate anemia in patients with cancer. Am J Clin Pathol 2014;142:506–12.

[5] Blouin AG, Hsu M, Fleisher M, Ramanathan LV, Pastores SM. Utility of procalcitonin as a predictor of bloodstream infections and supportive modality requirements in critically ill cancer patients. Clin Chim Acta 2020;510:181–5.

[6] Durnas B, Watek M, Wollny T, et al. Utility of blood procalcitonin concentration in the management of cancer patients with infections. Onco Targets Ther 2016;9:469–75.

[7] Carroll B, Richardson S, Murata K, Ramanathan L, Pessin M. Efficacy of calcium correction in a cancer population. Am J Clin Pathol 2017;147:S172–3.

[8] Burtis CA, Bruns DE, Sawyer BG, Tietz NW. Tietz fundamentals of clinical chemistry and molecular diagnostics. 7th ed. St. Louis, MO: Elsevier/Saunders; 2015.

[9] McCash SI, Goldfrank DJ, Pessin MS, Ramanathan LV. Reducing false-positive pregnancy test results in patients with cancer. Obstet Gynecol 2017;130:825–9.

[10] Gronowski AM, Fantz CR, Parvin CA, et al. Use of serum FSH to identify perimenopausal women with pituitary hCG. Clin Chem 2008;54:652–6.

[11] McPherson RA, Pincus MR. Henry's clinical diagnosis and management by laboratory methods. 24th ed. Philadelphia: Elsevier; 2021.

[12] Amgen. Filgrastim. Amgen; 2012. Reference ID: 3136501, https://www.accessdata.fda.gov/drugsatfda_docs/label/2012/103353s5147lbl.pdf.

[13] He RR. Abstracts of the American Society of Hematology 53rd Annual Meeting, December 10–13, 2011, San Diego, California, USA. Blood 2011;118:3–1820.

[14] Mansoor S, Holtzman NG, Emadi A. Reverse pseudohyperkalemia: an important clinical entity in chronic lymphocytic leukemia. Case Rep Hematol 2015;2015:930379.

[15] Meng QH, Wagar EA. Pseudohyperkalemia: a new twist on an old phenomenon. Crit Rev Clin Lab Sci 2015;52:45–55.

[16] Lippi G, Daves M, Mattiuzzi C. Interference of medical contrast media on laboratory testing. Biochem Med (Zagreb) 2014;24:80–8.

[17] Suresh J, Kyle BD. Clinical false positives resulting from recent intravenous immunoglobulin therapy: case report. BMC Infect Dis 2021;21:288.

[18] BH Corporation. Highlights of prescribing information: GAMMAGARD S/D. BH Corporation; 2014, https://www.fda.gov/media/78185/download.

[19] Oostendorp M, Lammerts van Bueren JJ, Doshi P, et al. When blood transfusion medicine becomes complicated due to interference by monoclonal antibody therapy. Transfusion 2015;55:1555–62.

[20] Murata K, McCash SI, Carroll B, et al. Treatment of multiple myeloma with monoclonal antibodies and the dilemma of false positive M-spikes in peripheral blood. Clin Biochem 2018;51:66–71.

[21] Sebia. HYDRASHIFT 2/4 daratumumab (20 tests). Sebia; Ref: 4639-2019, 2019.

[22] Nerenz RD, Song H, Gronowski AM. Screening method to evaluate point-of-care human chorionic gonadotropin (hCG) devices for susceptibility to the hook effect by hCG beta core fragment: evaluation of 11 devices. Clin Chem 2014;60:667–74.

[23] Lundberg GD. It is time to extend the laboratory critical (panic) value system to include vital values. MedGenMed. 2007;9:20.

[24] Lundberg G. When to panic over abnormal values. Med Lab Obs 1972;4:47–54.

CHAPTER

Thyroglobulin and thyroid cancer 4

William S. Phipps[1], Andrew N. Hoofnagle[1,2], Mara Y. Roth[2] and Christopher M. Shuford[3]

[1]*Department of Laboratory Medicine and Pathology, University of Washington, Seattle, WA, United States*

[2]*Department of Medicine, University of Washington, Seattle, WA, United States*

[3]*Laboratory Corporation of America Holdings, Burlington, NC, United States*

Objectives

- Describe the established roles of measuring thyroglobulin (Tg, and anti-Tg antibodies) in the management of differentiated thyroid cancer.
- Discuss the history of quantitative Tg measurement by immunoassay and the impact of Tg autoantibodies in clinical practice.
- Outline reported liquid chromatography-tandem mass spectrometry methods that use peptide immunoaffinity enrichment.
- Discuss challenges facing mass spectrometry-based assays, and ongoing efforts to improve quality of testing.

4.1 Introduction

Thyroglobulin (Tg) is a 660 kDa homodimeric glycoprotein central to thyroid hormone production [1,2]. Tg is synthesized by thyroid follicular cells (thyrocytes) and stored extracellularly as "colloid" within thyroid follicles. Through the process of iodination (by thyroid peroxidase), certain tyrosine residues become linked with iodine to form mono- and diiodotyrosine (MIT and DIT, respectively), which are then coupled to generate triiodothyronine

Cancer Biomarkers: Clinical Aspects and Laboratory Determination. DOI: https://doi.org/10.1016/B978-0-12-824302-2.00006-0
© 2022 Elsevier Inc. All rights reserved.

(T3, from MIT and DIT) and thyroxine (T4, from DIT and DIT) [3]. Upon stimulation by thyroid stimulating hormone (TSH), thyrocytes resorb and degrade iodinated Tg to release T4 and T3 for secretion into the bloodstream.

Unlike the TSH receptor, which is expressed in tissues throughout the body, Tg expression is unique to thyroid tissue, and leakage of Tg to the blood provides a metric for the total mass and activity of thyroid-differentiated tissue [4–7]. Tg measurement is primarily applied following total thyroidectomy for the treatment of differentiated thyroid cancer (DTC), and in monitoring for recurrence [8]. DTC arises from thyroid follicular cells and comprises papillary (predominant), follicular, and Hürthle cell types. Overall, thyroid cancer (of which DTC comprises 80%) is the most common endocrine malignancy, with estimated new cases in the United States exceeding 40,000 per year [9,10]. It is typically treatable, with a five-year survival rate of >98% (data from 2011 to 2017) [10]. Measurement of serum Tg is an essential tool for cancer surveillance. However, the presence of anti-Tg antibodies (TgAb), present in a substantial portion of the cancer population (up to 30%), poses challenges to Tg measurement and interpretation by traditional assays. Here, we review the current clinical application of Tg measurement, the challenge posed by anti-Tg antibodies to standard immunometric platforms, and the development of mass spectrometry-based approaches. Efforts to harmonize newer methods and further improve quantification at extremely low protein concentrations are discussed. Additionally, we provide clinical workflows for using mass spectrometry alongside standard immunoassays and an outlook on future directions in testing.

4.2 Thyroid nodule evaluation and the limited role of preoperative serum thyroglobulin measurements

The primary clinical manifestation of DTC is a thyroid nodule, either palpable during physical exam, or discovered incidentally by

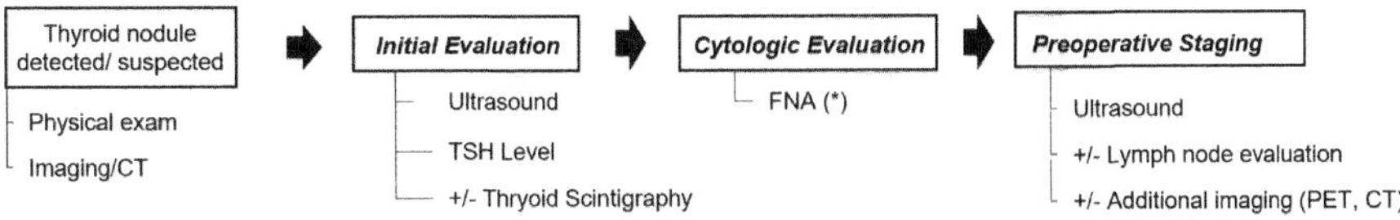

FIGURE 4.1

The diagnostic evaluation of differentiated thyroid cancer includes an initial evaluation by US and circulating thyroid stimulating hormone (TSH) level to assess the need for cytologic evaluation by fine needle aspiration (FNA). Nodules that are concerning for malignancy are treated with surgery, the extent of which depends on findings in preoperative staging. Although serum Tg is not used in diagnosis, very high levels may point to underlying metastasis. Repetitively nondiagnostic FNA (*) may be followed by a core biopsy or diagnostic surgery. *Tg*, thyroglobulin.

imaging. Only a small fraction (~5%–15%) of nodules harbor malignancy [11–13]. If palpable, malignant thyroid nodules tend to be hard and nonpainful, but patients may be entirely asymptomatic. Initial testing (Fig. 4.1) consists of ultrasound (US, to assess the nodule size, number, sonographic features, and adjacent structures) and TSH measurement (to gauge nodule functional activity). Reduced TSH levels suggests a nodule is hyperfunctioning—that is, producing an excess of thyroid hormones with resulting negative feedback on TSH production. Radionuclide thyroid scanning (scintigraphy) can confirm a nodule is hyperfunctioning (or "hot"), and therefore unlikely to be malignant [14,15]. In contrast, nodules associated with normal or increased TSH levels (or those identified as "cold" by scintigraphy) are more concerning for underlying carcinoma and must be further evaluated on the basis of sonographic features (e.g., size, shape, microcalcifications). If suspicious, cytologic evaluation is performed by fine needle aspiration (FNA), with estimated risk for malignancy based on diagnostic categories defined within one of several reporting systems (e.g., Bethesda System for Reporting Thyroid Cytopathology) [16–18]. Nondiagnostic aspirations can be followed with a repeat FNA, genetic testing, or diagnostic surgery, if appropriate [19]. Although more commonly applied to lymph nodes, the measurement of endocrine markers (including calcitonin and Tg) can be applied to the FNA washout fluids to aid in distinguishing DTC from medullary thyroid carcinoma or other pathology [18,20,21].

Due to elevations of Tg in nonmalignant conditions and thus a lack of specificity, the measurement of serum Tg does not play a clear role in routine nodule evaluation [18,22–24]. In a 2014 systematic review evaluating 10 studies of Tg as a diagnostic marker for carcinoma, only one study reported no significant difference in preoperative serum Tg levels between benign and malignant nodule groups [25,26]. However, proposed diagnostic cutoffs varied from 75 to 1000 ng/mL with modest-to-poor sensitivities and specificities, and frequently worse positive and/or negative predictive values. Further studies investigating the diagnostic utility of serum Tg continue to demonstrate mixed performance (Table 4.1), including one in which serum Tg appeared significantly higher in patients with underlying benign pathology [24,27–29].

Surgery is the first-line treatment for DTC, with the choice of procedure (e.g., lobectomy vs total thyroidectomy) dependent on the extent of disease [30]. US is foundational in this assessment, both for characterizing the primary tumor and detecting involvement of cervical lymph nodes (present in half, or potentially more of newly diagnosed cases) [23,31]. Lymph node sonographic features concerning for malignancy (e.g., calcifications) are well described, but offer variable sensitivity and specificity. Thus confirmation of lateral neck lymph node involvement (and beyond) may be performed by US-guided FNA for cytology and Tg measurements of saline rinses of FNA needles [23]. The latter can provide additional clarity, particularly in the setting of indeterminate cytologic findings [18]. However, despite reported "high accuracy" of FNA-Tg, there are challenges with this technique, for example, methods for sampling, collection, processing, and measurement have not been standardized [20,32]. As a result, broadly validated FNA-Tg cutoff levels are not available and may need to be established within a particular institution or practice [18]. Interpretation of intermediate levels can be challenging and must consider the possibility of serum contamination, particularly in patients who have not yet undergone thyroidectomy [18,23,32]. Nonetheless,

Table 4.1 Recent investigations of serum thyroglobulin (Tg) as a diagnostic marker for thyroid carcinoma.

Reference (year)	Study population	Tg testing platform	Findings (Tg in ng/mL)[a]	Suggested cutoff
[28] (2020)[b]	*N* = 92 total (47 malignant)	Beckman Access-2	Benign: 0.1–500 (median 29) Malignant: 0.3 to 816 (median 162)	53 ng/mL (sensitivity, 72%; specificity, 73%)
[29] (2017)[c]	*N* = 198 total (32 malignant)	Cisbio Bioassay (IRMA)	Benign: median 34 Malignant: median 449	165 ng/mL, (sensitivity, 69%; specificity, 84%)
[24] (2016)[d]	*N* = 202 (68 malignant)	Roche COBAS, e411	Benign: 0.25–1286 (median 64) Malignant: median 0.07–567 (median 20)	None
[27] (2014)[e]	*N* = 611	Not specified	Preoperative levels not associated with cancer	5000 ng/mL (metastatic disease)[f]

IRMA, *Immunoradiometric assay.*
[a]*Baseline/preoperative Tg levels.*
[b]*Prospective study (PMID 32275647) of patients ages 18–70 presenting with "thyroid swelling" and who underwent thyroidectomy. The study excluded patients with recurrent disease, diffuse thyroid enlargement, clinical hypothyroidism, thyrotoxicosis, proven extrathyroidal disease, and other factors/ cancers.*
[c]*Retrospective study (PMID 27936964) of 854 consecutive patients who underwent thyroid surgery, 198 of which presented with a thyroid nodule.*
[d]*Study (PMID 27376428) of patients who underwent thyroid surgery. 202 of 254 consecutive patients were included. Excluded patients with medullary thyroid cancer, high thyroid stimulating hormone level, or "elevated" anti-Tg antibodies.*
[e]*Patients undergoing total thyroidectomy or thyroid lobectomy without previous thyroid surgery were included (PMID 24411304,* n = *1014). Only 611 had preoperative Tg levels for review.*
[f]*All patients with Tg greater than 5000 ng/mL had metastatic disease* (n = *6).*

extreme concentrations may be universally informative. For example, undetectable Tg (<1 ng/mL) can be considered reassuring, while very high levels (1000s–10,000s ng/mL) may conclusively point to presence of metastatic tumor cells. Of note, the measurements of Tg in FNA needle rinses may be subject to many of the same analytical factors impacting measurement in serum. These include "hook effect" and Tg autoantibodies as described later in the chapter [32,33].

Although serum Tg measurement is not a routine component in the preoperative workup of DTC and is not recommended by clinical guidelines, several uses have been described. Because circulating Tg levels correlate with the "burden" of thyroid follicular cells present in the body, above certain thresholds, preoperative elevations could potentially signify initial metastatic disease, a rare situation associated with increased mortality [34]. In one larger of study of 4735 individuals with DTC, 57 (1.2%) had initial distant metastases, comprising 26 cases with micronodular lung metastasis, 5 with macronodular lung metastasis, 2 with single bone metastasis, 15 with multiple bone metastases, 9 with combined lung and bone metastases. The initial distant metastasis group had higher median Tg (328.4 ng/mL, compared to 10.0 ng/mL in the non-DM group), with highest elevations appearing in patients with bone metastases. Nonetheless, a modest cutoff of 63.4 ng/mL was selected based on receiver operating characteristic curve analysis in differentiating between the presence of initial distant metastases and nondistant metastases groups with 84.2% sensitivity, 90.6% specificity, 96.4% negative predictive value, and 9.29% positive predictive value. Notably, the optimal cutoff shifted when limiting analyses to cases ultimately diagnosed as follicular thyroid carcinoma (as opposed to papillary thyroid carcinoma, and the group as a whole) to 1192.0 ng/mL (75.0% sensitivity, 92.5% specificity, 68.3% positive predictive value, and 94.5% negative predictive value). In another study, all patients with preoperative Tg >5000 ng/mL had metastases to lung or bone, or extensive lateral cervical disease [27]. However, it is not clear from these published examples that measuring serum Tg preoperatively could identify metastatic disease that would not otherwise be detected using standard procedures.

Apart from potential identification of initial distant metastases, some authors have suggested that achieving optimal utilization of Tg in the postoperative state depends on knowing the preoperative level [27,35]. For example, postoperative Tg levels may be misleading in patients for whom tumors do not produce Tg or for whom preoperative levels appear already low or undetectable [27,36]. Overall, the evidence is limited for such cases, and the value of performing preoperative testing to identify these patients may be assay-specific [23]. For example, one study of 436 patients with DTC reported that 17 patients had undetectable or reduced Tg levels using a given assay [36]. The majority of these patients had measurable levels when retesting by a different immunoassay (despite exclusion of cases that had tested positive for TgAb).

4.3 Postoperative evaluation of differentiated thyroid cancer

By American Joint Committee on Cancer (AJCC) standards, pathological staging of DTC considers tumor size, extrathyroidal extension, involvement of specific surrounding anatomy (e.g., carotid artery), and nearby and distant lymph node metastases [23]. Staging alone is predictive of cancer-related mortality, but not of persistent or recurrent disease [37]. Nonetheless, many of the characteristics identified in staging are important determinants in stratifying risk of recurrent or persistent disease when considered *in combination* with additional histologic, genetic, and clinical features (including Tg levels) [23,37]. Based on these factors the 2009 American Thyroid Association (ATA) Risk Stratification system places patients into "low," "intermediate," and "high"-risk groups, corresponding to validated likelihoods of achieving remission (so-called "no evidence of disease," or NED) following total thyroidectomy and radioactive iodine (RAI) ablation—78%–91%, 52%–63%, and 14%–31%, respectively (see Tables 11 and 12 in 2015 ATA Guidelines). The postoperative risk level aids, for example, in the initial decision to perform RAI, which is recommended for high-risk and select intermediate-risk cases. It also impacts target the TSH range when administering thyroid replacement therapy. At sufficient doses, levothyroxine can—in addition to preventing hypothyroidism—suppress TSH below the reference range and thus reduce potential stimulation of residual carcinoma. For low-risk patients with undetectable Tg, a target range in the lower half of the reference interval may be more appropriate (0.5–2 mU/L), whereas for intermediate- and high-risk cases, as well as low-risk with still detectable Tg following total thyroidectomy, TSH suppression below this range may be indicated ("mild suppression" of 0.1–0.5 mU/L, or "full suppression" <0.1 mU/L).

Although the ATA Risk Categories offer initial guidance in management, these preliminary "static" risk levels do not adequately account for variation in longer term outcomes [37,38]. As a result, standard of care in the management of DTC now includes continual re-evaluation of risk as new information is acquired. Tuttle and colleagues conceived a dynamic "response to therapy" categorization scheme by which management is adjusted in accordance to biochemical and structural investigations performed at follow-up visits [37,39]. Routine monitoring for DTC at treatment follow-up constitutes imaging (typically US) and serum Tg measurements performed several times each in the first 1–2 years following total thyroidectomy, and potentially less frequently over time. Biochemical response to therapy is assessed via measurement of serum Tg, with a "biochemical incomplete" response constituting serum Tg concentrations either above a specified threshold, or "rising" (Table 4.2). More intermediate Tg levels may support a so-called "indeterminate" response, if stable or declining. The presence of TgAb also contributes to categorization, reflecting its potential as a surrogate tumor marker (see Section 4.5). However, structural evidence of disease (based on imaging or other approaches) supersedes any Tg or TgAb findings leading to placement in a "structural incomplete" category. The response to therapy categorization has potential management implications with respect to frequency of follow-up, intensity of TSH suppression, and performing additional imaging.

Table 4.2 Thyroglobulin (Tg) levels in treatment response categories (posttotal thyroidectomy and RAI).[a]

Category	Tg-suppressed (ng/mL)[b]	Tg-stimulated (ng/mL)[b]	TgAb	Other requirement (s)
Excellent[c]	<0.2	<1	Absent	AND negative imaging
Indeterminate	≥0.2 but <1.0	Detectable, <10	Stable or declining	OR imaging nonspecific, OR RAI scanning faint uptake in thyroid bed
Biochemical incomplete	≥1	≥10	Rising	AND negative imaging
Structural incomplete	Any	Any	Any	AND structural or functional evidence of disease

RAI, *Radioactive iodine.*
[a]*Adapted from Table 13 in PMID 26462967.*
[b]*In expert guidelines, these cutoffs are stated "In the absence of anti-Tg antibodies."*
[c]*"Excellent" disease status is essentially equivalent to "No Evidence of Disease" (defined as stimulated Tg <1 ng/mL with no other radiological or clinical evidence of disease).*

4.4 Thyroglobulin in the monitoring of differentiated thyroid cancer

Serum Tg has a half-life of 30–60 h [40,41], and is typically measured starting at least 4 weeks postoperatively to begin monitoring. Levels are influenced by both residual thyroid tissue mass (benign or malignant) and its TSH-dependent stimulation. Thus, in the postoperative state, both Tg and TSH are modulated through the administration of thyroid hormone replacement therapy, which reduces TSH levels through negative feedback at the anterior pituitary and hypothalamus. Tg evaluations performed while taking suppressive doses of thyroid hormones are termed "nonstimulated" (TSH-suppressed), whereas those performed following withdrawal of thyroid hormone or the administration of recombinant TSH (rTSH) are termed "stimulated." Increasingly sensitive assays for serum Tg ("2nd generation," described further in Section 4.6) have enabled broader use of nonstimulated measurements, where feasible, avoiding the inconvenience of interrupting TSH suppression and the expense of rTSH. Of note, the definition of "excellent" biochemical response in Table 4.2 corresponds specifically to patients treated with both total thyroidectomy and RAI. For this population, the listed serum Tg thresholds correspond to empirically determined recurrence rates of 1%–4% (over 5–10 years) for patients with ATA-low or intermediate risk disease and somewhat higher risk (5%–15%) in patients with ATA-high risk disease [37]. Given that not all patients are treated with RAI [42], Tuttle et al. have suggested modifications to the "excellent" response serum Tg thresholds for patients treated with thyroidectomy-only, or lobectomy (Table 4.3). However, Tg surveillance postlobectomy is not routinely recommended due to variation in this population and difficulties in interpretation [43].

By existing guidelines, there is no specific level of increase in Tg that is considered significant during follow-up. Additionally, although detectable Tg implies biochemically persistent disease (following total thyroidectomy), many, if not most affected patients have good long-term outcomes [37,44]. Nonetheless, the trend of the serum Tg over time is considered important in clinical practice. One approach to evaluating change in serum Tg is to consider the "doubling time" (TgDT) as a measure tumor growth rate. Miyauchi and colleagues investigated the kinetics of Tg changes in a group of 1515 patients who underwent total thyroidectomy for papillary thyroid cancer [38,45]. Four hundred twenty-six (426) patients had at least four Tg measurements performed over a median of 87 months, and were negative for TgAb (cases positive for TgAb were excluded due to presumed interference). One-hundred and thirty-seven (137) of these had detectable levels that increased, remained stable or decreased with time. Under TSH suppression (<0.1 mIU/L), observed changes in Tg

Table 4.3 Excellent biochemical response based on treatment and initial risk category.

Treatment	Tg cutoff (ng/mL)		ATA-risk (initial category)	Recurrence risk[a]
	Nonstimulated	Stimulated		
TT/RAI	0.2	1	Low or Intermediate	1%–4%[b]
			High	5%–15%
TT only	0.2	2	Low	<1%–3%[c]
Lobectomy	30[d]	n/a	Low	3.8%[e]

ATA, *American Thyroid Association;* n/a, *not applicable;* RAI, *radioactive iodine;* Tg, *thyroglobulin;* TT, *total thyroidectomy.*
[a]*Based on studies cited in PMID 24891169 (TT/RAI and TT only).*
[b]*Specified as over 5–10 years.*
[c]*For the TT only group, the stimulated cutoff is based on investigations of patient risk of with preablation Tg measurements. Stimulated values less than 2.0 ng/mL are associated with a risk of persistent or recurrent disease of less than 1%–3%, increasing to more than 20% for results greater than 10.0 ng/mL.*
[d]*Expected "excellent response" to lobectomy is based on an extrapolation from expected Tg levels in patients with intact thyroid and no thyroid disease (approximately 60 ng/mL), halved to reflect half of normal tissue. Serum Tg measurements postlobectomy are not routinely recommended.*
[e]*PMID 28089996, over median 8.5 years.*

were exponential (in line with the concept that the growth of tumors is exponential [46]), and assessment of doubling time was based on nonlinear least square regression, $\log(y) = b^*x + \log a$, where y equals Tg concentration and x is the time in years postsurgery. TgDT was then calculated as $\log 2/b$. In the multivariate analyses, when considered among many "classical" prognostic factors (e.g., tumor node metastasis stage), the authors found TgDT to be the only independent predictor of survival, distant metastases, and locoregional recurrence over the studied time period. Survival in patients with TgDT <1 year was particularly poor (50%), improving to over 90% with TgDT > 1 year. A subsequent study focusing on patients with progressive disease ($N = 99$) did not confirm the same simple correlation between TgDT and survival (this was ascribed to differences in study design) but did note more than double the mortality rate in individuals with

TgDT <5 months when compared with individuals with TgDT >14 months [47]. Interestingly, the differences in survival were limited to individuals with significant tumor burden (defined as serum Tg >100 ng/mL). More recently, Zhang et al. [48] assessed the utility of TgDT in early detection of progressive disease occurring in patients with known metastatic DTC. Although the study was small ($N=21$), the authors observed doubling times less than 1 year for all cases ultimately demonstrating progressive disease ($N=11$), and doubling times greater than 1 year for all cases associated with non-progressive disease. The most accurate differentiation was observed when limiting Tg values used in TgDT calculation to the four most recent Tg measurements.

Other investigations have explored the relationship between TgDT and findings in patients who undergo imaging using combined 18F-2-fluoro-2-deoxy-D-glucose (18F-FDG)-PET and X-ray computed tomography (18FDG-PET/CT). 18FDG-PET/CT is one of several whole-body imaging techniques used in the postoperative evaluations of DTC. It has particular utility in detecting metastatic foci negative by ^{131}I whole-body scanning (^{131}I-WBS) and has prognostic implications similar to TgDT [49,50]. In advanced DTC in particular, cancer cells lose the ability to take up iodine, but can still produce Tg. Unsurprisingly then, such cases have been found to be associated with shorter TgDT [51]. In turn, determination of TgDT has been reported as a potential tool in improving selection of patients for 18FDG-PET/CT scanning, or discerning its potential accuracy in a given patient [52,53].

4.5 Utilization of thyroglobulin autoantibodies as surrogate marker

TgAbs are immunoglobulin G antibodies directed against Tg. These occur in nearly a third of patients with DTC, and are also found at significant levels in patients with benign thyroid disorders [54]. TgAbs bind Tg in clinical samples and can interfere with traditional immunoassays, but also appear to offer a potential surrogate marker for DTC recurrence and progression (Table 4.4) [7,55–58]. Similar

Table 4.4 Select investigations of TgAb as a prognostic marker.

Reference (year)[a]	Study Details	Findings
[55] (2002)	• Comparison of persistence/recurrence in patients with or without TgAb • DTC patients s/p TT and RAI, with undetectable Tg[b] (*N* = 226) • Patients followed for average 14 months	• 51 cases were TgAb positive. 25 of 51 (49%) had persistent or recurrent disease in follow-up. • 6 of 175 (3.4%) TgAb-negative cases had persistent or recurrent disease in follow-up.
[56] (2008)	• Evaluation of change in TgAb as a predictor of clinical recurrence • DTC patients s/p TT and RAI, with undetectable Tg (*N* = 824)	• 56 had positive TgAb at 6–12 after treatment. • 10 of 56 (18%) TgAb-positive had recurrence. • 10 of 768 (1%) TgAb-negative had recurrence. • Recurrence was highest (37%) in patients with increasing TgAb.
[57] (2018)	• Low- or intermediate-risk PTC patients s/p TT and RAI, with undetectable Tg and TgAb below reference limit (*N* = 576)	• TgAb detectable: 5/156 (3.2%) developed recurrence. • TgAb undetectable: 11/420 (2.6%) developed recurrence.[c]
[58] (2020)	• Evaluation of the significance of de novo detection of TgAb • DTC patients (*N* = 1545) from national registry, 1325 initially TgAb-negative, 812 included in final analyses with at least 3 years follow-up	• 40 of 812 became TgAb-positive. • Recurrence in TgAb-persistent negative group (9.6%) was lower but not significantly than in the de novo TgAb-positive group (15.0%).

DTC, *Differentiated thyroid cancer;* PTC, *papillary thyroid cancer;* RAI, *radioactive iodine;* Tg, *thyroglobulin;* TT, *total thyroidectomy.*
[a]*PMID: 12153600, 18812478, 29325506, and 32228151.*
[b]*By immunoradiometric assay.*
[c]*Conclusion was that low- and intermediate-risk PTC patients with borderline TgAb are not at higher risk or recurrence.*

to serum Tg, existing guidelines do not recommend routine measurement of anti-Tg antibodies in preoperative assessments, including in the evaluation of thyroid nodules [23]. However, following surgery, assessment of trend of TgAb is included in guidelines as a prognostic factor to consider.

4.6 The measurement of thyroglobulin by immunoassays and the challenge of autoantibodies

Quantitative measurement of serum Tg originated in the 1960s and 1970s with the development of radioimmunoassay (RIA) and immunoradiometic (IRM) methods [59–64]. Assay "sensitivity" in early reports ranged from to 1.0 to 10 ng/mL. Since this period, Tg immunoassays have undergone improvements with respect to format and analytical performance [65]. Modern automated platforms employ a sandwich chemiluminescent readout, eliminating the use of radioactive "tracer" reagents (used by RIAs) or radiolabeled antibodies (used in IRM assays). The suitability of an assay in reporting low serum Tg levels is described by its so-called "functional sensitivity," that is, the lowest analyte concentration that has between-run variability less than 20% of the measured value over the typical interval of measurements (6–12 months for Tg) [66,67]. This terminology contrasts somewhat with the "lower limit of quantitation" (or lower limit of the measuring interval), typically defined through assessments of analytical precision at low concentrations over much shorter periods [68]. Current "second-generation" assays have a functional sensitivity of $\leq$0.10 ng/mL, representing an order-of-magnitude improvement over previous, first-generation platforms ($\sim$1.0 ng/mL). By improving detection of persistent or recurrent disease, advancements in Tg assay sensitivity have obviated the need for stimulated Tg measurements and have facilitated the shift in treatment of DTC from standardized therapy (constituting total thyroidectomy and RAI ablation for all patients) to more individualized care based on risk [67].

Despite these developments, significant challenges continue to plague serum Tg measurement, particularly with respect to assay interference. The presence of Tg autoantibodies in human sera was reported prior to the development of the first Tg immunoassays [69–71]. Investigators quickly recognized that TgAb could interfere with immunoassay-based serum Tg measurement, leading to spuriously increased or decreased results depending on assay format and other factors. In testing by RIA, which relies on competition between the endogenous analyte and radioactive reagent antigen ("tracer") for a fixed quantity of reagent antibody, bidirectional interference from TgAb was observed (constituting false elevations at low or absent Tg concentrations, and false decreases at higher concentrations) [63]. In testing by IRM assay, by contrast, unidirectional, poor recovery of Tg was observed in the presence TgAb (false-negative results). Modern sandwich immunoassays continue to demonstrate a similar negative interference, reflecting steric hindrance of antibody binding [67]. Early proponents of the IRM format suggested that falsely depressed Tg values could be corrected by means of Tg fractional recovery studies [59,63], in which exogeneous Tg is added to anti-Tg containing sera to assess the degree to which TgAbs suppress Tg readout. However, we know now that recovery experiments are fundamentally unreliable and can fail to detect inferring TgAb [67]. Of note, in addition to TgAb, serum Tg immunoassays are also impacted by other interferents (e.g., heterophile antibodies) reviewed in Chapter 1 of this book, and elsewhere [66,72].

Another critical challenge impacting Tg immunoassays is harmonization. As an analyte, Tg presents particular challenges in measurement related to its size and heterogeneity and is more accurately described as a protein family comprising various truncated and post-translationally modified states with different conformations (rather than a single analyte). Compounding this challenge is that posttranslational modifications (glycosylation, phosphorylation) are often altered in malignancy, which can impact epitope recognition by reagent antibodies [65,67,73,74]. Given the use of proprietary antibodies and other reagents in modern commercial assays, it is unlikely that the disparities between immunoassays related to antibody specificity can be resolved. This reality has been underscored by the continual failure of assay harmonization despite widespread traceability of immunoassays to an international standard reference material derived from human tissue (BCR 457) [75,76]. Use of the

BCR 457 material, which is now depleted, improved agreement between immunoassays but still left significant disparities in many cases [65,67]. Variability in TgAb testing itself has made comparison between Tg immunoassays even more challenging. In one prominent study [77], TgAb status, when measured by a range of assays, was concordant in only 65% of cases. Regardless, all Tg methods tested in the same study were prone to underestimate Tg concentration in the presence of any given positive TgAb result. Guidelines recommend a common method be used for Tg monitoring and TgAb, but this does not solve the challenge that patient care could be altered, at least theoretically, depending on which assay is selected for use at the inception of Tg monitoring.

4.7 Thyroglobulin measurement by mass spectrometry using peptide immunoaffinity enrichment

Methods for the quantification of Tg by mass spectrometry have been developed using liquid chromatography-tandem mass spectrometry (LC-MS/MS)-based analyses of surrogate peptides. In this approach, biological samples are treated with trypsin (or other protease) to release peptides representative of the target protein. Peptides, being smaller, are more reliably detected by mass spectrometers found in clinical laboratories (e.g., triple quadrupole analyzers). Additionally, the problem of analyte heterogeneity is avoided, since the same peptide can be released from many forms (e.g., multiple glycoforms) of the intact protein. One or more "proteotypic" peptides are selected as a proxy for the protein analyte on the basis of sequence specificity for the target protein, amino acid length, and other factors [78]. An MS method is then constructed using one or more "transitions" specific for the selected peptide. These consist of pairs of mass-to-charge (m/z) ratios corresponding to the intact peptide ion (Q1), and a corresponding fragment ion (Q3). The mass spectrometer is set up to monitor for these transitions over the course of an analysis. Coupling of MS with chromatography enables separation of the target peptide(s) from other peptides and interferents present in human sera (e.g., lipids), improving selectivity and sensitivity of the analysis.

The utilization of the tryptic digest in sample preparation addresses the main challenges of interference affecting immunoassays by simultaneously destroying autoantibodies and other proteins [66]. However, LC-MS/MS assays are not free from other issues. The most immediate hurdle facing the application of LC-MS/MS to the measurement of serum Tg is sensitivity. Quantitative protein LC-MS/MS has been greatly facilitated through the application of electrospray ionization, which enables nondestructive ionization of proteins/peptides and is easily coupled with chromatography. However, when using this technique, coelution of multiple peptides creates competition for charge and prevents ionization of all species—particularly low abundance peptides. This phenomenon is called "ion suppression." Tg differs from other serum proteins by more than several orders of magnitude, such that proteolytic digestion of serum will generate one peptide from Tg for every ~40,000,000 peptides from albumin (when Tg is 1 ng/mL and albumin is 4 g/dL). Consequently, some form of sample purification or enrichment is required to analyze Tg by LC-MS/MS. For other applications, investigators have attempted to decrease the limit of detection of MS analyses by immunodepleting abundant proteins (e.g., albumin, immunoglobulins). However, removing immunoglobulins could also deplete the analyte of interest in patients with autoantibodies against that target. Instead, to bring Tg measurement within range of modern immunoassays, the technique of proteolysis-peptide immunoenrichment has been applied [79,80], with several assays reported in the literature [81–84] (Table 4.5). The first reported LC-MS/MS-based Tg assay was not adequately sensitive for clinical use, with a limit of detection of 2.6 ng/mL [82]. Multiple commercial laboratories have since deployed LC-MS/MS-based Tg quantification with varying sample preparation methods and more adequate sensitivity. One approach addresses the problem of TgAb by adding rabbit anti-Tg antibody to serum or plasma prior to processing [84,85]. This "levels the playing field" for the analyte, such that all Tg molecules carry Ab across all samples (and calibrators).

Despite multiple demonstrations that LC-MS/MS can detect low levels of Tg in TgAb-positive samples, additional reports have suggested persistent false-negative Tg results occurring in patients

Table 4.5 Assays for thyroglobulin measurement by LC-MS/MS peptide immunoaffinity purification.

Reference (year)[a]	Antibody	Peptide(s)[b,c]	Reported transitions[d]	Additional features[e]
[82] (2008)	Polyclonal (rabbit)	VIF LGD FPL VIF*	636.36 → 1059.6 (541.3) 475.25 → 836.4 (779.4) 604.34 → 850.47 (963.6) 639.75 → 1065.56 (547.4)	Protein precipitation: No Internal standard: tSIL peptide LOD: 2.6 ng/mL
[83] (2012)	Polyclonal (chicken)	VIF VIF** VIF*	636.36 → 1059.6 642.32 → 1071.4	Protein precipitation: No Internal standard: tSIL Peptide LOD: 0.3 ng/mL LOQ: 0.4 ng/mL
[84] (2013)	Polyclonal	VIF VIF*	636.36 → 1059.56 639.34 → 1065.56	Protein precipitation: Yes Internal standard: cSIL peptide LOQ: 0.5 ng/mL
[81] (2020)	Monoclonal[f]	FSP FSP (IS)	703.8490 → 586.7989 703.8490 → 687.4148	Protein precipitation: No Internal standard: cSIL peptide LOQ: 0.02 ng/mL

LC-Ms/MS, *Liquid chromatography-tandem mass spectrometry;* LOD, *limit of detection;* LOQ, *limit of quantification.*
[a]*PMID 18801935, 23138009, 23396140; DOI: 10.1016/j.clinms.2020.01.001.*
[b]*Endogenous peptides: VIF, VIFDANAPVAVR; LGD, LGDQEFIK; FPLGESFLVAK, FPL.*
[c]*Isotopically labeled ($^{13}C_5$ ± ^{15}N valine labeling) internal standard (IS) peptides: VIF*, one labeled valine; VIF**, two labeled valines.*
[d]*Alternative product ion in parentheses (if provided).*
[e]*All methods utilize protein digestion and peptide immunoaffinity enrichment as well as multipoint external calibration. However, not all methods employ initial protein precipitation or the same IS approach. Most assays employ tryptic peptide stabile isotope labeled (tSIL) peptide as IS. One alternative is a cleavable (or "winged") stabile isotope labeled (cSIL) peptide, described further in PMID 28605192.*
[f]*Clinical Proteomic Tumor Analysis Consortium (CPTAC) antibody against FSPDDSAGASALLR peptide.*

with TgAb despite using an LC-MS/MS-based assay [86–89]. One possible explanation for these findings is that LC-MS/MS methods utilized had inadequate lower limit of quantification (LLOQ)/sensitivity (lower cutoff of 0.5 ng/nL was used in [86,87,89]). Even at a LLOQ of 0.1 ng/mL [88], sensitivity may not be adequate in all cases and even better analytical sensitivity may be needed. Given these findings, Shuford, et al. have described an approach to further maximize sensitivity beyond that of "second-generation" immunoassays using microflow chromatography coupled with MS/MS [81]. Standard LC-MS/MS analyses in the clinical laboratory utilize flow rates in the hundreds of microliters per minute to low milliliters per minute range. While these higher flow rates generally correlate with more robust assays and higher throughput, this comes at a cost of sensitivity. The Tg method published by Shuford et al. demonstrated that the LLOQ can be reduced approximately an order of magnitude (to 0.02 ng/mL) when analyses are performed at 3 μL per minute, without significant loss of robustness. Nonetheless, further study is required to verify if improved analytical sensitivity will correspondingly improve the clinical sensitivity of LC-MS/MS methods.

4.8 Thyroglobulin measurement by mass spectrometry—toward harmonization

As noted above, despite the existence of a certified reference material for Tg (BCR 457, now depleted), harmonization has remained out of reach for immunoassays, and it is not clear that this can ever be achieved given differences in reagent antibody specificity [77]. Preliminary investigation into the harmonization of existing LC-MS/MS-based Tg assays has demonstrated that, despite differences sample preparation, quantifying peptide, and other features, laboratory developed tests for the measurement of Tg more strongly correlate and achieve similar results compared with commercial immunoassays. Results from an example study are shown in Table 4.6 ($N = 40$ patient samples, including 18 with elevated TgAb levels) [90]. Two of the four methods used protein precipitation to enrich

Table 4.6 Pilot study comparing LC-MS/MS methods.[a]

				Comparison to Avg		Comparison to Method A		Comparison to Method A (adjusted)[c]	
Method	**PP**	**Quantifying peptide**	**Calibrator[b]**	R^2	**Slope**	R^2	**Slope**	R^2	**Slope**
A	Yes	FSPDDSAGASALLR	BCR-457	0.988	1.102	N/A	N/A	NA	N/A
B	No	FSPDDSAGASALLR	Human serum	0.981	1.043	0.970	0.942	0.970	1.16
C	No	FSPDDSAGASALLR	DxI (native)	0.996	1.027	0.983	0.920	0.983	1.05
D	Yes	VIFDANAPVAVR	DxI (native)	0.961	0.833	0.932	0.740	0.932	0.895

BCR, *Community Bureau of Reference;* LC-MS/MS, *liquid chromatography-tandem mass spectrometry;* N/A, *not applicable;* PP, *protein precipitation.*

[a]*Study described in PMID 26430076. Comparison was performed for 40 samples spanning* <*0.1–10 ng/mL (concentration assigned by DxI, Beckman Coulter).*

[b]*Calibrators consisted of BCR-457 spiked into Tg-negative serum, human serum with Tg concentrations assigned by Beckman DxI assay, or the Beckman DxI calibrator itself.*

[c]*Adjustment was based on each assay's calibration bias versus the Method A's calibrators (assessed via separate analyses in which Methods B, C, and D were used to evaluate the calibrators of Method A).*

for Tg prior to protein digestion and peptide immunoaffinity enrichment. For quantification, three of four assays utilized an antibody generated by Clinical Proteomic Tumor Analysis Consortium (CPTAC, an NIH-funded initiative) for immunoaffinity precipitation and quantification of the "FSP" tryptic peptide (amino acid sequence FSPDDSAGASALLR), whereas the fourth assay used an in house−generated antibody targeting the "VIF" peptide (VIFDANAPVAVR). Notably, the latter assay (Method D in Table 4.6) corresponded to the lowest slope (0.833) when its results were compared to the average result across all four assays. Method A utilized the BCR-457 calibrator, which may explain the differences observed with Methods B and C, both of which were calibrated against an immunoassay (Beckman DxI).

Of note, all four laboratories utilized an *intact protein calibrant* from various sources, each ultimately traceable to BCR 457. Although the high degree of agreement between LC-MS/MS methods cannot be solely attributable to their common metrology, it is noteworthy because protein quantification by mass spectrometry has a long history of using surrogate calibrants such as spiked isotope-labeled peptides rather than intact proteins [91,92]. While spike-in surrogate peptide calibrants are more convenient for "omics"-scale quantification in basic research, they infrequently provide an accurate estimate of protein concentration due to variability in the digestion process [93,94]. Indeed, Shuford, et al. specifically demonstrated the shortcomings of such surrogate calibrants for the quantification of Tg [95]. It is important to recognize, however, that even an intact protein calibrant is fundamentally a surrogate of its native protein counterpart, given the heterogeneity (differences in glycosylation, etc.) of many proteins may affect efficiency in digestion and other processes leading up to ions entering the mass spectrometer. This work demonstrated that poor commutability of protein reference materials among mass spectrometry-based measurement procedures will be problematic, similar to immunometric assays. As such, this emphasizes the need for harmonization of Tg measurements across the clinical community. Such programs should ideally include immunoassay manufacturers and LC-MS/MS test developers because both platforms are expected to persist in the care of patients.

4.9 Thyroglobulin measurement by mass spectrometry—other challenges

An important caveat in the quantification of a protein using a surrogate peptide is that if the mass of the peptide is altered due to a mutation, then the peptide cannot be detected. This can be addressed by the inclusion of nonpolymorphic peptides as targets in LC-MS/MS assays, which can be guided by large databases of human polymorphisms [66,96]. Another challenge in utilizing LC-MS/MS is the cost to the laboratory [66].

4.10 Practical application of thyroglobulin measurement by mass spectrometry: a suggested approach

Given the higher cost of LC-MS/MS-based Tg measurement and the speed and convenience of immunoassay, testing can include reflexing to LC-MS/MS when appropriate. For example, for reference laboratories in a conventional setting, reflex could include: immunoassay for TbAb-Neg specimens and LC-MS/MS for TgAb-Pos specimens. To further reduce utilization of LC-MS/MS, another testing approach as shown in Fig. 4.2. Testing is first performed using an automated immunoassay platform with simultaneous testing for TgAb. Below a specified threshold (e.g., for TgAb = 1.0 IU/mL), TgAb might be assumed to not interfere with Tg measurement. As a result, the immunoassay result could be accepted and reported. Above this threshold,

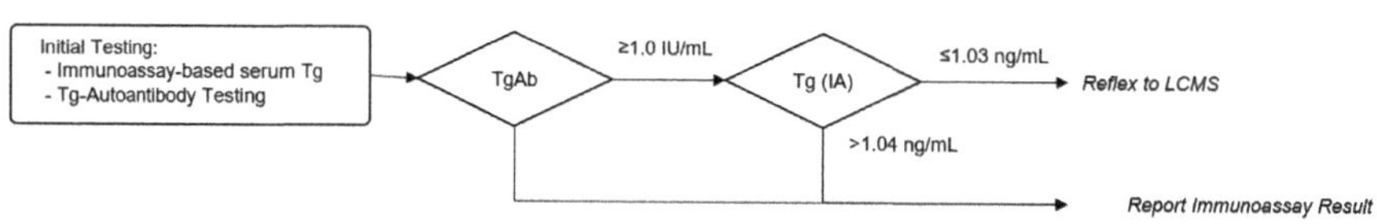

FIGURE 4.2

A potential workflow for measuring low thyroglobulin levels is shown. Above certain antibody level thresholds, concern for false-negative results may more warrant additional testing by spectrometry to confirm the absence of thyroglobulin.

negative interference in the serum Tg measurement might be presumed, and only low results would be confirmed by LC-MS/MS. Although TgAb might interfere with measurements of Tg greater than 1 ng/mL, the Tg concentration in these cases is usually detectable. Thus, repeat testing of samples >1.0 ng/mL by LC-MS/MS is probably not needed unless otherwise clinically indicated. Of note, thresholds used from lab to lab will be dependent upon the TgAb assay and will need to be lab-specific.

An example LC-MS/MS assay to use in this workflow is detailed in Table 4.7. Sample preparation includes protein digestion, immunoenrichment, and LC-MS/MS analysis and can be completed in a single day. Magnetic beads for immunoenrichment must be prepared beforehand but can be used multiple times using a validated cleaning protocol. The method incorporates internal normalization to an isotope-labeled tryptic peptide and calibration using 5-point external standard calibration. Calibrators and quality control samples are matrix-matched when feasible. Given the complexity of Tg, including the variability of its posttranslational modifications and the variability of human serum, using pooled human samples as calibrators and QC materials may help control for these differences in the patient population. Surrogate matrices of similar quality (e.g., chicken serum) can be used for dilution, if needed.

This method reflects a refinement of the first method listed in Table 4.5 and has a LLOQ of 0.15 ng/mL and LOD of 0.08 ng/mL. One conspicuous change from the original version of the assay is the shift from monitoring the VIF peptide (full sequence VIFDANAPVAVR) to the FSP peptide (FSPDDSAGASALLR) as a surrogate for Tg. This change coincided with the development of an anti-FSP antibody through the efforts of the CPTAC [97]. Of note, due its size, Tg generates ~160 theoretical fully tryptic peptides. Many can be eliminated as candidates for monitoring based on size (too small to be sequence specific or too large to be detectable), because they contain so-called "ragged edges" (leading to inconsistent digestion), because they have amino acids prone to modification (Cys, Met, His), and other factors [78]. Of the peptides that remain, some have higher and more precise signal in LC-MS/MS than others. In addition, the antibody used to enrich peptides, specifically its affinity and ability to enrich the target peptide, is also a critical

Table 4.7 Example liquid chromatography-tandem mass spectrometry method.

Assay component	Specifications				
Specimen	400 μL serum (1 mL minimum collection, serum separator tube preferred)				
Antibody (IPE)	Anti-FSP (rabbit, monoclonal)[a]				
Bead preparation[b]	Magnetic beads (tosyl activated) undergo washing (3 ×, in a borate buffer), antibody coupling (per sample, 2 μg Ab is coupled to 255 μg bead in 100 mM sodium borate), and second washing (e.g., in Tris, with a protein blocking agent such as bovine serum albumin). Beads stored in PBS containing 0.1% CHAPS/0.02% NaN_3.				
Digestion	**Step**	**Reagent addition**[c]	**Incubation**[d]		
	1—Denature/Reduce	100 μL 20% deoxycholate/0.19 M TCEP	1 h, 40°C, shaking 1400 RPM		
	2—Alkylate	100 μL 0.06 M iodoacetamide	30 min, RT, in the dark		
	3—Add IS	100 μL of 0.2 M Tris + 20 μL I.S.[e]	None		
	4—Digest	100 μL trypsin (10 mg/mL) in 10 mM HCl	30 min, 37°C, shaking 1400 RPM		
	5—Quench	20 μL TLCK (3 mg/mL) in 10 mM HCl	5 min, RT, shaking 1400 RPM		
IPE[f]	Protein digest is mixed with 10 μL IA bead slurry and incubated for 1 h; beads are washed with PBS and 0.1% CHAPS, and peptide is eluted (50 μL) using 2.5% acetic acid/0.1% CHAPS.				
Gradient	20%–67% B over 3 min (solvents: % DMSO/0.1% formic acid in A, water, or B, methanol)				
MS acquisition (MRM)[g]	**Target**	**Q1 (m/z)**	**Q3 (m/z)**	**Cone voltage** (V)	**Collision energy** (eV)
	FSP2	704.1	587.0	40	20
	FSP2 (heavy)	707.1	590.0	40	20
Calibration	5-point, external calibration (0.1, 0.3, 3.5, 10, 15 ng/mL)				
Internal standard	Tryptic peptide, FSPDDSAGASALL (leucine 13C(6)-labeled)				
QC	Negative, "Ultralow" (target, 0.1 ng/mL), low, high QC				

CHAPS, 3-[(3-Cholamidopropyl)dimethylammonio]-1-propanesulfonate; DMSO, *dimethylsulfoxide;* IA, *immunoaffinity;* IPE, *immunoaffinity peptide enrichment;* I.S., *internal standard;* MRM, *multiple reaction monitoring;* MS, *mass spectrometry;* PBS, *phosphate-buffered saline;* QC, *quality control;* RT, *room temperature;* TCEP, *tris(2-carboxyethyl)phosphine;* TLCK, *tosyl-L-lysyl-chloromethane hydrochloride (trypsin inhibitor).*

[a]*Clinical Proteomic Tumor Analysis Consortium (CPTAC)-39c, commercially available.*

[b]*Detailed instructions are available from bead manufacturer.*

[c]*Each followed by light vortex.*

[d]*Each followed by brief centrifugation.*

[e]*Prepared as 0.5 nM in 5% acetonitrile/0.1% formic acid. Further preparation detail is available upon request.*

[f]*225 μg beads/reaction.*

[g]*Additional MS settings (Waters TQS Mass Spectrometer): Cone voltage, 26 V; Capillary voltage 1.4 kV; desolvation temp 500C; desolvation gas 1000 L/Hr; Cone gas 150 L/Hr.*

determinant of LLOQ. Of note, three of four methods in Ref. [90] quantify by the FSP peptide, indicating a shift in preference to toward this peptide as the "proteotypic peptide" surrogate of Tg.

4.11 Thyroglobulin measurement—future directions

Efforts are ongoing to improve standardization between LC-MS/MS approaches, to maximize commutability of reference materials, and to further improve assay sensitivity. As described above, the size and complexity of Tg pose significant challenges to method harmonization. As a result, harmonization of Tg measurements will undoubtedly inform harmonization efforts for other protein biomarkers. Additionally, as measurement of Tg continues to improve, the door may open to new approaches in the evaluation of patients treated for DTC. For example, algorithms that tie Tg antibody titer and low levels of measurable Tg have not been possible previously, given the interference of TgAb in immunoassays at low concentrations. Additionally, many if not most studies involving measurement of serum Tg have excluded patients with positive TgAb levels [34]. LC-MS/MS can enable inclusion of these patients in future investigations. It is worth noting that the determined utility of all Tg measures largely revolves around the imperfect methods for identifying TgAb. Ultimately, if we cannot expect to do this robustly in clinical practice, we should strive for a future in which TgAb measurements are obsolete—namely a future in which only high-sensitivity (hs) LC-MS/MS measurements are standard of care.

4.12 Conclusion

Measurement of Tg by immunoassay has become increasingly sensitive, which has transformed the evaluation and care of patients with DTC. However, the success of Tg immunoassays in clinical practice demands that results deemed below the lower limit of quantification (or functional sensitivity) are accurately negative. This cannot be achieved for many patients due to Tg autoantibodies and many patients may not benefit from this individualized approach to care. Instead, the introduction of LC-MS/MS-based assays offers a complementary approach, providing clarity in these cases. For now, immunoassays which offer speed and convenience may be used alongside LC-MS/MS to manage testing for larger patient populations. However, as assay harmonization continues to improve and LC-MS/MS becomes more prevalent in clinical laboratories, we anticipate the evaluation of patients of with DTC will migrate to this more sensitive and specific platform.

4.13 Summary points

- Thyroglobulin (Tg) plays a critical role in detecting persistent and recurrent differentiated thyroid cancer.
- Increasingly sensitive Tg measurements have enabled more individualized approaches to the management of thyroid cancer, but application in clinical practice is stifled by the high prevalence of naturally occurring Tg autoantibodies.
- Endogenous anti-Tg antibodies can cause both falsely low and falsely high serum Tg results when measured by radioimmunoassay, and falsely low results by sandwich immunoassays.
- Liquid chromatography-tandem mass spectrometry-based assays circumvent autoantibody interference and have achieved equivalent sensitivity to immunoassays via peptide immunoaffinity purification.
- The size and complexity of Tg impose challenges related to harmonization across all method types. However, preliminary investigations demonstrate that harmonization of LC-MS/MS exceeds longstanding Food and Drug Administration–approved immunoassay platforms.

References

[1] Coscia F, Taler-Verčič A, Chang VT, Sinn L, O'Reilly FJ, Izoré T, et al. The structure of human thyroglobulin. Nature 2020;578:627–30. Available from: https://doi.org/10.1038/s41586-020-1995-4.

[2] Citterio CE, Morishita Y, Dakka N, Veluswamy B, Arvan P. Relationship between the dimerization of thyroglobulin and its ability to form triiodothyronine. J Biol Chem 2018;293:4860–9. Available from: https://doi.org/10.1074/jbc.RA118.001786.

[3] Shahid M, Ashraf M, Sharma S. Thyroid hormone. In: StatPearls [Internet]. Treasure Island, FL: StatPearls Publishing; 2021. <https://www.ncbi.nlm.nih.gov/books/NBK500006/>.

[4] Smith TJ, Padovani-Claudio DA, Lu Y, Raychaudhuri N, Fernando R, Atkins S, et al. Fibroblasts expressing the thyrotropin receptor overarch thyroid and orbit in Graves' disease. J Clin Endocrinol Metab 2011;96:3827–37. Available from: https://doi.org/10.1210/jc.2011-1249.

[5] Ma S, Jing F, Xu C, Zhou L, Song Y, Yu C, et al. Thyrotropin and obesity: increased adipose triglyceride content through glycerol-3-phosphate acyltransferase 3. Sci Rep 2015;5:7633. Available from: https://doi.org/10.1038/srep07633.

[6] Lin J-D. Thyroglobulin and human thyroid cancer. Clin Chim Acta 2008;388:15–21. Available from: https://doi.org/10.1016/j.cca.2007.11.002.

[7] Spencer CA, Lopresti JS. Measuring thyroglobulin and thyroglobulin autoantibody in patients with differentiated thyroid cancer. Nat Clin Pract Endocrinol Metab 2008;4:223–33. Available from: https://doi.org/10.1038/ncpendmet0757.

[8] Giovanella L, Clark PM, Chiovato L, Duntas L, Elisei R, Feldt-Rasmussen U, et al. Thyroglobulin measurement using highly sensitive assays in patients with differentiated thyroid cancer: a clinical position paper. Eur J Endocrinol 2014;171:R33–46. Available from: https://doi.org/10.1530/EJE-14-0148.

[9] Nguyen QT, Lee EJ, Huang MG, Park YI, Khullar A, Plodkowski RA. Diagnosis and treatment of patients with thyroid cancer. Am Heal Drug Benefits 2015;8:30–40. Available from: http://www.ncbi.nlm.nih.gov/pubmed/25964831.

[10] Cancer.gov. Cancer Stat Facts: Thyroid. <https://seer.cancer.gov/statfacts/html/thyro.html>; n.d.

[11] Hegedüs L. Clinical practice. The thyroid nodule. N Engl J Med 2004;351:1764–71. Available from: https://doi.org/10.1056/NEJMcp031436.

[12] Kwong N, Medici M, Angell TE, Liu X, Marqusee E, Cibas ES, et al. The Influence of patient age on thyroid nodule formation, multinodularity, and thyroid cancer risk. J Clin Endocrinol Metab 2015;100:4434–40. Available from: https://doi.org/10.1210/jc.2015-3100.

[13] Yassa L, Cibas ES, Benson CB, Frates MC, Doubilet PM, Gawande AA, et al. Long-term assessment of a multidisciplinary approach to thyroid nodule diagnostic evaluation. Cancer 2007;111:508–16. Available from: https://doi.org/10.1002/cncr.23116.

[14] Noto B, Eveslage M, Pixberg M, Gonzalez Carvalho JM, Schäfers M, Riemann B, et al. Prevalence of hyperfunctioning thyroid nodules among those in need of fine needle aspiration cytology according to ATA 2015, EU-TIRADS, and ACR-TIRADS. Eur J Nucl Med Mol Imaging 2020;47:1518–26. Available from: https://doi.org/10.1007/s00259-020-04740-y.

[15] Rosario PW, Rocha TG, Mourão GF, Calsolari MR. Is radioiodine scintigraphy still of value in thyroid nodules with indeterminate cytology?: a prospective study in an iodine-sufficient area. Nucl Med Commun 2018;39:1059–60. Available from: https://doi.org/10.1097/MNM.0000000000000896.

[16] Wong LQ, Baloch ZW. Analysis of the bethesda system for reporting thyroid cytopathology and similar precursor thyroid cytopathology reporting schemes. Adv Anat Pathol 2012;19:313–19. Available from: https://doi.org/10.1097/PAP.0b013e3182666398.

[17] Cibas ES, Ali SZ. The 2017 bethesda system for reporting thyroid cytopathology. Thyroid 2017;27:1341–6. Available from: https://doi.org/10.1089/thy.2017.0500.

[18] Gharib H, Papini E, Garber JR, Duick DS, Harrell RM, Hegedüs L, et al. AACE/ACE/AME Task Force on Thyroid Nodules, American Association of Clinical Endocrinologists, American College of Endocrinology, and Associazione Medici Endocrinologi Medical guidelines for clinical practice for the diagnosis and management of thyroid nodules–2016 update. Endocr Pract 2016;22:622–39. Available from: https://doi.org/10.4158/EP161208.GL.

[19] Jug RC, Datto MB, Jiang XS. Molecular testing for indeterminate thyroid nodules: performance of the Afirma gene expression classifier and ThyroSeq panel. Cancer Cytopathol 2018;126:471–80. Available from: https://doi.org/10.1002/cncy.21993.

[20] Trimboli P, D'Aurizio F, Tozzoli R, Giovanella L. Measurement of thyroglobulin, calcitonin, and PTH in FNA washout fluids. Clin Chem Lab Med 2017;55:914–25. Available from: https://doi.org/10.1515/cclm-2016-0543.

[21] Wells SA, Asa SL, Dralle H, Elisei R, Evans DB, Gagel RF, et al. American Thyroid Association Guidelines Task Force on Medullary Thyroid Carcinoma, Revised American Thyroid Association

guidelines for the management of medullary thyroid carcinoma. Thyroid 2015;25:567–610. Available from: https://doi.org/10.1089/thy.2014.0335.

[22] Okamoto T, Kanbe M, Iihara M, Yamazaki K, Okamoto J, Yamashita T, et al. Measuring serum thyroglobulin in patients with follicular thyroid nodule: its diagnostic implications. Endocr J 1997;44:187–93. Available from: https://doi.org/10.1507/endocrj.44.187.

[23] Haugen BR, Alexander EK, Bible KC, Doherty GM, Mandel SJ, Nikiforov YE, et al. 2015 American Thyroid Association Management Guidelines for Adult Patients with Thyroid Nodules and Differentiated Thyroid Cancer: The American Thyroid Association guidelines task force on thyroid nodules and differentiated thyroid cancer. Thyroid 2016;26:1–133. Available from: https://doi.org/10.1089/thy.2015.0020.

[24] Yazici P, Mihmanli M, Bozkurt E, Ozturk FY, Uludag M. Which is the best predictor of thyroid cancer: thyrotropin, thyroglobulin or their ratio? Hormones (Athens) 2016;15:256–63. Available from: https://doi.org/10.14310/horm.2002.1677.

[25] Trimboli P, Treglia G, Giovanella L. Preoperative measurement of serum thyroglobulin to predict malignancy in thyroid nodules: a systematic review. Horm Metab Res 2015;47:247–52. Available from: https://doi.org/10.1055/s-0034-1395517.

[26] Suh I, Vriens MR, Guerrero MA, Griffin A, Shen WT, Duh Q-Y, et al. Serum thyroglobulin is a poor diagnostic biomarker of malignancy in follicular and Hurthle-cell neoplasms of the thyroid. Am J Surg 2010;200:41–6. Available from: https://doi.org/10.1016/j.amjsurg.2009.08.030.

[27] Oltmann SC, Leverson G, Lin SH-I, Schneider DF, Chen H, Sippel RS. Markedly elevated thyroglobulin levels in the preoperative thyroidectomy patient correlates with metastatic burden. J Surg Res 2014;187:1–5. Available from: https://doi.org/10.1016/j.jss.2013.12.017.

[28] Hulikal N, Re A, Banoth M, Chowhan AK, Yutla M, Sachan A. Can preoperative serum thyroglobulin levels predict the risk of malignancy? Results from prospective analysis of biochemical predictors of malignancy in thyroid nodules. Acta Otorhinolaryngol Ital 2020;40:33–7. Available from: https://doi.org/10.14639/0392-100X-N0276.

[29] Kim HJ, Mok J-O, Kim CH, Kim YJ, Kim SJ, Park HK, et al. Preoperative serum thyroglobulin and changes in serum thyroglobulin during TSH suppression independently predict follicular thyroid carcinoma in thyroid nodules with a cytological diagnosis of follicular

lesion. Endocr Res 2017;42:154–62. Available from: https://doi.org/10.1080/07435800.2016.1262395.

[30] King AD. Imaging for staging and management of thyroid cancer. Cancer Imaging 2008;8:57–69. Available from: https://doi.org/10.1102/1470-7330.2008.0007.

[31] Yeh MW, Bauer AJ, Bernet VA, Ferris RL, Loevner LA, Mandel SJ, et al. American Thyroid Association Surgical Affairs Committee Writing Task Force, American Thyroid Association statement on preoperative imaging for thyroid cancer surgery. Thyroid 2015;25:3–14. Available from: https://doi.org/10.1089/thy.2014.0096.

[32] Grani G, Fumarola A. Thyroglobulin in lymph node fine-needle aspiration washout: a systematic review and meta-analysis of diagnostic accuracy. J Clin Endocrinol Metab 2014;99:1970–82. Available from: https://doi.org/10.1210/jc.2014-1098.

[33] Li QK, Nugent SL, Straseski J, Cooper D, Riedel S, Askin FB, et al. Thyroglobulin measurements in fine-needle aspiration cytology of lymph nodes for the detection of metastatic papillary thyroid carcinoma. Cancer Cytopathol 2013;121:440–8. Available from: https://doi.org/10.1002/cncy.21285.

[34] Kim H, Kim YN, Kim HI, Park SY, Choe J-H, Kim J-H, et al. Preoperative serum thyroglobulin predicts initial distant metastasis in patients with differentiated thyroid cancer. Sci Rep 2017;7:16955. Available from: https://doi.org/10.1038/s41598-017-17176-6.

[35] Indrasena BSH. Use of thyroglobulin as a tumour marker. World J Biol Chem 2017;8:81–5. Available from: https://doi.org/10.4331/wjbc.v8.i1.81.

[36] Giovanella L, Ceriani L, Ghelfo A, Maffioli M, Keller F. Preoperative undetectable serum thyroglobulin in differentiated thyroid carcinoma: incidence, causes and management strategy. Clin Endocrinol (Oxf) 2007;67:547–51. Available from: https://doi.org/10.1111/j.1365-2265.2007.02922.x.

[37] Momesso DP, Tuttle RM. Update on differentiated thyroid cancer staging. Endocrinol Metab Clin North Am 2014;43:401–21. Available from: https://doi.org/10.1016/j.ecl.2014.02.010.

[38] Pacini F, Sabra MM, Tuttle RM. Clinical relevance of thyroglobulin doubling time in the management of patients with differentiated thyroid cancer. Thyroid 2011;21:691–2. Available from: https://doi.org/10.1089/thy.2011.2107.ed1.

[39] Tuttle RM, Tala H, Shah J, Leboeuf R, Ghossein R, Gonen M, et al. Estimating risk of recurrence in differentiated thyroid cancer after total

thyroidectomy and radioactive iodine remnant ablation: using response to therapy variables to modify the initial risk estimates predicted by the new American Thyroid Association staging system. Thyroid 2010;20:1341–9. Available from: https://doi.org/10.1089/thy.2010.0178.

[40] Hocevar M, Auersperg M, Stanovnik L. The dynamics of serum thyroglobulin elimination from the body after thyroid surgery. Eur J Surg Oncol 1997;23:208–10. Available from: https://doi.org/10.1016/s0748-7983(97)92292-7.

[41] Giovanella L, Ceriani L, Maffioli M. Postsurgery serum thyroglobulin disappearance kinetic in patients with differentiated thyroid carcinoma. Head Neck 2010;32:568–71. Available from: https://doi.org/10.1002/hed.21214.

[42] Kwon H, Jeon MJ, Kim WG, Park S, Kim M, Song DE, et al. A comparison of lobectomy and total thyroidectomy in patients with papillary thyroid microcarcinoma: a retrospective individual risk factor-matched cohort study. Eur J Endocrinol 2017;176:371–8. Available from: https://doi.org/10.1530/EJE-16-0845.

[43] Park S, Jeon MJ, Oh H-S, Lee Y-M, Sung T-Y, Han M, et al. Changes in serum thyroglobulin levels after lobectomy in patients with low-risk papillary thyroid cancer. Thyroid 2018;28:997–1003. Available from: https://doi.org/10.1089/thy.2018.0046.

[44] Vaisman F, Momesso D, Bulzico DA, Pessoa CHCN, Dias F, Corbo R, et al. Spontaneous remission in thyroid cancer patients after biochemical incomplete response to initial therapy. Clin Endocrinol (Oxf) 2012;77:132–8. Available from: https://doi.org/10.1111/j.1365-2265.2012.04342.x.

[45] Miyauchi A, Kudo T, Miya A, Kobayashi K, Ito Y, Takamura Y, et al. Prognostic impact of serum thyroglobulin doubling-time under thyrotropin suppression in patients with papillary thyroid carcinoma who underwent total thyroidectomy. Thyroid 2011;21:707–16. Available from: https://doi.org/10.1089/thy.2010.0355.

[46] Collins VP, Loeffler RK, Tivey H. Observations on growth rates of human tumors. Am J Roentgenol Radium Ther Nucl Med 1956;76:988–1000. Available from: http://www.ncbi.nlm.nih.gov/pubmed/13362715.

[47] Rössing RM, Jentzen W, Nagarajah J, Bockisch A, Görges R. Serum thyroglobulin doubling time in progressive thyroid cancer. Thyroid 2016;26:1712–18. Available from: https://doi.org/10.1089/thy.2016.0031.

[48] Zhang X, Higuchi T, Tomonaga H, Lamid-Ochir O, Bhattarai A, Nguyen-Thu H, et al. Early detection of progressive disease

using thyroglobulin doubling-time in metastatic differentiated thyroid carcinoma treated with radioactive iodine. Nucl Med Commun 2020;41:350–5. Available from: https://doi.org/10.1097/MNM.0000000000001154.

[49] Kelders A, Kennes LN, Krohn T, Behrendt FF, Mottaghy FM, Verburg FA. Relationship between positive thyroglobulin doubling time and 18F-FDG PET/CT-positive, ^{131}I-negative lesions. Nucl Med Commun 2014;35:176–81. Available from: https://doi.org/10.1097/MNM.0000000000000025.

[50] Robbins RJ, Wan Q, Grewal RK, Reibke R, Gonen M, Strauss HW, et al. Real-time prognosis for metastatic thyroid carcinoma based on 2-[18 F]fluoro-2-deoxy-D-glucose-positron emission tomography scanning. J Clin Endocrinol Metab 2006;91:498–505. Available from: https://doi.org/10.1210/jc.2005-1534.

[51] Araz M, Soydal Ç, Özkan E, Akkuş P, Nak D, Küçük NÖ, et al. Role of thyroglobulin doubling time in differentiated thyroid cancer and its relationship with demographic-histopathologic risk factors and 18F-fluorodeoxyglucose positron emission tomography/computed tomography parameters. Cancer Biother Radiopharm 2021;36:425–32. Available from: https://doi.org/10.1089/cbr.2019.3203.

[52] Giovanella L, Trimboli P, Verburg FA, Treglia G, Piccardo A, Foppiani L, et al. Thyroglobulin levels and thyroglobulin doubling time independently predict a positive 18F-FDG PET/CT scan in patients with biochemical recurrence of differentiated thyroid carcinoma. Eur J Nucl Med Mol Imaging 2013;40:874–80. Available from: https://doi.org/10.1007/s00259-013-2370-6.

[53] Albano D, Tulchinsky M, Dondi F, Mazzoletti A, Lombardi D, Bertagna F, et al. Thyroglobulin doubling time offers a better threshold than thyroglobulin level for selecting optimal candidates to undergo localizing [18 F]FDG PET/CT in non-iodine avid differentiated thyroid carcinoma. Eur J Nucl Med Mol Imaging 2021;48:461–8. Available from: https://doi.org/10.1007/s00259-020-04992-8.

[54] Jo K, Lim D-J. Clinical implications of anti-thyroglobulin antibody measurement before surgery in thyroid cancer. Korean J Intern Med 2018;33:1050–7. Available from: https://doi.org/10.3904/kjim.2018.289.

[55] Chung J-K, Park YJ, Kim TY, So Y, Kim S-K, Park DJ, et al. Clinical significance of elevated level of serum antithyroglobulin antibody in patients with differentiated thyroid cancer after thyroid ablation. Clin Endocrinol (Oxf) 2002;57:215–21. Available from: https://doi.org/10.1046/j.1365-2265.2002.01592.x.

[56] Kim WG, Yoon JH, Kim WB, Kim TY, Kim EY, Kim JM, et al. Change of serum antithyroglobulin antibody levels is useful for prediction of clinical recurrence in thyroglobulin-negative patients with differentiated thyroid carcinoma. J Clin Endocrinol Metab 2008;93:4683–9. Available from: https://doi.org/10.1210/jc.2008-0962.

[57] Côrtes MCS, Rosario PW, Oliveira LFF, Calsolari MR. Clinical impact of detectable antithyroglobulin antibodies below the reference limit (borderline) in patients with papillary thyroid carcinoma with undetectable serum thyroglobulin and normal neck ultrasonography after ablation: a prospective study. Thyroid 2018; 28:229–35. Available from: https://doi.org/10.1089/thy.2017.0350.

[58] Yin N, Sherman SI, Pak Y, Litofsky DR, Gianoukakis AG. The de novo detection of anti-thyroglobulin antibodies and differentiated thyroid cancer recurrence. Thyroid 2020;30:1490–5. Available from: https://doi.org/10.1089/thy.2019.0791.

[59] Bayer MF, Kriss JP. Immunoradiometric assay for serum thyroglobulin: semiquantitative measurement of thyroglobulin in antithyroglobulin-positive sera. J Clin Endocrinol Metab 1979;49:557–64. Available from: https://doi.org/10.1210/jcem-49-4-557.

[60] Schneider AB, Pervos R. Radioimmunoassay of human thyroglobulin: effect of antithyroglobulin autoantibodies. J Clin Endocrinol Metab 1978;47:126–37. Available from: https://doi.org/10.1210/jcem-47-1-126.

[61] Roitt IM, Torrigiani G. Identification and estimation of undegraded thyroglobulin in human serum. Endocrinology 1967;81:421–9. Available from: https://doi.org/10.1210/endo-81-3-421.

[62] Bayer MF, Kriss JP. A solid phase, sandwich-type radioimmunoassay for antithyroglobulin: elimination of false positive results and semi-quantitative measurement of antithyroglobulin in the presence of elevated thyroglobulin. J Clin Endocrinol Metab 1979;49:565–71. Available from: https://doi.org/10.1210/jcem-49-4-565.

[63] Mariotti S, Cupini C, Giani C, Lari R, Rolleri E, Falco A, et al. Evaluation of a solid-phase immunoradiometric assay (IRMA) for serum thyroglobulin: effect of anti-thyroglobulin autoantibody. Clin Chim Acta 1982;123:347–55. Available from: https://doi.org/10.1016/0009-8981(82)90181-4.

[64] Van Herle AJ, Uller RP, Matthews NI, Brown J. Radioimmunoassay for measurement of thyroglobulin in human serum. J Clin Invest 1973;52:1320–7. Available from: https://doi.org/10.1172/JCI107303.

[65] Giovanella L, Feldt-Rasmussen U, Verburg FA, Grebe SK, Plebani M, Clark PM. Thyroglobulin measurement by highly sensitive

assays: focus on laboratory challenges. Clin Chem Lab Med 2015;53:1301–14. Available from: https://doi.org/10.1515/cclm-2014-0813.

[66] Hoofnagle AN, Roth MY. Clinical review: improving the measurement of serum thyroglobulin with mass spectrometry. J Clin Endocrinol Metab 2013;98:1343–52. Available from: https://doi.org/10.1210/jc.2012-4172.

[67] Spencer C, LoPresti J, Fatemi S. How sensitive (second-generation) thyroglobulin measurement is changing paradigms for monitoring patients with differentiated thyroid cancer, in the absence or presence of thyroglobulin autoantibodies. Curr Opin Endocrinol Diabetes Obes 2014;21:394–404. Available from: https://doi.org/10.1097/MED.0000000000000092.

[68] CLSI. Evaluation of detection capability for clinical laboratory measurement procedures; approved guideline—second edition, CLSI Doc. EP17-A2; 2012.

[69] Doniach D, Roitt IM. Auto-immunity in Hashimoto's disease and its implications. J Clin Endocrinol Metab 1957;17:1293–304. Available from: https://doi.org/10.1210/jcem-17-11-1293.

[70] Owen CA. A review of auto-immunization in Hashimoto's disease. J Clin Endocrinol Metab 1958;18:1015–23. Available from: https://doi.org/10.1210/jcem-18-9-1015.

[71] Roitt IM, Doniach D. Human autoimmune thyroiditis: serological studies. Lancet (London, England) 1958;2:1027–33. Available from: https://doi.org/10.1016/s0140-6736(58)90530-0.

[72] Hoofnagle AN, Wener MH. The fundamental flaws of immunoassays and potential solutions using tandem mass spectrometry. J Immunol Methods 2009;347:3–11. Available from: https://doi.org/10.1016/j.jim.2009.06.003.

[73] Spencer CA, Takeuchi M, Kazarosyan M. Current status and performance goals for serum thyroglobulin assays. Clin Chem 1996;42:164–73. Available from: http://www.ncbi.nlm.nih.gov/pubmed/8565221.

[74] Peixoto A, Relvas-Santos M, Azevedo R, Santos LL, Ferreira JA. Protein glycosylation and tumor microenvironment alterations driving cancer hallmarks. Front Oncol 2019;9:380. Available from: https://doi.org/10.3389/fonc.2019.00380.

[75] Feldt-Rasmussen U, Profilis C, Colinet E, Black E, Bornet H, Bourdoux P, et al. Human thyroglobulin reference material (CRM 457). 1st part: assessment of homogeneity, stability and immunoreactivity. Ann Biol Clin (Paris) 1996;54:337–42. Available from: http://www.ncbi.nlm.nih.gov/pubmed/9092300.

[76] Feldt-Rasmussen U, Profilis C, Colinet E, Black E, Bornet H, Bourdoux P, et al. Human thyroglobulin reference material (CRM 457). 2nd part: physicochemical characterization and certification. Ann Biol Clin (Paris) 1996;54:343–8. Available from: http://www.ncbi.nlm.nih.gov/pubmed/9092301.

[77] Spencer CA, Bergoglio LM, Kazarosyan M, Fatemi S, LoPresti JS. Clinical impact of thyroglobulin (Tg) and Tg autoantibody method differences on the management of patients with differentiated thyroid carcinomas. J Clin Endocrinol Metab 2005;90:5566–75. Available from: https://doi.org/10.1210/jc.2005-0671.

[78] Kim H, Sohn A, Yeo I, Yu SJ, Yoon J-H, Kim Y. Clinical assay for AFP-L3 by using multiple reaction monitoring–mass spectrometry for diagnosing hepatocellular carcinoma. Clin Chem 2018;64:1230–8. Available from: https://doi.org/10.1373/clinchem.2018.289702.

[79] Whiteaker JR, Paulovich AG. Peptide immunoaffinity enrichment coupled with mass spectrometry for peptide and protein quantification. Clin Lab Med 2011;31:385–96. Available from: https://doi.org/10.1016/j.cll.2011.07.004.

[80] Kiernan UA. Quantitation of target proteins and post-translational modifications in affinity-based proteomics approaches. Expert Rev Proteomics 2007;4:421–8. Available from: https://doi.org/10.1586/14789450.4.3.421.

[81] Shuford CM, Johnson JS, Thompson JW, Holland PL, Hoofnagle AN, Grant RP. More sensitivity is always better: measuring subclinical levels of serum thyroglobulin on a μLC–MS/MS system. Clin Mass Spectrom 2020;15:29–35. Available from: https://doi.org/10.1016/j.clinms.2020.01.001.

[82] Hoofnagle AN, Becker JO, Wener MH, Heinecke JW. Quantification of thyroglobulin, a low-abundance serum protein, by immunoaffinity peptide enrichment and tandem mass spectrometry. Clin Chem 2008;54:1796–804. Available from: https://doi.org/10.1373/clinchem.2008.109652.

[83] Clarke NJ, Zhang Y, Reitz RE. A novel mass spectrometry-based assay for the accurate measurement of thyroglobulin from patient samples containing antithyroglobulin autoantibodies. J Investig Med 2012;60:1157–63. Available from: https://doi.org/10.2310/JIM.0b013e318276deb4.

[84] Kushnir MM, Rockwood AL, Roberts WL, Abraham D, Hoofnagle AN, Meikle AW. Measurement of thyroglobulin by liquid chromatography-tandem mass spectrometry in serum and plasma in the presence of antithyroglobulin autoantibodies. Clin

Chem 2013;59:982–90. Available from: https://doi.org/10.1373/clinchem.2012.195594.

[85] Kushnir MM, Rockwood AL, Straseski JA, Meikle AW. Comparison of LC-MS/MS to immunoassay for measurement of thyroglobulin in fine-needle aspiration samples. Clin Chem 2014;60:1452–3. Available from: https://doi.org/10.1373/clinchem.2014.227504.

[86] Spencer C, Petrovic I, Fatemi S, LoPresti J. Serum thyroglobulin (Tg) monitoring of patients with differentiated thyroid cancer using sensitive (second-generation) immunometric assays can be disrupted by false-negative and false-positive serum thyroglobulin autoantibody misclassifications. J Clin Endocrinol Metab 2014;99:4589–99. Available from: https://doi.org/10.1210/jc.2014-1203.

[87] Netzel BC, Grebe SKG, Carranza Leon BG, Castro MR, Clark PM, Hoofnagle AN, et al. Thyroglobulin (Tg) testing revisited: Tg assays, TgAb assays, and correlation of results with clinical outcomes. J Clin Endocrinol Metab 2015;100:E1074–83. Available from: https://doi.org/10.1210/jc.2015-1967.

[88] Azmat U, Porter K, Senter L, Ringel MD, Nabhan F. Thyroglobulin liquid chromatography-tandem mass spectrometry has a low sensitivity for detecting structural disease in patients with antithyroglobulin antibodies. Thyroid 2017;27:74–80. Available from: https://doi.org/10.1089/thy.2016.0210.

[89] Jindal A, Khan U. Is thyroglobulin level by liquid chromatography tandem-mass spectrometry always reliable for follow-up of DTC after thyroidectomy: a report on two patients. Thyroid 2016;26:1334–5. Available from: https://doi.org/10.1089/thy.2015.0556.

[90] Netzel BC, Grant RP, Hoofnagle AN, Rockwood AL, Shuford CM, Grebe SKG. First steps toward harmonization of LC-MS/MS thyroglobulin assays. Clin Chem 2016;62:297–9. Available from: https://doi.org/10.1373/clinchem.2015.245266.

[91] Barr JR, Maggio VL, Patterson DG, Cooper GR, Henderson LO, Turner WE, et al. Isotope dilution–mass spectrometric quantification of specific proteins: model application with apolipoprotein A-I. Clin Chem 1996;42:1676–82. Available from: http://www.ncbi.nlm.nih.gov/pubmed/8855153.

[92] Gerber SA, Rush J, Stemman O, Kirschner MW, Gygi SP. Absolute quantification of proteins and phosphoproteins from cell lysates by tandem MS. Proc Natl Acad Sci U S A 2003;100:6940–5. Available from: https://doi.org/10.1073/pnas.0832254100.

[93] Arnold SL, Stevison F, Isoherranen N. Impact of sample matrix on accuracy of peptide quantification: assessment of calibrator and internal standard selection and method validation. Anal Chem 2016;88:746–53. Available from: https://doi.org/10.1021/acs.analchem.5b03004.

[94] Brun V, Dupuis A, Adrait A, Marcellin M, Thomas D, Court M, et al. Isotope-labeled protein standards: toward absolute quantitative proteomics. Mol Cell Proteomics 2007;6:2139–49. Available from: https://doi.org/10.1074/mcp.M700163-MCP200.

[95] Shuford CM, Walters JJ, Holland PM, Sreenivasan U, Askari N, Ray K, et al. Absolute protein quantification by mass spectrometry: not as simple as advertised. Anal Chem 2017;89:7406–15. Available from: https://doi.org/10.1021/acs.analchem.7b00858.

[96] 1000 Genomes Project Consortium, Abecasis GR, Auton A, Brooks LD, DePristo MA, Durbin RM, Handsaker RE, et al. An integrated map of genetic variation from 1,092 human genomes. Nature 2012;491:56–65. Available from: https://doi.org/10.1038/nature11632.

[97] Rodriguez H, Pennington SR. Revolutionizing precision oncology through collaborative proteogenomics and data sharing. Cell 2018;173:535–9. Available from: https://doi.org/10.1016/j.cell.2018.04.008.

Further reading

Cabanillas et al., 2016Cabanillas ME, McFadden DG, Durante C. Thyroid cancer. Lancet (London, England) 2016;388:2783–95. Available from: https://doi.org/10.1016/S0140-6736(16)30172-6.

McLeod et al., 2014McLeod DSA, Cooper DS, Ladenson PW, Ain KB, Brierley JD, Fein HG, et al.The National Thyroid Cancer Treatment Cooperative Study Group Prognosis of differentiated thyroid cancer in relation to serum thyrotropin and thyroglobulin antibody status at time of diagnosis. Thyroid 2014;24:35–42. Available from: https://doi.org/10.1089/thy.2013.0062.

Miyauchi et al., 2013Miyauchi A, Kudo T, Kihara M, Higashiyama T, Ito Y, Kobayashi K, et al. Relationship of biochemically persistent disease and thyroglobulin-doubling time to age at surgery in patients with papillary thyroid carcinoma. Endocr J 2013;60:415–21. Available from: http://www.ncbi.nlm.nih.gov/pubmed/23182919.

Piccardo et al., 2021Piccardo A, Siri G, Ugolini M, Fiz F, Puntoni M, Bottoni G, et al. A three-domain scoring system to customize the risk of relapse of differentiated thyroid carcinoma. Cancers (Basel) 2021;13. Available from: https://doi.org/10.3390/cancers13174335.

Roman et al., 2017Roman BR, Morris LG, Davies L. The thyroid cancer epidemic, 2017 perspective. Curr Opin Endocrinol Diabetes Obes 2017;24:332–6. Available from: https://doi.org/10.1097/MED.0000000000000359.

Stevic et al., 2015Stevic I, Dembinski TC, Pathak KA, Leslie WD. Transient early increase in thyroglobulin levels post-radioiodine ablation in patients with differentiated thyroid cancer. Clin Biochem 2015;48:658–61. Available from: https://doi.org/10.1016/j.clinbiochem.2015.04.009.

Tuttle and Alzahrani, 2019Tuttle RM, Alzahrani AS. Risk stratification in differentiated thyroid cancer: from detection to final follow-up. J Clin Endocrinol Metab 2019. Available from: https://doi.org/10.1210/jc.2019-00177.

Wu et al., 2020Wu D, Gomes-Lima CJ, Kulkarni K, Burman KD, Wartofsky L, Van Nostrand D. 30 mCi exploratory scan for two-step dosimetric ^{131}I therapy in differentiated thyroid cancer patients: a novel approach and case report. World J Nucl Med 2020;19:266–70. Available from: https://doi.org/10.4103/wjnm.WJNM_35_19.

Yang et al., 2021Yang SP, Koh LCW, Kong KW, Parameswaran R, Loke KSH, Ngiam KY, et al. A Multiplex thyroid-specific assay for quantification of circulating thyroid cell-free RNA in plasma of thyroid cancer patients. Front Genet 2021;12:721832. Available from: https://doi.org/10.3389/fgene.2021.721832.

Ylli et al., 2021Ylli D, Soldin SJ, Stolze B, Wei B, Nigussie G, Nguyen H, et al. Biotin interference in assays for thyroid hormones, thyrotropin and thyroglobulin. Thyroid 2021;31:1160–70. Available from: https://doi.org/10.1089/thy.2020.0866.

CHAPTER

PSA: role in screening and monitoring patients with prostate cancer

5

Sigrid V. Carlsson[1,2], Kazunori Murata[3], Daniel C. Danila[4] and Hans Lilja[3,5,6]

[1]*Departments of Surgery (Urology Service) and Epidemiology and Biostatistics, Memorial Sloan Kettering Cancer Center, New York, NY, United States*

[2]*Department of Urology, Institute of Clinical Sciences, Sahlgrenska Academy at University of Gothenburg, Gothenburg, Sweden*

[3]*Department of Pathology and Laboratory Medicine, Memorial Sloan Kettering Cancer Center, New York, NY, United States*

[4]*Department of Medicine, Memorial Sloan Kettering Cancer Center, New York, NY, United States*

[5]*Departments of Surgery (Urology Service) and Medicine, Memorial Sloan Kettering Cancer Center, New York, NY, United States*

[6]*Department of Translational Medicine, Lund University, Malmö, Sweden*

Objectives

- To present the physiological function and analytical assay of the prostate-specific antigen (PSA) test
- To present the role of PSA as a tumor marker and screening tool for prostate cancer and recommendations from national guideline groups
- To discuss the evidence from randomized PSA screening trials
- To review studies on the use of baseline PSA and long-term risk of lethal prostate cancer
- To present PSA derivatives to improve the specificity of the PSA test and reduce unnecessary prostate biopsies
- To present the important role of the PSA-test in monitoring the disease and detecting recurrence of the disease

Cancer Biomarkers: Clinical Aspects and Laboratory Determination. DOI: https://doi.org/10.1016/B978-0-12-824302-2.00001-1
© 2022 Elsevier Inc. All rights reserved.

5.1 Introduction

The discovery in 1970s and early 1980s of a unique tissue-specific antigen that enabled detection of prostate cancer at an early localized stage and contributed means to monitor recurrence and progression of the disease in blood was remarkable [1–4]. Since the discovery, testing for prostate-specific antigen (PSA) in blood has become one of the most widely and frequently used tumor markers in medicine today [5]. This chapter presents the role of PSA in screening men for prostate cancer and monitoring patients with the disease.

5.2 PSA: physiological function and analytical assay

PSA is encoded by the kallikrein-related peptidase 3 (*KLK3*) gene located in a cluster of 15 kallikrein-related related genes at chromosome 19q.13.3–13.4 [2]. Abundant expression of *KLK3* is regulated by the androgen receptor (*AR*) [3] and highly restricted to the prostate epithelial cells generating a 261-amino acid prepro-protein that is converted to a 237-residue 28.4 kDa single-chain active serine proteinase glycoprotein with chymotryptic-like substrate specificity [6,7]. Functional orthologs to *KLK3* are lacking in most mammals (e.g., rodents) and unique to primates due to recent duplication of a progenitor to *KLK3/KLK2*. *KLK2* is a closely related gene encoding a 237-residue single-chain active serine proteinase glycoprotein called human kallikrein-related peptidase 2 (hK2), which is 80% identical to PSA in amino acid sequence but manifests tryptic-like serine proteinase activity. Like PSA, abundant expression and secretion of hK2 are regulated by *AR* and specific to the epithelial cells of the prostate, a gland unique to males. At miniscule levels however, expression of PSA or hK2 has been detected in a few other tissues including breast tissue although prostate breast tissue appears to be the only source of PSA/hK2 detected in blood [8].

The biologic function and catalytic action of PSA and hK2 have relevance to reproductive function as it catalyzes the sol-to-liquid transformation of ejaculated semen through cleavage of gel-forming semenogelin (*SEMG1* [9] and *SEMG2* [10]) that are predominant in the seminal vesicle secretion, thereby influencing the motility of the sperm [11].

Normally, PSA (and hK2) occurs at picomolar concentrations in peripheral blood, which is about a million-fold lower than the levels of PSA found in semen [12]. Most of PSA detected in blood occurs in covalently stable $\approx$90-kDa 1:1 molar ratio with α-1-antichymotrypsin (ACT, *SERPINA3*), an extracellular serine proteinase inhibitor present at $\approx 10^4$-fold molar excess to PSA in blood [13]. However, a minority of PSA remains in unbound and noncatalytic free forms. α-2-Macroglobulin (A2M) is another complexing ligand that largely encapsulates PSA making it difficult to immunodetect [14] and no current data support clinical utility for assays specific to the PSA–A2M complex. Analytical laboratory assays used to measure PSA in blood are based on equimolar immunodetection of both free PSA and PSA–ACT complex forms (total PSA) using high-affinity monoclonal antibodies (MAbs) enabling functional detection limits for total PSA $\leq$0.10 μg/L [15]. After puberty, PSA levels remain low for several decades whereas median levels increase from $\approx$0.6 to 1.1 μg/L for men aged 40–65 [16–18], largely due to higher frequency of benign prostate enlargement among the older men.

The PSA test was first approved by the US Food and Drug Administration (FDA) in 1986 for monitoring prostate cancer recurrence and in 1994 it was approved for screening for prostate cancer [19]. Although PSA levels in the hundreds or thousands μg/L in blood may be viewed as pathognomonic of very advanced stages of prostate cancer, it should be emphasized that diagnosis of prostate cancer typically requires histopathological examination of prostate tissue from transrectal ultrasound (TRUS)-guided systematic biopsy, magnetic resonance imaging (MRI)-targeted biopsy, or transperineal biopsy. Other examples are histopathological examination of tissue from transurethral resection of the prostate (TURP), radical prostatectomy (RP), and cystoprostatectomy.

There is no universally agreed upon PSA cut-off above which to consider additional risk stratification using reflex biomarkers and/or MRI-imaging as well as when to refer men for a clinical workup with digital rectal examination (DRE) and a prostate biopsy procedure. Despite lack of rigorously examined or reported empirical evidence, it became common practice for very long time to use PSA as a dichotomous marker: "positive" leading to prostate biopsy when men had PSA ≥4.0 μg/L versus "negative" or "normal" if below that level and no biopsy unless DRE was suspicious. However, Thompson and colleagues reported in 2004 a landmark study based on the end-of-study biopsy data among men in the control arm of the Prostate Cancer Prevention Trial (PCPT), which challenged this paradigm. Based on 2950 men with PSA <4.0 μg/L who were subject to end-of-study biopsies after 7 years, they found that both risk of any grade and risk of high-grade (Gleason score ≥7) prostate cancer were associated with the levels of PSA, but there was no lower limit of PSA at which prostate cancer was not detected, and 15.2% of all men were diagnosed with any-grade prostate cancer of which 67 of the 449 men with cancer at biopsy (14.9%) were high grade [20]. As many men (26.9%) were diagnosed with prostate cancer at PSA levels of 3.0–4.0 μg/L, and similar data emerged from interim analyses of ERSPC, the European Randomized Screening for Prostate Cancer, several guideline groups subsequently lowered the PSA cut-off to ≥3.0 g/L. Hence, PSA ≥3 μg/L as a cut-off is still commonly used in clinical practice, whereas others advocate age-specific reference ranges and yet others have suggested a strategy whereby the clinician has further discussions or initiates workup with, for example, additional markers, MRI, DRE, and/or biopsy for men with PSA levels ≥1.5 μg/L [8,19,21].

While the PSA test is a highly sensitive blood biomarker to detect increased risk of prostate cancer at a very early stage, it is imperfect tool due to lack of specificity, as the majority of modestly increased PSA levels are due to causes other than prostate cancer [22]. In fact, most men with a modestly elevated PSA level (3–10 μg/L) do *not* have prostate cancer on subsequent biopsy (false positives), but more commonly a benign enlargement of the prostate gland, benign prostatic hyperplasia (BPH), or inflammation that causes the elevation [22]. The range of false positive PSA tests in repeated screening programs varies between about 1 in 5 men and 2 in 3 men [23,24]. DRE, prostate massage, and ejaculation may cause minor short-term changes in PSA level, but rarely of clinical importance [8]. There is no robust evidence that bicycle riding increases PSA levels [25–28]. Instrumentation and invasive procedures such as rigid cystoscopy, prostate biopsy, and TURP may provide increases in PSA levels that are predominantly attributed to short-term elevation of the free PSA forms, which eliminates with a half-life of 12–18 hours, whereas any additional influx of complexed PSA is eliminated at a much slower rate [29], which may persist for up to 6 weeks [8]. Medications for BPH such as 5-α-reductase-inhibitors (finasteride and dutasteride) reduce PSA levels by approximately 50% [8].

5.3 PSA screening trials

No tumor marker has had such a significant impact on stage-migration that enables early detection, and monitoring after treatment for prostate cancer as the blood test PSA. In the late 1980s to early 1990s, several groups including those of Catalona, Wang, Cooner, Brawer, and Stamey demonstrated that PSA could be used as a tumor marker to screen for prostate cancer and distinguish cancer from BPH [30–35]. This set the stage for a widespread implementation of PSA testing in the United States, with much slower adoption in worldwide use of the test [36]. Before the introduction of PSA testing, about half of men with newly diagnosed prostate cancer had incurable, metastatic disease. At the time, DRE was the only screening strategy for prostate cancer, that is, the doctor palpating the patient's prostate via a finger in the rectum. However, DRE as a screening tool has poor sensitivity [37] and finds many tumors to be incurable by the time they are palpable. In contrast, only 5 years after the introduction of PSA, the majority of men were diagnosed with early-stage, curable disease [30–32].

The observations that the PSA test could be used as a tumor marker to screen for prostate cancer became the impetus, which prompted the planning of large prospective randomized controlled trials to investigate the benefits and harms of PSA-screening. To date, four large-scale, randomized controlled trials of PSA screening versus usual care have been completed and provide evidence regarding the effectiveness of PSA screening: the US Prostate Lung Colorectal and Ovarian (PLCO) cancer screening trial [38]; the ERSPC trial [39]; the Göteborg Randomized Population-Based Prostate Cancer Screening trial (Göteborg-1 [40]); and the UK Cluster Randomized Trial of PSA Testing for Prostate Cancer (CAP) [41], which was an extension of the Prostate Testing for Cancer and Treatment (ProtecT) trial [42] (Table 5.1).

In brief, the collective evidence from these trials has demonstrated that regular PSA screening every 2–4 years in men aged 50–70 years detects the vast majority of prostate cancers at an early, curable stage, reducing men's risk of developing metastatic prostate cancer and dying from the disease by approximately 20%–35% at 13–18 years of follow-up [38–40,44–46]. The evidence also supports the principle that PSA screening is more effective if it is pursued as a systematic program, that is, repeated at regular intervals over time, and conversely, that there is little, if any, benefit to one-time PSA screening only [41,47]. In the United States, where PSA-testing has been used ubiquitously, the age-adjusted prostate cancer mortality has decreased by 52% since 1993 and men diagnosed with prostate cancer generally have a good prognosis [36].

In detail, the European trials, ERSPC, and Göteborg-1 showed that regular PSA screening every 2–4 years of men aged 50–70 was associated with a statistically significant relative reduction in prostate cancer mortality by 20%–35% at 16–18 years of follow-up [39,40]. In contrast, the US PLCO trial and the UK CAP trials found no evidence of a statistically significant prostate cancer mortality benefit associated with PSA testing. However, the PLCO trial suffered from a very high contamination rate as there was as many men in the control arm who received opportunistic PSA-testing as compared with the number of men accepting to participate in the screening arm [43,48]. As a result, at 17 years of follow-up, there was no statistically significant difference in prostate cancer mortality rates between the randomized arms [relative risk (RR), 0.93, 95% confidence interval (CI), 0.81–1.08; P-value = 0.38) [38]. Furthermore, there were also important differences in the design of the ERSPC and PLCO trials and participant adherence to the study protocols; the PLCO utilized annual screening, while the re-

Table 5.1 Contemporary, large-scale, randomized controlled trials of PSA-screening versus usual care.

Trial	PLCO [38,43]	ERSPC [39]	Göteborg-1 [40]	CAP/ProtecT [41,42]
Population	10 US centers	8 European countries	Göteborg, Sweden	573 UK primary care centers
Enrollment years	1993–2001	1991–2003	1994	2001–09
Sample size	76,685	162,389	20,000	CAP: 419,582 men; 189,386 men invited; 64,436 men with a valid PSA testing result ProtecT: 1643 men with prostate cancer
Age at screening	55–74	55–69	50–64	50–69
Intervention (vs. usual care)	PSA & DRE	PSA (± DRE in some countries)	PSA	CAP: single PSA ProtecT: random selection between three arms: active monitoring; radical prostatectomy; radiotherapy
Screening interval	Annual PSA for 6 years, DRE 4 years	4 years (most centers)	2 years	One-time PSA
Indication for prostate biopsy	PSA $\geq$ 4 ng/mL or suspicious DRE	PSA $\geq$ 3 ng/mL	PSA $\geq$2.5–3.0 ng/mL	PSA $\geq$ 3.0 ng/mL

(Continued)

Table 5.1 Contemporary, large-scale, randomized controlled trials of PSA-screening versus usual care. *Continued*

Trial	PLCO [38,43]	ERSPC [39]	Göteborg-1 [40]	CAP/ProtecT [41,42]
Biopsy	Followup of abnormal tests was at the discretion of the patients and their primary care physicians	Sextant TRUS biopsy, later 10–12 cores	Sextant TRUS biopsy, later 10–12 cores	10-core TRUS biopsy
Participation rate (%)	85	83	77	34
Biopsy compliance (%)	30–40	86	93	85
Followup to date	17 years	16 years	18 years	10 years
Number of screening rounds	Mean 5.0 rounds in screening arm Mean 2.7 rounds in control arm	Mean 1.9 rounds	Median 3 rounds	One-time PSA
RR of prostate cancer mortality	RR 0.93 (95% CI: 0.81–1.08; $P = 0.38$)	RR 0.80 (95% CI: 0.72–0.89; $P < 0.001$)	RR 0.65 (95% CI: 0.49–0.87; $P = 0.003$)	CAP: RR 0.96 (95% CI: 0.85–1.08; $P = 0.50$) ProtecT: No difference among the randomized treatment groups ($P = 0.48$)

screening interval was every 2–4 years in the ERSPC. The recommended biopsy threshold was PSA ≥4.0 μg/L in the PLCO trial but follow-up of abnormal tests was at the discretion of the patients and their primary care physicians, resulting in less than half of men who were screening positives were subject to prostate biopsy (30%–40%), as compared to adherence to a study protocol that recommended biopsy when PSA was ≥3.0 μg/L at dedicated urology clinics and screening centers at the ERSPC sites, resulting in a very high biopsy compliance rate (86%) [46]. Accounting for these differences in implementation and settings, Tsodikov and colleagues used methods based on mean lead time (i.e., the length of time by which screening advances the time of diagnosis as compared to its clinical presentation [49]), to reconcile the effects of screening on prostate cancer mortality in the two trials. This analysis showed that both the ERSPC and PLCO trials provided compatible evidence, finding that regular PSA screening reduces prostate cancer mortality relative to no screening at 11 years of follow-up (ERSPC: 25%–31%; PLCO: 27%–32%) [46].

Finally, the CAP trial in the United Kingdom found no difference in prostate cancer mortality at 10 years of follow-up compared to usual practice (RR 0.96, 95% CI, 0.85–1.08, *P*-value = 0.50). There was considerable contamination rates of PSA testing in the CAP trial [50], which complicates the interpretation of this report. However, the study provided no support for using a one-time only PSA test as a population-based screening program [41], in contrast to the important features of the PLCO, ERSPC, and Göteborg-1 trial protocols where men were re-screened at regular intervals of either 1, 2, or 4 years.

5.4 Baseline PSA studies and long-term risk of lethal prostate cancer

5.4.1 PSA variation

The levels of PSA and related markers measured in separate tubes of blood collected within few hours show high reproducibility, both biological and technical [51]. Still, the tendency of a man's PSA level to fluctuate may on average be 10%–15% from 1 week to another [51], which includes analytical laboratory assay variation of no more than 5%–10% [51,52], and is associated with the levels of PSA in prostatic fluid that are about million-fold higher than those in blood. Transient year-to-year fluctuations in PSA levels may pose a problem in clinical practice as they negatively impact the utility of PSA kinetics (e.g., PSA velocity or PSA doubling time) [53], and may provoke unnecessary antibiotic use. Widespread use of antibiotics prescribed to asymptomatic men with elevated PSA and normal DRE—assuming that the patient has bacterial prostatitis—risks contributing to the development of antibiotic-resistant organisms that increase the risk of sepsis after biopsy [54–56]. The number of previous biopsies is associated with a significant risk of infectious complications and each previous biopsy increases the risk [57].

Eastham and colleagues documented year-to-year variation in PSA-levels in a landmark study based on the use of archived serum samples collected consecutively 1 year apart over 4 years from 972 men participating in the Polyp Prevention Trial [58], a randomized study to evaluate the effect of diet in preventing colon polyps. The serum samples were collected between 1991 and 1994, cryopreserved and subsequently used to measure PSA levels between 1999 and 2000. Contingent on the definition of an "abnormal" value, the study found that about a third (21%–37%) of these men who had "abnormal" PSA (e.g., ≥4.0 ng/mL) in at least one of their five samples had "normal" levels of PSA (e.g., <4.0 ng/mL) in subsequent measurements. Once "normalized," the PSA level remained "normal" for most men (65%–83%) during the 4 years without any intervention [59]. A smaller study of 149 men confirmed the presence of a low short-term variation of approximately ±10% for total PSA repeated within 2 weeks. Measurements repeated on the same day varied less than measurements taken 2 weeks apart [51]. A report from the Göteborg-1 trial estimated the intraindividual variability for total PSA to be less than 14% for men with "normal" PSAs <2.0 μg/L [60].

The clinical implication of these studies is generally that an isolated PSA elevation should be confirmed several weeks later before further follow-up. This recommendation to repeat the PSA is included in many urology guidelines [61,62]. For instance, the National Comprehensive Cancer Network (NCCN) guidelines recommend that: "any man with a PSA ≥ 3.0 ng/mL undergo workup for benign disease, a repeat PSA and a DRE to inform decisions about whether to proceed with TRUS-guided biopsy" [62].

5.4.2 PSA as a prognostic marker

Although elevations of PSA levels in blood are strongly associated with increased risk of prostate cancer, the major limitation with the PSA test is the limited specificity, that is, the most men with modestly elevated PSA do not have prostate cancer and the elevated PSA is due to reasons other than prostate cancer such as BPH [22]. Because of this, many men undergo unnecessary prostate biopsy procedures to rule out cancer. In addition to the anxiety induced by an elevated PSA [63], there are many downstream harms from invasive prostate biopsy procedures, which may lead to complications, including urinary tract infections/sepsis (1%–18%), hematuria (2%–84%), urinary retention (0.5%–6%), and rectal bleeding (1%–45%) [59,64]. The incidence of biopsy-related hospitalization, primarily due to infectious complications, varies between 0.5% and 7% and has increased dramatically over time [59,64]. Critically, lack of stringent criteria to perform systematic 10+ core prostate biopsies also carry a substantial risk of overdiagnosis, that is, detecting small, insignificant prostate cancers that would never have led to clinical symptoms or reduced life-expectancy during a man's lifetime [57,59,65]. Notably, the majority of US men with such low-risk cancers choose to undergo immediate treatment (surgery or radiation) without evidence of oncologic benefit (overtreatment), but with substantial risks of harms that include urinary incontinence and erectile dysfunction [66,67].

Prostate cancer is a heterogeneous disease with natural history and prognosis that is highly variable and difficult to predict accurately at the time of diagnosis [68]. The fact that prostate cancer metastases mainly present later in life does not mean that the systemic dissemination of the disease does not begin much earlier. Mounting evidence supports the notion that a baseline PSA level during early middle-life (ages 45–60) is a remarkably strong prognostic marker of lethal prostate cancer, that is, can be used to inform men about their the long-term risk of developing prostate cancer with propensity to progress and disseminate to locally advanced stages, distant metastasis or death from disease 20–25 years later [16,17,69–72]. Hence, men with PSA levels at or below age-median (e.g., <1.1 μg/L at age 60) have very low long-term risk for a prostate cancer with propensity to become lethal, and for whom re-screening intervals can be either extended or exempted [16,17,73,74]. In other words, these studies have consistently shown that most men who developed lethal prostate cancer had increased PSA levels in the blood decades before their cancer diagnosis. As such, the major strength with the PSA test lies in its ability to stratify risk and provide a remarkably high negative predictive value for long-term risk of lethal prostate cancer. These findings have been demonstrated in a series of studies of cohorts of men followed conservatively without screening and with PSA measured retrospectively in archived blood samples collected in midlife, such as the Malmö Preventive Project (MPP), Malmö Diet and Cancer (MDC) study, Västerbotten Intervention Project (VIP) in Sweden, and the Danish Diet, Cancer and Health cohort. The strong association between a midlife PSA and lethal prostate cancer has subsequently been confirmed in cohorts with participants who were subject to PSA-testing [PLCO, Physicians Health Study (PHS), and Southern Community Cohort Study (SCCS)] (Table 5.2) [16,17,70–76].

Evaluating a test's performance characteristics, or discrimination or predictive accuracy, is traditionally done by comparing the area under the receiver operating characteristics curve (AUC), or concordance statistic when outcomes are measured as time to event (e.g., Harrel's C-index). AUC gives us the probability that, for a randomly selected pair of individuals, one with the disease and one without, a model will give a higher score to the patient with the disease [79,80]. Tests or models with poor discrimination similar to a coin flip result in an AUC of 0.5 whereas a test with perfect discrimination has an AUC of 1 [81].

Based on access to samples and data from a highly representative population-based cohort of 60-year-old men not subject to PSA testing but giving archived blood when they participated in the MPP in 1981,

Table 5.2 Baseline PSA studies and risk of aggressive prostate cancer.

Cohort	*N*	Age at blood draw	Followup (years)	Outcome	Results
MPP [73]	810	44–50	17	Locally advanced prostate cancer (≥cT3) or distant metastasis at diagnosis	AUC 0.791; 66% of advanced cancers occurred among the top 20% (PSA ≥0.9 μg/L)
MPP [17]	1167	60	25	Death from prostate cancer	AUC 0.90; 90% of prostate cancer deaths occurred among the top 25% (PSA >2 μg/L); 0.5% risk of death from prostate cancer if PSA ≤1 μg/L (below median)
MPP [74]	5040	44–50	23	Locally advanced prostate cancer (≥cT3) or distant metastasis at diagnosis	AUC 0.751; 69% of advanced cancers occurred among the top tertile (≥0.83 μg/L)
MPP [16]	10,357	45–49	27	Death from prostate cancer	Ages 45–49: AUC 0.718; 44% of prostate cancer deaths were in men in the top 10% (PSA ≥1.6 μg/L)
	4063	51–55			Ages 51–55: AUC 0.755; similar proportion in the top 10% (PSA ≥2.4 μg/L)
MDC [77]	4251	45–73	21	Death from prostate cancer	C-index 0.862
VIP [78]	43,692	50, 60	20	Death from prostate cancer	Age 50: C-index 0.859
					Age 60: C-index 0.840
					~75% of prostate cancer deaths occurred in men with PSA in the top quartile (>1.3 μg/L at age 50 and >2.3 μg/L at age 60)

(*Continued*)

Table 5.2 Baseline PSA studies and risk of aggressive prostate cancer. *Continued*

Cohort	*N*	Age at blood draw	Followup (years)	Outcome	Results
Danish Diet, Cancer and Health cohort [75]	1822	50–64	14	Aggressive cancer (≥cT3 or Gleason score ≥7 or N1 or M1)	Risk of aggressive cancer increased with each PSA quintile: incidence rate ratio (cases vs. controls) 100 in the top quintile (PSA >5.1 μg/L)
PHS [71]	945	40–59	30	Death from prostate cancer	40–49: 82% of prostate cancer deaths were in men with PSA above the median (PSA ≥0.68 μg/L); 50–54: 71% of deaths (PSA ≥0.88), 55–69: 86% of deaths (PSA ≥0.96)
SCCS [72]	766	40–64	9	Aggressive prostate cancer (Gleason ≥7, stage III/IV or prostate cancer death)	40–54: 100% of aggressive prostate cancers occurred in men above the median (PSA >0.72 for 40–49, >0.80 for 50–54); 55–64: 95% of deaths (PSA >0.94 for 55–69, >1.03 for 60–64)
PLCO [70]	10,968	55–60	13	Clinically significant prostate cancer (≥cT2b, Gleason ≥7 or prostate cancer death; or ≥pT3, Gleason ≥7 or N1 among men undergoing radical prostatectomy)	60% of prostate cancer deaths occurred in men with PSA ≥ 2 μg/L. 13-year risk of significant prostate cancer: 0.3% if PSA ≤ 0.49 μg/L and 1.1% if PSA 0.50–0.99 μg/L

PSA was found to be very strongly predictive of subsequent death from prostate cancer by age 85 with an AUC of 0.90 (95% CI: 0.84–0.96) [17]. Although only a minority of the men with a PSA level in the top quartile (≥2.0 μg/L) at age 60 developed lethal prostate cancer, 90% of all deaths from prostate cancer occurred among these men. These findings implicate an extremely favorable relationship between benefits and harms from continuing biennial screening of middle-aged men with PSA levels ≥2.0 μg/L when data from men regularly screened in the Göteborg-1 trial were compared to those from unscreened men in the MPP cohort in Malmö, suggesting that 72% of men aged 60 had little if any benefit from repeat screening in the Göteborg-1 trial [69]. Conversely, men with a PSA concentration at the median or lower at age 60 (PSA <1.1 μg/L) were highly unlikely to die (risk of 0.2%) or develop distant metastasis (risk of 0.5%) from prostate cancer by age 85. Hence, regular screening of the top quartile of 60-year-old men with PSA levels ≥2.0 μg/L at regular intervals was associated with a highly significant reduction in metastatic disease and prostate cancer mortality [69]. Conversely, no further screening was recommended for men aged 60 years with PSA levels <1.1 μg/L [17,61,69].

In another study based on archived samples and data from the MPP cohort in Malmö, Sweden, that focused on men early middle-age (ages 51–55) at baseline, long-term risk of death or distant metastasis from prostate cancer was strongly associated with PSA levels at baseline with 20-year risk of death from prostate cancer being 5.7% (95% CI: 3.7%–8.6%) for those with PSA levels in the top decile (PSA ≥2.4 μg/L), accounting for 44% of prostate cancer deaths; whereas, the corresponding 20-year risk for 51–55-year-old men with PSA levels below age-median (PSA <0.9 μg/L) was about 10-fold lower [absolute risk of 0.5% (95% CI: 0.2%–0.9%)] [16]. While the European cohorts [16,17,73–76] and the PHS [71] comprised >95% white men, the ability of baseline PSA to capture aggressive disease equally as well in black men has subsequently been confirmed in the SCCS (100% black men) [72] and PLCO (4% black men) [70,82].

These studies have helped inform the development of risk-adapted screening guidelines, such as those from the EAU (European

Association of Urology) and those in place at Memorial Sloan Kettering Cancer Center (MSK), with re-screening intervals based on a man's age, general health, and prior PSA [61,83].

5.5 PSA as a tumor marker and screening tool for prostate cancer and recommendations from national guideline groups

Who, when, and how often to screen men for prostate cancer remains controversial. Recommendations from various guideline groups have changed considerably over time and the age at which to begin and stop PSA screening as well as the re-screening interval vary between guidelines depending on how the guideline groups reviewed the evidence basis [84]. In a dramatic turnaround, and based on concerns that the harms from overdiagnosis and overtreatment of tumors that were unlikely to give rise to any symptoms during a man's lifetime may outweigh the benefits of screening, the United States Preventive Services Task Force (USPSTF) issued a grade D recommendation in 2012, that is, a recommendation against PSA testing in all men [85]. The justification was based on a rapid surge in the incidence of diagnosis of any grade prostate cancer among US males after the introduction of the PSA test and that many men treated with surgery or radiation harbored very low risk disease and may have been unnecessarily treated [36,66]. However, in the years following the USPSTF's D recommendation in 2012, there were trends reported toward higher stage and grade at diagnosis [86,87], and more recent reports on a definitive and sustained increase in distant metastatic stage at diagnosis, consistent with decreased PSA-testing rates in the wake of the USPSTF recommendation in 2012 [87–91]. Similarly, the death rate from prostate cancer in the United States, which declined by 4% per year from the mid-1990s through 2010 mainly due to early detection by PSA-testing and advancements in treatment, plateaued and did not continue to decline between 2013 and 2018 [36].

In a second turnaround in 2018, the USPSTF changed their recommendations regarding PSA-testing to "informed decision making" for men 55–69 years, stating that "screening offers a small potential benefit of reducing the chance of death from prostate cancer" [92] and noting that the rate of active surveillance for men with low risk prostate cancer in the United States increased, an important management strategy to reduce the harms of overtreatment [66]. Overtreatment would not be a problem if the treatment did not have associated morbidity. Because the prostate is anatomically located in close proximity to the urinary bladder, the rectum and the nerves that control erections, surgical removal of, or radiation, to the gland may lead to long-term side-effects including urinary leakage, urinary urgency, rectal bleeding, and erectile dysfunction, which may have a significant impact on the man's quality of life [93,94]. For this reason, active surveillance, as a modified and more pro-active version of the watchful waiting approach investigated by Johansson and colleagues [95,96], was first proposed by Klotz and colleagues in 2001 [97] and has developed over the past decades as a management strategy that implies postponing immediate curative treatment and allowing for close monitoring for signs of disease progression through repeat clinical workup and biopsy. In contrast, watchful waiting is a strategy for administering hormonal therapy when patients develop symptoms consistent with local or systemic progression, and is therefore fundamentally different from active surveillance.

Several active surveillance cohorts worldwide have demonstrated that active surveillance is a safe strategy over intermediate-to-long-term follow-up for carefully selected men with low-risk prostate cancer who are regularly monitored in an active surveillance protocol [98]. The 2018 update of the USPSTF [99] recommendation aligns with the American Urological Association (AUA) guideline from 2013 [100], which recommends that men between the ages of 55 and 69 years engage in shared decision-making about the benefits and disadvantages of PSA screening. Other guideline groups, including the NCCN guidelines (v.2.2021) for early detection of prostate cancer [21] recommends a baseline evaluation and discussion about risk and benefit to start at age 45 years (age 40 years if African ancestry or those with a family history of prostate cancer), which is similar to recommendations by the EAU [101] and the guidelines in place at MSK [61].

The PSA cutoff for recommending prostate biopsy also varies between guidelines. The USPSTF cites the ERSPC and Göteborg-1 trials used a PSA threshold ranging between 2.5 and 4 μg/L [99]. The

AUA an EAU guidelines do not specify a particular PSA cutoff for biopsy [100,101], and the NCCN recommends further evaluation of men ages 40–75 years with a PSA level of $\geq 4.0\ \mu g/L$ [21]. The MSK guideline recommends clinical workup and consideration of biopsy for men with PSA levels $\geq 3.0\ \mu g/L$ [61]. Recommendations for re-screening intervals also vary between the guidelines, ranging from every 1, 2, 4 years to risk-adapted strategies with intervals personalized based on a man's age, PSA level, risk factors, and general health.

5.6 PSA-related blood biomarkers to improve the specificity of PSA and rule out the need for prostate biopsy

5.6.1 PSA forms

What is generally termed the "PSA test" is actually not a single molecule but rather a "bag" of several molecularly distinct forms (Fig. 5.1) either arising from the catalytic serine proteinase PSA activity or due to lack of catalytic activity, for example, precursor forms or internally cleaved PSA, and which were first reported in 1990–91 [13,14]. The predominant form in blood is "complexed" PSA where PSA has formed a stable 1:1 molar ratio, covalently linked $\approx$90-kDa complex with α_1-antichymotrypsin (*SERPINA3*, ACT). Noncomplexed, "unbound," or "free" PSA comprise several distinct noncatalytic forms such as internally cleaved ("multichain" or "nicked") PSA, single-chain ("intact") PSA [102], and proPSA [103]. Although PSA has ability to also bind to α_1-antitrypsin (*SERPINA1*) and *A2M*, none of these complexes have been found to be of clinical value and are not detectable by currently available immunoassays ("invisible PSA") [8]. As discussed earlier in this chapter, hK2 is closely similar to PSA with 80% identity in primary structure but with a distinctly different trypsin-like substrate specificity [104] compared to the chymotrypsin-like substrate specificity of the catalytically active endoproteolytic PSA enzyme [105].

Laboratory assays to measure free PSA, hK2, and intact PSA are, similar to total PSA, based on immunodetection using high affinity MAbs that are independently binding to epitopes that are highly

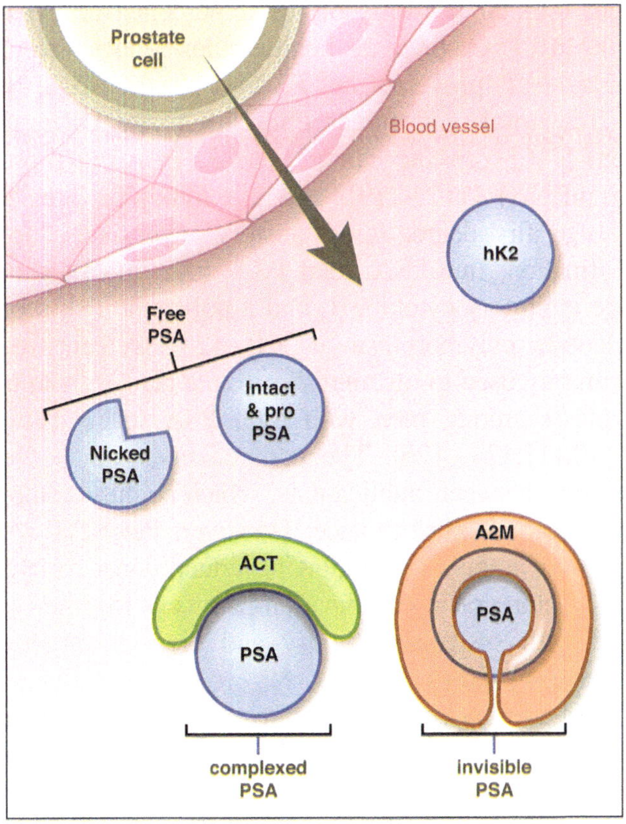

FIGURE 5.1 Different molecular forms of PSA in blood.

PSA, Prostate-specific antigen; *hK2*, human Kallikrein-related peptidase 2; *ACT*, α1-antichymotrypsin; *A2M*, α2-macroglobulin.

Schematic illustration: S. Carlsson. *©2022 Memorial Sloan Kettering. All rights reserved.*

unique to each of these extensively similar molecular structures. The functional detection limit is 0.04 μg/L for free PSA, 0.5 pg/mL for hK2, and 0.035 μg/L for intact PSA [15,106,107].

5.6.2 Biomarkers to improve the specificity of PSA

Several serum and urinary biomarkers are available to improve the specificity of PSA and for clinicians and patients to consider before prostate biopsy in order to reduce the number of unnecessary biopsies and overdiagnosis [108–109]. These tests are primarily based on different molecular forms of PSA: percent-free PSA (free/total PSA ratio) [110,111], prostate health index (PHI), calculated using the formula $(-2 \text{ proPSA/free PSA}) \times \sqrt{\text{total PSA}}$) [112–115], 4Kscore (free, total, and intact PSA, hK2, age DRE, prior biopsy information) [116], prostate cancer gene 3 (PCA3, a prostate-specific noncoding mRNA) [117–119], and the multiplex Stockholm-3 test (STHLM3, age, first-degree family history of prostate cancer, previous biopsy information, free PSA, total PSA, hK2, microseminoprotein-β, macrophage inhibitory cytokine-1, and a polygenic risk score based on single-nucleotide polymorphisms) [120–123]. Percent-free PSA and PHI are primarily used to distinguish prostate cancer from benign prostate hyperplasia among men with total PSA levels between 2 and 10 μg/L [110,111,124–126]. The components used in the STHLM3 test have varied between publications, which slightly complicate interpretations of the test performance. However, the STHLM3 has been shown in several large-scale studies to reduce unnecessary biopsy and overdiagnosis, alone or in combination pathways together with MRI of the prostate with systematic biopsy and biopsies targeted to MRI-suspicious lesions [120–123].

Other tests are based on urine testing, with or without serum PSA: Michigan Prostate Score (MiPS, total PSA plus urine TMPRSS2:ERG and PCA3) [127], ExoDx Prostate Intelliscore [EPI, exosomal RNA expression levels in urine of three genes: PCA3, V-ets erythroblastosis virus E26 oncogene homologs (ERG), and SAM pointed domain-containing Ets transcription factor (SPDEF)] [128], and SelectMDx (gene expression in urine, after DRE, of mRNA levels of two genes DLX1 and HOXC6) [129,130] and the urine test PCA3 that is approved for men ≥ 50 years with previous negative biopsy and other indications for repeat biopsy [127,131].

5.6.3 4-Kallikrein panel

Four kallikrein-related blood biomarker assays to measure levels of free, total, and intact PSA, as well as hK2, have been combined in what is known as the 4-kallikrein (4K) panel. The 4K panel has been tested in 13 studies involving nearly 22,000 men and shown to strongly predict risk of high-grade prostate cancer [Gleason Score ≥7, also known as International Society Urological Pathology [ISUP] Grade Group ≥2] on biopsy. The reported AUC has been in the range of 0.80–0.90, which is 0.05–0.10 higher than that of standard clinical predictors, such as age, DRE, and total PSA alone or a widely used prostate cancer risk calculator developed from the PCPT (Prostate Cancer Prevention Trial). These studies used archived samples and data from the large population-based screening trials in Europe (ERSPC and ProtecT) and the United States (PLCO) and were designed to systematically cover a range of clinical scenarios including men not previously subject to PSA-screening (ERSPC-Göteborg, ERSPC-Rotterdam, and ProtecT), men previously screened, biopsy-naïve men (ERSPC-Göteborg and ERSPC-Rotterdam), previously biopsied men (ERSPC-Rotterdam), men referred to biopsy on the basis of elevated PSA (PLCO), men referred to biopsy on the basis of clinical workup (ERSPC France, PLCO), men referred to biopsy on the basis of elevated PSA and low ratio of free-to-total PSA (Skåne University Hospital), and prospectively evaluated in multiinstitutional trials enrolling men referred to biopsy [106,116,132–141].

This prespecified statistical model based on the 4K panel has also been shown to be well-calibrated and use of the 4K panel to determine indication for biopsy has been shown to be able to decrease unnecessary prostate biopsies by 30%–50% without delaying the diagnosis of an undue number of high-grade cancers. Results have also been replicated in studies including men of diverse ethnicities (multiethnic cohort, PLCO) [106,116,132–141]. Improvement in the predictive accuracy was observed in each racial/ethnic group (African American, Japanese American, Latino, Native Hawaiian, and White) for any grade and aggressive prostate cancer as well as prostate cancer death [132].

On the basis of these studies, the prespecified statistical model based on the 4K panel has been licensed by OPKO Health and made commercially available as the 4Kscore in which the marker concentrations are combined with clinical variables, such as age, DRE, and prior biopsy to predict probability of high-grade prostate cancer on biopsy. As of December 2021, the 4Kscore was approved by FDA for use in men aged 45 and older who have not had a prior prostate biopsy or are biopsy negative and have an age-specific abnormal total PSA and/or abnormal DRE. Use of a tentative 4Kscore cut-off of 7.5% for biopsy decisions is associated with a 32% reduction in biopsies and a risk of missing or delaying the detection of high-grade prostate cancer (ISUP Grade Group 2 or 3) in 4.8% of men [142]. In a clinical utility study of 611 US patients with abnormal PSA and/or DRE, the use of the 4Kscore test reduced the number of biopsies by 65% [143].

In addition to the data on the use of the prespecified statistical model based on the 4K panel to predict high-grade prostate cancer on biopsy, the test has also been shown to predict aggressive prostate cancer using more robust outcome end-points, such as adverse pathology at RP (AUC 0.72–0.84) when added to a clinical model [144–147] and improving the specificity of the PSA test by further stratifying long-term risk of lethal prostate cancer; i.e., distant metastasis and death from prostate cancer with high predictive accuracy (AUC 0.75–0.88) (Table 5.3) [18,73,76–78]. For instance, in men with PSA ≥ 2.0 μg/L at age 50–60, the 4K panel was found to strongly improve accuracy in predicting risk of distant prostate cancer metastasis at 15 years followup with a C-index of 0.86–0.87 versus 0.76–0.81 for PSA alone [18]. Among the 19% of men with a PSA ≥ 1.5 μg/L at 45–59 years in the MDC study, 20-year risk of prostate cancer death was strongly associated with 4-kallikrein panel risk: 6.8% risk of death among those with a 4K panel risk score $\geq 7.5\%$ versus 1.2% risk of death for those with a 4K panel risk score $<7.5\%$ [77]. Similarly, among the 34% of men with a PSA ≥ 2.0 μg/L at ages 60–73, those with a 4K panel risk $\geq 7.5\%$ had a 13% risk of prostate cancer death at 20 years, while those with a 4K panel risk $<7.5\%$ had a 1.7% risk. These findings suggest that men with elevated PSA, but low 4K panel risk can be monitored and re-screened for many years instead of undergoing prostate biopsy, with little risk of developing life-threatening prostate cancer [77]. The association between the marker levels and risks of metastasis and prostate cancer death was separately validated in another large population-based cohort, the VIP in Sweden [18,78]. Men with elevated PSA ≥ 3.0 μg/L at age 60 and a 4K panel risk $<7.5\%$ had 0% risk of prostate cancer death at 10 years

Table 5.3 Summary of studies reporting on 4-kallikrein model predicting cancer outcomes beyond biopsy.

(A) Adverse pathology at radical prostatectomy or biochemical recurrence (BCR)			
Cohort	***N***	**Outcome**	**Results**
MSK [144]	1356	BCR	AUC 0.798 (clinical model + fPSA + hK2) vs. 0.786 (clinical model: clinical stage, biopsy Gleason Score, tPSA)
ERSPC Rotterdam [145]	392	Adverse pathology: pathologically aggressive prostate cancer (nonorgan confined disease, tumor volume >0.5 cm^3 or any Gleason grade ≥4)	AUC 0.84 (clinical model + 4kallikeins) vs. 0.81 (clinical model: age, clinical stage, tPSA, biopsy Gleason Score, number of positive cores, millimeters of cancer)
Hamburg [146]	2330	Adverse pathology: primary Gleason pattern 4, any pattern 5 or nonorgan-confined disease; BCR: postoperative PSA ≥0.20 μg/L	Adverse pathology: biopsy Gleason 3 + 3 and 3 + 4, respectively: AUC 0.718 and 0.659 (clinical model + 4-kallikreins) vs. 0.672 and 0.644 (clinical model: age, tPSA, biopsy Gleason Score, clinical stage); BCR: 0.660 and 0.616 vs. 0.630 and 0.620
Mayo Clinic [147] (very high risk population)	106	BCR or metastasis after radical prostatectomy	Poor discrimination of BCR and metastasis for clinical models, respectively (0.596 and 0.600 for preoperative nomogram and 0.576 and 0.655 for postoperative nomogram). Adding

(*Continued*)

Table 5.3 Summary of studies reporting on 4-kallikrein model predicting cancer outcomes beyond biopsy. *Continued*

(A) Adverse pathology at radical prostatectomy or biochemical recurrence (BCR)			
Cohort	***N***	**Outcome**	**Results**
			hK2 improved AUC for metastasis by 0.058. Adding intact PSA improved the AUC for BCR and metastasis by 0.011–0.057
(B) Long-term outcomes (prediction) of 4-kallikrein panel in cohorts of unscreened men			
Cohort	*N*	Outcome	Results
MPP [73]	21,277	Advanced prostate cancer: clinical stage ≥T3 or metastasis at diagnosis	AUC (4-kallikreins) 0.785 vs. 0.791 (tPSA) (No evidence that the 4-kallikrein panel added discriminatory accuracy among men ages 45–50 above tPSA alone in this study.)
MDC [76]	943	Clinically diagnosed prostate cancer; advanced prostate cancer (≥T3 or metastasis at diagnosis)	Ages 45–73 and PSA ≥3 μg/L: AUC 0.751 (age + 4-kallikreins) vs. 0.654 (age + tPSA) for clinically diagnosed cancer; AUC 0.824 vs. 0.716 for advanced cancer
MDC [77]	4251	Death from prostate cancer >20 years	Ages 45–73 and PSA ≥2 μg/L: C-index 0.803 (age + 4-kallikreins) vs. 0.735 (tPSA)

(*Continued*)

Table 5.3 Summary of studies reporting on 4-kallikrein model predicting cancer outcomes beyond biopsy. *Continued*

(A) Adverse pathology at radical prostatectomy or biochemical recurrence (BCR)			
Cohort	***N***	**Outcome**	**Results**
VIP [18]	12,542	Distant metastasis >15 years from blood draw	Age 50 and PSA >2 μg/L: C-index 0.863 (age + 4-kallikreins) vs. 0.751 (tPSA); age 60 and PSA >2 μg/L: C-index 0.875 vs. 0.805
VIP [78]	43,692	Death from prostate cancer >15 years	Age 50 and PSA ≥ 2 μg/L: C-index 0.828 (age + DRE + prior biopsy information + 4-kallikreins) vs. 0.767 (tPSA); age 60 and PSA ≥ 2 μg/L: 0.862 vs. 0.774

BCR, *Biochemical recurrence;* C-index, *Harrel's concordance index;* DRE, *digital rectal examination;* ECE, *extra capsular extension;* fPSA, *free PSA;* hK2, *human Kallikrein-related peptidase 2;* MDC, *Malmö Diet and Cancer;* MPP, *Malmö Preventive Project;* pT, *pathological tumor stage;* tPSA, *total PSA;* VIP, *Västerbotten Intervention Project.*

and 0.55% risk at 15 years, whereas those with a very high 4K panel risk of 20% had a high and similar risk of prostate cancer death as men with PSA levels of ≥ 10 μg/L [78].

5.7 Methods of detecting PSA

As of writing, there are multiple FDA-approved/510k-cleared commercial assays available for clinical use. Most major in Vitro diagnostics (IVD) manufacturers, including Abbott, Beckman, Roche, Siemens, and Tosoh, all have assays that are approved by FDA and available for use on their automated assay platforms. All currently available commercialized assays to detect total PSA are immunoassays, which utilize MAbs for the capture and detection of nonoverlapping epitopes on PSA covalently bound to ACT (*SERPINA3*).

The Hybritech PSA assay developed by the Hybritech Corporation was the first PSA assay to gain FDA approval for prostate cancer monitoring in 1986. Based on data using the Hybritech Tandem R method, a cutoff of 4.0 ng/mL was suggested and subsequently widely implemented for the detection of prostate cancer in men. However, widely discrepant results stemming from standardization of PSA assays with PSA from human semen, combined with the discovery that PSA occurs in multiple forms in blood [13], prompted considerable efforts to ensure comparability between commercially available assays. This was manifested through the first proposal for an International Standard (IS) by Thomas Stamey of Stanford [148]. This initial standard consisted of a mixture of 10% free PSA and 90% PSA complexed to ACT. This was followed by two World Health Organization standards, the 96/668 standard for free PSA, and the 96/670 standard for total PSA (i.e., free + complexed PSA) [149]. As of writing, stocks of WHO IS PSA 96/670 are depleted, and WHO IS PSA 96/668 are close to depletion, although assays that were based on these standards are still utilized. The current PSA standards are WHO IS PSA 17/100, which replaces 96/670, and the WHO IS PSA 17/102, which replaces 96/668 [150]. Both are available from the National Institute for Biological Standards and Control (NIBSC) in the United Kingdom, which serves as the designated distributor of the standard material. As of writing, most commercially available FDA-approved assays for PSA in clinical use are standardized to either the Hybritech or WHO standard. Clinicians should be aware that there are differences in PSA values obtained from the two standards, with values obtained from assays calibrated to the WHO standard showing a negative bias of approximately 25% relative to assays calibrated to the Hybritech standard [151].

Most commercially available FDA-approved assays for PSA are now what is referred to as the third-generation ultrasensitive PSA. These assays were developed in the 1990s, and have the ability to quantitate PSA into the 0.04 ng/mL range [152]. In recent years, there has been significant emphasis on the ability of PSA assays to detect lower and lower concentrations of PSA. This is because PSA assays play a key role in the detection of disease recurrence or metastases following either RP or radiation therapy for prostate cancer. Indeed, after RP, PSA levels may be particularly low.

5.8 Role of PSA in monitoring disease recurrence and progression

The rise in PSA following RP or radiation therapy is referred to as "biochemical recurrence" (BCR). The most commonly cited criteria for BCR following RP are those published by the American Urological Association, and is defined as a serum PSA ≥ 0.2 ng/mL, which is confirmed by a repeat PSA value ≥ 0.2 ng/mL [153]. Different criteria exist for the definition of biochemical progression following radiation therapy. This is partly due to differences in the types of radiation therapies available for the treatment of prostate cancer, such as external beam methods, brachytherapy, or a combination of these methods. In addition, radiation therapies only rarely result in PSA levels below the detectable limits unless combined with androgen deprivation therapy.

The most recent development in ultrasensitive PSA detection has come from Quanterix, a biomarker manufacturer, which has released a PSA assay that they refer to as a "fifth-generation" PSA assay. These immunoassays are available as RUO (Research Use Only) kits to be analyzed on either the proprietary Simoa HD-X or Simoa SR-X immunoassay platforms. The manufacturer claims a lower limit of quantitation of 0.024 pg/mL, with a dynamic range of 0–400 pg/mL. It is important to note, however, that these assays are not approved by FDA as of writing in 2021 and will need to be validated as laboratory developed tests by the performing laboratory prior to use in a clinical setting. Early studies have demonstrated that the assay can be used in a clinical setting with acceptable precision and accuracy [154]. At this point, it remains to be seen if use of a fifth-generation sensitive assay can result in significantly improved outcomes for patients being monitored following prostate cancer treatment.

PSA also remains important in monitoring response to systemic therapy for men with advanced or metastatic prostate cancer alongside imaging modalities such as prostate-specific membrane antigen (PSMA) PET imaging. Although not universally accepted or standardized, PSA changes are often interpreted as an early indicator of response to treatment or progression of disease on current treatments. PSA changes by 50% or 90% with treatment have been reported, although percent change from baseline at 8–12 weeks, and separately, the maximal change (rise or fall) at any time using a waterfall plot is most informative. Three patterns of PSA changes have been reported to a specific treatment. Sensitivity to a treatment may be considered when substantial and durable PSA declines are consistent with biological activity of the treatment [155]. An initial decline followed by a rise may represent only partial response to treatment, and selection or development of resistant disease. Primary resistance to treatment may result in a small decline followed by rapid rise and consistent rise of PSA [155].

To date, PSA changes are not considered a reliable surrogate end point, as they do not always correlate with clinical benefit or overall survival [156], and FDA has not used PSA changes for the approval of a novel treatment. Tumor dedifferentiation or dissociation between PSA changes and tumor growth may result in unreliable PSA declines, when true clinical progression is noted, as well as a small PSA rise may not indicate lack of treatment benefit from current treatment [156]. Several retrospective studies have proposed PSA changes as a surrogate for survival [157], however, prospective validation in dedicated clinical trials is still needed.

In selected patients, a rise in PSA is the only evidence of progression. In clinical studies, Prostate Cancer Working Group 3 (PCWG3) recommends reporting the time from start of therapy resulting in an initial decline to the first PSA increase that is $\geq 25\%$ and ≥ 2 ng/mL above the nadir, and which is confirmed by a second value ≥ 3 weeks later. If there was no decline in PSA from baseline, progression is defined as a $\geq 25\%$ increase and ≥ 2 ng/mL increase from baseline beyond 12 weeks [156]. In here, continuing treatment for at least 12 weeks is suggested before considering interpreting PSA changes, due to delayed effects of treatment on PSA and possible initial spikes in PSA measurements, despite clinical benefit.

For eligibility to clinical studies based solely on PSA rises, PCWG3 recommends progression to be documented by two consecutive increases in PSA, with the first rise at least 1 week from the reference value, and it should be confirmed by another value. This criterion recognizes fluctuations in PSA, which may require documentation of an additional PSA rise, and recommends considerations of PSA of at least 2 ng/mL, given variability and difficulty of interpreting lower values. Minimal starting value of 1.0 ng/mL, if confirmed, is recommended for PSA as the only indication of progression in pure small-cell carcinoma.

5.9 Conclusions and future directions

In conclusion, the PSA test remains an accurate and frequently used tumor marker for early detection of prostate cancer, staging and monitoring after treatment. Molecular subforms of PSA can further enhance risk stratification prior to prostate biopsy and in prognosticating risk of future life-threatening prostate cancer. Current research endeavors are focused on evaluating various screening and early strategies of combining the PSA-test with other biomarkers, with or without MRI, to achieve a better balance between the benefits and harms of early detection. This way, the risks of unnecessary biopsy and overdiagnosis could be mitigated while keeping the reduction in prostate cancer metastasis and death. The EAU recently proposed a risk-adapted strategy for prostate cancer early detection following a pathway based on PSA testing, risk calculators, and MRI [158]. In addition to the recently published results from the STHLM3-MRI trial [120–123], several large-scale trials (e.g., Göteborg-2 trial in Sweden, ProScreen in Finland) are currently evaluating such pathways [159]. The Göteborg-2 trial evaluates PSA followed by MRI with targeted and systematic biopsy versus targeted biopsy alone, as well as one arm utilizing a lower cut-off for biopsy ($\geq$1.8 ng/mL vs. 3 ng/mL). The ProScreen trial evaluates PSA followed by the 4-kallikrein panel followed by MRI and only targeted biopsy in case of positive MRI. Both pilot studies of these two trials showed promising results regarding feasibility and a signal that these pathways may keep the detection of clinically significant prostate cancer while reducing the detection of insignificant disease (overdiagnosis) [160,161]. Results regarding prostate cancer mortality are expected over the next decade.

Conflict of interest declaration

Hans Lilja is named on patents on assays for intact PSA (antibody, immunoassay, and method for prostate cancer detection. US Patent No. US 7,872,104 B2) and patents on a statistical method to detect prostate cancer (Methods and Apparatuses for Predicting Risk of Prostate Cancer and Prostate Gland Volume. US Patent No.US. 9,672,329 B2) held by Arctic Partners and licensed to and commercialized by OPKO Health as the 4Kscore test. Dr. Lilja receives royalties from sales of this test and has stock in Arctic Partners, OPKO Health, Diaprost, and Acousort. Kazunori Murata reports Research/contract funding from Abbott Laboratories. All other authors have no conflicts of interest to declare.

References

[1] Hara M, et al. Some physico-chemical characteristics of "-seminoprotein," an antigenic component specific for human seminal plasma. Forensic immunological study of body fluids and secretion. VII. Nihon Hoigaku Zasshi 1971;25(4):322–4.

[2] Wang MC, et al. Purification of a human prostate specific antigen. Invest Urol 1979;17(2):159–63.

[3] Kuriyama M, et al. Quantitation of prostate-specific antigen in serum by a sensitive enzyme immunoassay. Cancer Res 1980;40(12):4658–62.

[4] Papsidero LD, et al. A prostate antigen in sera of prostatic cancer patients. Cancer Res 1980;40(7):2428–32.

[5] Duffy MJ. Biomarkers for prostate cancer: prostate-specific antigen and beyond. Clin Chem Lab Med 2020;58(3):326–39.

[6] Denmeade SR, et al. Specific and efficient peptide substrates for assaying the proteolytic activity of prostate-specific antigen. Cancer Res 1997;57(21):4924–30.

[7] Lundwall A, Lilja H. Molecular cloning of human prostate specific antigen cDNA. FEBS Lett 1987;214(2):317–22.

[8] Duffy, M.J. and P. McGing, Guidelines for the use of tumour markers. Association of Clinical Biochemists in Ireland Scientific Committee Guidelines. 5th ed.: November 2018.
[9] Lilja H, Abrahamsson PA, Lundwall A. Semenogelin, the predominant protein in human semen. Primary structure and identification of closely related proteins in the male accessory sex glands and on the spermatozoa. J Biol Chem 1989;264(3):1894–900.
[10] Lilja H, Lundwall A. Molecular cloning of epididymal and seminal vesicular transcripts encoding a semenogelin-related protein. Proc Natl Acad Sci U S A 1992;89(10):4559–63.
[11] Balk SP, Ko YJ, Bubley GJ. Biology of prostate-specific antigen. J Clin Oncol 2003;21(2):383–91.
[12] Savblom C, et al. Genetic variation in KLK2 and KLK3 is associated with concentrations of hK2 and PSA in serum and seminal plasma in young men. Clin Chem 2014;60(3):490–9.
[13] Lilja H, et al. Prostate-specific antigen in serum occurs predominantly in complex with alpha 1-antichymotrypsin. Clin Chem 1991;37(9):1618–25.
[14] Christensson A, Laurell CB, Lilja H. Enzymatic activity of prostate-specific antigen and its reactions with extracellular serine proteinase inhibitors. Eur J Biochem 1990;194(3):755–63.
[15] Mitrunen K, et al. Dual-label one-step immunoassay for simultaneous measurement of free and total prostate-specific antigen concentrations and ratios in serum. Clin Chem 1995;41(8 Pt 1):1115–20.
[16] Vickers AJ, et al. Strategy for detection of prostate cancer based on relation between prostate specific antigen at age 40–55 and long term risk of metastasis: case-control study. BMJ 2013;346:f2023.
[17] Vickers AJ, et al. Prostate specific antigen concentration at age 60 and death or metastasis from prostate cancer: case-control study. BMJ 2010;341:c4521.
[18] Stattin P, et al. Improving the specificity of screening for lethal prostate cancer using prostate-specific antigen and a panel of Kallikrein markers: a nested case-control study. Eur Urol 2015;68(2):207–13.
[19] Crawford ED, et al. An approach using PSA levels of 1.5 ng/mL as the cutoff for prostate cancer screening in primary care. Urology 2016;96:116–20.
[20] Thompson IM, et al. *Prevalence of prostate cancer among men with a prostate-specific antigen level* < or = 4.0 ng per milliliter. N Engl J Med 2004;350(22):2239–46.
[21] National Comprehensive Cancer Network. NCCN Clinical Practice Guidelines in Oncology: Prostate Cancer Early Detection Version

2.2021. <https://www.nccn.org/professionals/physician_gls/pdf/prostate_detection.pdf.>; 2021 [accessed 09.08.21].

[22] Schröder FH, et al. Screening and prostate-cancer mortality in a randomized European study. N Engl J Med 2009;360(13):1320–8.

[23] Kilpeläinen TP, et al. False-positive screening results in the European randomized study of screening for prostate cancer. Eur J Cancer 2011;47(18):2698–705.

[24] Pinsky PF, Parnes HL, Andriole G. Mortality and complications after prostate biopsy in the Prostate, Lung, Colorectal and Ovarian Cancer Screening (PLCO) trial. BJU Int 2014;113(2):254–9.

[25] Jiandani D, et al. The effect of bicycling on PSA levels: a systematic review and meta-analysis. Prostate Cancer Prostatic Dis 2015;18 (3):208–12.

[26] Heger Z, et al. Influence of long-distance bicycle riding on serum/ urinary biomarkers of prostate cancer. Int J Mol Sci 2016;17(3):377.

[27] Mejak SL, Bayliss J, Hanks SD. Long distance bicycle riding causes prostate-specific antigen to increase in men aged 50 years and over. PLoS One 2013;8(2):e56030.

[28] Herrmann M, et al. Long-distance mountain biking does not disturb the measurement of total, free or complexed prostate-specific antigen in healthy men. Clin Chem Lab Med 2004;42(3):347–9.

[29] Bjork T, et al. Rapid exponential elimination of free prostate-specific antigen contrasts the slow, capacity-limited elimination of PSA complexed to alpha 1-antichymotrypsin from serum. Urology 1998;51(1):57–62.

[30] Catalona WJ, et al. Detection of organ-confined prostate cancer is increased through prostate-specific antigen—based screening. JAMA 1993;270(8):948–54.

[31] Catalona WJ, et al. Measurement of prostate-specific antigen in serum as a screening test for prostate cancer. N Engl J Med 1991;324(17):1156–61.

[32] Stamey TA, et al. Prostate-specific antigen as a serum marker for adenocarcinoma of the prostate. N Engl J Med 1987;317 (15):909–16.

[33] Wang TY, Kawaguchi TP. Preliminary evaluation of measurement of serum prostate-specific antigen level in detection of prostate cancer. Ann Clin Lab Sci 1986;16(6):461–6.

[34] Cooner WH, et al. Prostate cancer detection in a clinical urological practice by ultrasonography, digital rectal examination and prostate specific antigen. J Urol 1990;143(6):1146–52.

[35] Brawer MK, et al. Screening for prostatic carcinoma with prostate specific antigen. J Urol 1992;147(3 Pt 2):841–5.
[36] Siegel RL, et al. Cancer statistics, 2021. CA Cancer J Clin 2021;71 (1):7–33.
[37] Schröder FH, et al. Evaluation of the digital rectal examination as a screening test for prostate cancer. J Natl Cancer Inst 1998;90 (23):1817–23.
[38] Pinsky PF, et al. Extended follow-up for prostate cancer incidence and mortality among participants in the Prostate, Lung, Colorectal and Ovarian randomized cancer screening trial. BJU Int 2019;123 (5):854–60.
[39] Hugosson J, et al. A 16-yr follow-up of the European randomized study of screening for prostate cancer. Eur Urol 2019;76(1):43–51.
[40] Hugosson J, et al. Eighteen-year follow-up of the Goteborg randomized population-based prostate cancer screening trial: effect of sociodemographic variables on participation, prostate cancer incidence and mortality. Scand J Urol 2018;52(1):27–37.
[41] Martin RM, et al. Effect of a low-intensity PSA-based screening intervention on prostate cancer mortality: the CAP Randomized Clinical Trial. JAMA 2018;319(9):883–95.
[42] Hamdy FC, et al. 10-Year outcomes after monitoring, surgery, or radiotherapy for localized prostate cancer. N Engl J Med 2016;375 (15):1415–24.
[43] Pinsky PF, et al. Assessing contamination and compliance in the prostate component of the Prostate, Lung, Colorectal, and Ovarian (PLCO) Cancer Screening Trial. Clin Trials 2010;7 (4):303–11.
[44] Hugosson J, et al. Prostate specific antigen based biennial screening is sufficient to detect almost all prostate cancers while still curable. J Urol 2003;169(5):1720–3.
[45] Pinsky PF, et al. Metastatic prostate cancer at diagnosis and through progression in the Prostate, Lung, Colorectal, and Ovarian (PLCO) Cancer Screening Trial. Cancer 2019;125(17):2965–74.
[46] Tsodikov A, et al. Reconciling the effects of screening on prostate cancer mortality in the ERSPC and PLCO trials. Ann Intern Med 2017;167(7):449–55.
[47] Arnsrud Godtman R, et al. Opportunistic testing vs organized prostate-specific antigen screening: outcome after 18 years in the Goteborg randomized population-based prostate cancer screening trial. Eur Urol 2015;68(3):354–60.

[48] Shoag JE, Mittal S, Hu JC. Reevaluating PSA testing rates in the PLCO Trial. N Engl J Med 2016;374(18):1795–6.

[49] Draisma G, et al. Lead times and overdetection due to prostate-specific antigen screening: estimates from the European Randomized Study of Screening for Prostate Cancer. J Natl Cancer Inst 2003;95 (12):868–78.

[50] Clift AK, Coupland CA, Hippisley-Cox J. Prostate-specific antigen testing and opportunistic prostate cancer screening: a cohort study in England, 1998–2017. Br J Gen Pract 2021;71(703):e157–65.

[51] Christensson A, et al. Intra-individual short-term variability of prostate-specific antigen and other kallikrein markers in a serial collection of blood from men under evaluation for prostate cancer. BJU Int 2011;107(11):1769–74.

[52] Ulmert D, et al. Reproducibility and accuracy of measurements of free and total prostate-specific antigen in serum vs plasma after long-term storage at -20 degrees C. Clin Chem 2006;52(2):235–9.

[53] Vickers AJ, et al. An empirical evaluation of guidelines on prostate-specific antigen velocity in prostate cancer detection. J Natl Cancer Inst 2011;103(6):462–9.

[54] Eggener SE, et al. Empiric antibiotics for an elevated prostate-specific antigen (PSA) level: a randomised, prospective, controlled multi-institutional trial. BJU Int 2013;112(7):925–9.

[55] Scardino PT. The responsible use of antibiotics for an elevated PSA level. Nat Clin Pract Urol 2007;4(1):1.

[56] Habermacher GM, Chason JT, Schaeffer AJ. Prostatitis/chronic pelvic pain syndrome. Annu Rev Med 2006;57:195–206.

[57] Ehdaie B, et al. The impact of repeat biopsies on infectious complications in men with prostate cancer on active surveillance. J Urol 2014;191(3):660–4.

[58] Eastham JA, et al. Variation of serum prostate-specific antigen levels: an evaluation of year-to-year fluctuations. JAMA 2003;289 (20):2695–700.

[59] Loeb S, et al. Systematic review of complications of prostate biopsy. Eur Urol 2013;64(6):876–92.

[60] Bruun L, et al. Assessment of intra-individual variation in prostate-specific antigen levels in a biennial randomized prostate cancer screening program in Sweden. Prostate 2005;65(3):216–21.

[61] Vickers AJ, et al. The Memorial Sloan Kettering Cancer Center recommendations for prostate cancer screening. Urology 2016;91:12–18.

[62] Carroll, P.R, National Comprehensive Cancer Network (NCCN Guidelines) v.1.2021 Prostate Cancer Early Detection.
[63] Carlsson S, et al. Anxiety associated with prostate cancer screening with special reference to men with a positive screening test (elevated PSA) - results from a prospective, population-based, randomised study. Eur J Cancer 2007;43(14):2109–16.
[64] Borghesi M, et al. Complications after systematic, random, and image-guided prostate biopsy. Eur Urol 2017;71(3):353–65.
[65] Heijnsdijk EA, et al. Quality-of-life effects of prostate-specific antigen screening. N Engl J Med 2012;367(7):595–605.
[66] Cooperberg MR, Carroll PR. Trends in management for patients with localized prostate cancer, 1990–2013. JAMA 2015;314 (1):80–2.
[67] Donovan JL, et al. Patient-reported outcomes after monitoring, surgery, or radiotherapy for prostate cancer. N Engl J Med 2016;375 (15):1425–37.
[68] Mitchell JA, et al. Ability of 2 pretreatment risk assessment methods to predict prostate cancer recurrence after radical prostatectomy: data from CaPSURE. J Urol 2005;173(4):1126–31.
[69] Carlsson S, et al. Influence of blood prostate specific antigen levels at age 60 on benefits and harms of prostate cancer screening: population based cohort study. BMJ 2014;348:g2296.
[70] Kovac E, et al. Association of baseline prostate-specific antigen level with long-term diagnosis of clinically significant prostate cancer among patients aged 55 to 60 years: a secondary analysis of a cohort in the Prostate, Lung, Colorectal, and Ovarian (PLCO) Cancer Screening Trial. JAMA Netw Open 2020;3(1): e1919284.
[71] Preston MA, et al. Baseline prostate-specific antigen levels in midlife predict lethal prostate cancer. J Clin Oncol 2016;34 (23):2705–11.
[72] Preston MA, et al. Baseline prostate-specific antigen level in midlife and aggressive prostate cancer in Black men. Eur Urol 2019;75 (3):399–407.
[73] Ulmert D, et al. Prostate-specific antigen at or before age 50 as a predictor of advanced prostate cancer diagnosed up to 25 years later: a case-control study. BMC Med 2008;6:6.
[74] Lilja H, et al. Prediction of significant prostate cancer diagnosed 20 to 30 years later with a single measure of prostate-specific antigen at or before age 50. Cancer 2011;117(6):1210–19.

[75] Sjoberg DD, et al. Twenty-year risk of prostate cancer death by mid-life prostate-specific antigen and a panel of Four Kallikrein markers in a large population-based cohort of healthy men. Eur Urol 2018;73 (6):941–8.

[76] Vertosick EA, et al. Prespecified 4-Kallikrein Marker Model at age 50 or 60 for early detection of lethal prostate cancer in a large population based cohort of asymptomatic men followed for 20 years. J Urol 2020;204(2):281–8.

[77] Larsen SB, et al. Baseline prostate-specific antigen measurements and subsequent prostate cancer risk in the Danish Diet, Cancer and Health cohort. Eur J Cancer 2013;49(14):3041–8.

[78] Vickers AJ, et al. A panel of kallikrein marker predicts prostate cancer in a large, population-based cohort followed for 15 years without screening. Cancer Epidemiol Biomark Prev 2011;20(2):255–61.

[79] Vickers AJ. Markers for the early detection of prostate cancer: some principles for statistical reporting and interpretation. J Clin Oncol 2014;32(36):4033–4.

[80] Assel M, Vickers A. Biomarker evaluation and clinical development. SIU-ICUD ConsultatiMol Biomark Urol Oncol 2020;1(1).

[81] Steyerberg EW, et al. Assessing the incremental value of diagnostic and prognostic markers: a review and illustration. Eur J Clin Invest 2012;42(2):216–28.

[82] Andriole GL, et al. Mortality results from a randomized prostate-cancer screening trial. N Engl J Med 2009;360(13):1310–19.

[83] Mottet N, et al. EAU-EANM-ESTRO-ESUR-SIOG Guidelines on Prostate Cancer-2020 Update. Part 1: screening, diagnosis, and local treatment with curative intent. Eur Urol 2021;79(2):243–62.

[84] Kohestani K, Chilov M, Carlsson SV. Prostate cancer screening-when to start and how to screen? Transl Androl Urol 2018;7(1):34–45.

[85] Moyer VA, US Preventive Services Task Force. Screening for prostate cancer: U.S. Preventive Services Task Force recommendation statement. Ann Intern Med 2012;157(2):120–34.

[86] Fleshner K, Carlsson SV. The USPSTF screening recommendation: a swinging pendulum. Nat Rev Urol 2018;15(9):532–4.

[87] Fleshner K, Carlsson SV, Roobol MJ. The effect of the USPSTF PSA screening recommendation on prostate cancer incidence patterns in the USA. Nat Rev Urol 2017;14(1):26–37.

[88] Jemal A, et al. Prostate cancer incidence 5 years after US preventive services task force recommendations against screening. J Natl Cancer Inst 2021;113(1):64–71.

[89] Patel NA, et al. Definitive and sustained increase in prostate cancer metastases in the United States. Urol Oncol 2019;37(12):988–90.
[90] Negoita S, et al. Annual report to the nation on the status of cancer, part II: recent changes in prostate cancer trends and disease characteristics. Cancer 2018;124(13):2801–14.
[91] Dalela D, et al. Contemporary trends in the incidence of metastatic prostate cancer among US men: results from nationwide analyses. Eur Urol Focus 2019;5(1):77–80.
[92] US Preventive Services Task Force. Screening for prostate cancer: US preventive services task force recommendation statement. JAMA 2018;319(18):1901–13.
[93] Sanda MG, et al. Quality of life and satisfaction with outcome among prostate-cancer survivors. N Engl J Med 2008;358 (12):1250–61.
[94] Hoffman KE, et al. Patient-reported outcomes through 5 years for active surveillance, surgery, brachytherapy, or external beam radiation with or without androgen deprivation therapy for localized prostate cancer. JAMA 2020;323(2):149–63.
[95] Johansson JE, et al. Fifteen-year survival in prostate cancer. A prospective, population-based study in Sweden. JAMA 1997;277 (6):467–71.
[96] Johansson JE, et al. Natural history of early, localized prostate cancer. JAMA 2004;291(22):2713–19.
[97] Choo R, et al. PSA doubling time of prostate carcinoma managed with watchful observation alone. Int J Radiat Oncol Biol Phys 2001;50(3):615–20.
[98] Shill DK, et al. Active surveillance for prostate cancer. Transl Androl Urol 2021;10(6):2809–19.
[99] US Preventive Services Task Force, et al. Screening for prostate cancer: US preventive services task force recommendation statement. JAMA 2018;319(18):1901–13.
[100] Carter HB, et al. Early detection of prostate cancer: AUA guideline. J Urol 2013;190(2):419–26.
[101] Mottet N, et al., EAU–ESTRO–ESUR–SIOG guidelines on prostate cancer. Edn. Presented at the EAU Annual Congress Milan 2021, EAU Guidelines Office, Arnhem, The Netherlands.
[102] Nurmikko P, et al. Production and characterization of novel anti-prostate-specific antigen (PSA) monoclonal antibodies that do not detect internally cleaved Lys145-Lys146 inactive PSA. Clin Chem 2000;46(10):1610–18.

[103] Mikolajczyk SD, et al. Proenzyme forms of prostate-specific antigen in serum improve the detection of prostate cancer. Clin Chem 2004;50(6):1017–25.

[104] Lovgren J, Airas K, Lilja H. Enzymatic action of human glandular kallikrein 2 (hK2). Substrate specificity and regulation by Zn2 + and extracellular protease inhibitors. Eur J Biochem 1999;262 (3):781–9.

[105] Lilja H, Ulmert D, Vickers AJ. Prostate-specific antigen and prostate cancer: prediction, detection and monitoring. Nat Rev Cancer 2008;8(4):268–78.

[106] Vickers A, et al. Reducing unnecessary biopsy during prostate cancer screening using a four-kallikrein panel: an independent replication. J Clin Oncol 2010;28(15):2493–8.

[107] Nurmikko P, et al. Discrimination of prostate cancer from benign disease by plasma measurement of intact, free prostate-specific antigen lacking an internal cleavage site at Lys145-Lys146. Clin Chem 2001;47(8):1415–23.

[108] Kim JH, Hong SK. Clinical utility of current biomarkers for prostate cancer detection. Investig Clin Urol 2021;62(1):1–13.

[109] Visser WCH, et al. Commercialized blood-, urinary- and tissue-based biomarker tests for prostate cancer diagnosis and prognosis. Cancers (Basel) 2020;12(12).

[110] Catalona WJ, et al. Use of the percentage of free prostate-specific antigen to enhance differentiation of prostate cancer from benign prostatic disease: a prospective multicenter clinical trial. JAMA 1998;279(19):1542–7.

[111] Christensson A, et al. Serum prostate specific antigen complexed to alpha 1-antichymotrypsin as an indicator of prostate cancer. J Urol 1993;150(1):100–5.

[112] Filella X, Gimenez N. Evaluation of [-2] proPSA and prostate health index (phi) for the detection of prostate cancer: a systematic review and meta-analysis. Clin Chem Lab Med 2013;51 (4):729–39.

[113] Jansen FH, et al. Prostate-specific antigen (PSA) isoform p2PSA in combination with total PSA and free PSA improves diagnostic accuracy in prostate cancer detection. Eur Urol 2010;57(6):921–7.

[114] Lazzeri M, et al. Serum isoform [-2]proPSA derivatives significantly improve prediction of prostate cancer at initial biopsy in a total PSA range of 2–10 ng/mL: a multicentric European study. Eur Urol 2013;63(6):986–94.

[115] Stephan C, et al. Multicenter evaluation of [-2]proprostate-specific antigen and the prostate health index for detecting prostate cancer. Clin Chem 2013;59(1):306–14.

[116] Parekh DJ, et al. A multi-institutional prospective trial in the USA confirms that the 4Kscore accurately identifies men with high-grade prostate cancer. Eur Urol 2015;68(3):464–70.

[117] Bussemakers MJ, et al. DD3: a new prostate-specific gene, highly overexpressed in prostate cancer. Cancer Res 1999;59(23):5975–9.

[118] Groskopf J, et al. APTIMA PCA3 molecular urine test: development of a method to aid in the diagnosis of prostate cancer. Clin Chem 2006;52(6):1089–95.

[119] Hessels D, et al. DD3(PCA3)-based molecular urine analysis for the diagnosis of prostate cancer. Eur Urol 2003;44(1):8–15.

[120] Eklund M, et al. MRI-targeted or standard biopsy in prostate cancer screening. N Engl J Med 2021;.

[121] Eklund M, et al. The Stockholm-3 (STHLM3) Model can improve prostate cancer diagnostics in men aged 50–69 yr compared with current prostate cancer testing. Eur Urol Focus 2018;4(5):707–10.

[122] Gronberg H, et al. Prostate cancer screening in men aged 50–69 years (STHLM3): a prospective population-based diagnostic study. Lancet Oncol 2015;16(16):1667–76.

[123] Nordstrom T, et al. Prostate cancer screening using a combination of risk-prediction, MRI, and targeted prostate biopsies (STHLM3-MRI): a prospective, population-based, randomised, open-label, non-inferiority trial. Lancet Oncol 2021.

[124] Loeb S, Lilja H, Vickers A. Beyond prostate-specific antigen: utilizing novel strategies to screen men for prostate cancer. Curr Opin Urol 2016;26(5):459–65.

[125] de la Calle C, et al. Multicenter evaluation of the prostate health index to detect aggressive prostate cancer in biopsy naive men. J Urol 2015;194(1):65–72.

[126] Catalona WJ, et al. A multicenter study of [-2]pro-prostate specific antigen combined with prostate specific antigen and free prostate specific antigen for prostate cancer detection in the 2.0 to 10.0 ng/mL prostate specific antigen range. J Urol 2011;185(5):1650–5.

[127] Tomlins SA, et al. Urine TMPRSS2:ERG Plus PCA3 for individualized prostate cancer risk assessment. Eur Urol 2016;70(1):45–53.

[128] McKiernan J, et al. A novel urine exosome gene expression assay to predict high-grade prostate cancer at initial biopsy. JAMA Oncol 2016;2(7):882–9.

[129] Van Neste L, et al. Detection of high-grade prostate cancer using a urinary molecular biomarker-based risk score. Eur Urol 2016;70(5):740–8.
[130] Leyten GH, et al. Identification of a candidate gene panel for the early diagnosis of prostate cancer. Clin Cancer Res 2015;21(13):3061–70.
[131] Wei JT, et al. Can urinary PCA3 supplement PSA in the early detection of prostate cancer? J Clin Oncol 2014;32(36):4066–72.
[132] Darst BF, et al. The four-kallikrein panel is effective in identifying aggressive prostate cancer in a multiethnic population. Cancer Epidemiol Biomark Prev 2020;29(7):1381–8.
[133] Kim EH, et al. Detection of high grade prostate cancer among PLCO participants using a prespecified 4-kallikrein marker panel. J Urol 2017;197(4):1041–7.
[134] Braun K, et al. A four-kallikrein panel predicts high-grade cancer on biopsy: independent validation in a community cohort. Eur Urol 2016;69(3):505–11.
[135] Carlsson SV, et al. Can one blood draw replace transrectal ultrasonography-estimated prostate volume to predict prostate cancer risk? BJU Int 2013;112(5):602–9.
[136] Benchikh A, et al. A panel of kallikrein markers can predict outcome of prostate biopsy following clinical work-up: an independent validation study from the European Randomized Study of Prostate Cancer Screening, France. BMC Cancer 2010;10:635.
[137] Gupta A, et al. A four-kallikrein panel for the prediction of repeat prostate biopsy: data from the European Randomized Study of Prostate Cancer Screening in Rotterdam, Netherlands. Br J Cancer 2010;103(5):708–14.
[138] Vickers AJ, et al. A four-kallikrein panel predicts prostate cancer in men with recent screening: data from the European Randomized Study of Screening for Prostate Cancer, Rotterdam. Clin Cancer Res 2010;16(12):3232–9.
[139] Vickers AJ, et al. Impact of recent screening on predicting the outcome of prostate cancer biopsy in men with elevated prostate-specific antigen: data from the European Randomized Study of Prostate Cancer Screening in Gothenburg, Sweden. Cancer 2010;116(11):2612–20.
[140] Vickers AJ, et al. A panel of kallikrein markers can reduce unnecessary biopsy for prostate cancer: data from the European Randomized Study of Prostate Cancer Screening in Goteborg, Sweden. BMC Med 2008;6:19.

[141] Bryant RJ, et al. Predicting high-grade cancer at ten-core prostate biopsy using four kallikrein markers measured in blood in the ProtecT study. J Natl Cancer Inst 2015;107(7):djv095.

[142] Bhattu AS, et al. A 4Kscore cut-off of 7.5% for prostate biopsy decisions provides high sensitivity and negative predictive value for significant prostate cancer. Urology 2021;148:53–8.

[143] Konety B, et al. The 4Kscore® Test reduces prostate biopsy rates in community and academic urology practices. Rev Urol 2015;17 (4):231–40.

[144] Wenske S, et al. Evaluation of molecular forms of prostate-specific antigen and human kallikrein 2 in predicting biochemical failure after radical prostatectomy. Int J Cancer 2009;124(3):659–63.

[145] Carlsson S, et al. Predictive value of four kallikrein markers for pathologically insignificant compared with aggressive prostate cancer in radical prostatectomy specimens: results from the European Randomized Study of Screening for Prostate Cancer section Rotterdam. Eur Urol 2013;64(5):693–9.

[146] Haese A, et al. A pre-specified model based on four kallikrein markers in blood improves predictions of adverse pathology and biochemical recurrence after radical prostatectomy. Br J Cancer 2020;123(4):604–9.

[147] Assel MJ, et al. Kallikrein markers performance in pretreatment blood to predict early prostate cancer recurrence and metastasis after radical prostatectomy among very high-risk men. Prostate 2020;80(1):51–6.

[148] Stamey TA. *Second stanford conference on international standardization of prostate-specific antigen immunoassays: September 1 and 2, 1994*. Urology 1995;45(2):173–84.

[149] Rafferty B, et al. Reference reagents for prostate-specific antigen (PSA): establishment of the first international standards for free PSA and PSA (90:10). Clin Chem 2000;46(9):1310–17.

[150] Ferguson J, et al. Continued provision of WHO International Standards for total and free PSA: content and commutability of replacement preparations. Clin Biochem 2019;71:58–66.

[151] Vignati G, Giovanelli L. Standardization of PSA measures: a reappraisal and an experience with WHO calibration of Beckman Coulter Access Hybritech total and free PSA. Int J Biol Mark 2007;22(4):295–301.

[152] Mione R, et al. Third-generation PSA: ultrasensitive or ultraprecise assay? Int J Biol Markers 1995;10(4):229–33.

[153] Cookson MS, et al. Variation in the definition of biochemical recurrence in patients treated for localized prostate cancer: the American Urological Association Prostate Guidelines for Localized Prostate Cancer Update Panel report and recommendations for a standard in the reporting of surgical outcomes. J Urol 2007;177 (2):540–5.

[154] Wilson DH, et al. Fifth-generation digital immunoassay for prostate-specific antigen by single molecule array technology. Clin Chem 2011;57(12):1712–21.

[155] Rathkopf D, Scher HI. Androgen receptor antagonists in castration-resistant prostate cancer. Cancer J 2013;19(1):43–9.

[156] Scher HI, et al. Trial design and objectives for castration-resistant prostate cancer: updated recommendations from the Prostate Cancer Clinical Trials Working Group 3. J Clin Oncol 2016;34 (12):1402–18.

[157] Hussain M, et al. Prostate-specific antigen progression predicts overall survival in patients with metastatic prostate cancer: data from Southwest Oncology Group Trials 9346 (Intergroup Study 0162) and 9916. J Clin Oncol 2009;27(15):2450–6.

[158] Van Poppel H, et al. Prostate-specific antigen testing as part of a risk-adapted early detection strategy for prostate cancer: European Association of Urology Position and Recommendations for 2021. Eur Urol 2021;80(6):703–11.

[159] Kohestani K, et al. The GOTEBORG prostate cancer screening 2 trial: a prospective, randomised, population-based prostate cancer screening trial with prostate-specific antigen testing followed by magnetic resonance imaging of the prostate. Scand J Urol 2021;55 (2):116–24.

[160] Grenabo Bergdahl A, et al. Role of magnetic resonance imaging in prostate cancer screening: a pilot study within the Goteborg Randomised Screening Trial. Eur Urol 2016;70(4):566–73.

[161] Rannikko A, et al. Population-based randomized trial of screening for clinically significant prostate cancer ProScreen: pilot study. BJU Int 2021;.

CHAPTER

6 Monoclonal gammopathy detection and current technologies

Shelby M. Hutcherson and Katie L. Thoren

Department of Pathology and Laboratory Medicine, Memorial Sloan Kettering Cancer Center, New York, NY, United States

Objectives

- Describe the different types of monoclonal gammopathies and the role of the M-protein in disease diagnosis and monitoring
- Discuss current methods for routine detection of M-proteins
- Summarize new technologies being developed for M-protein detection and their potential applications

6.1 Introduction

A monoclonal gammopathy is a disorder characterized by abnormal proliferation of a single clone of B lymphocytes, usually a plasma cell. This clonal proliferation leads to excess production of a monoclonal immunoglobulin or M-protein. As we will discuss in more detail, monoclonal gammopathies range from premalignant disorders such as monoclonal gammopathy of undetermined significance (MGUS) and smoldering multiple myeloma (SMM) to life-threatening plasma cell malignancies such as multiple myeloma (MM), Waldenström's macroglobulinemia (WM), and amyloid light chain (AL) amyloidosis.

Regardless of the specific disease, the M-protein produced by the patient's abnormal clone of plasma cells serves as a useful tool for diagnosing and monitoring their condition. The M-protein was initially identified by Dr. Henry Bence Jones in 1847, making it one of

Cancer Biomarkers: Clinical Aspects and Laboratory Determination. DOI: https://doi.org/10.1016/B978-0-12-824302-2.00005-9
© 2022 Elsevier Inc. All rights reserved.

the oldest cancer biomarkers. Techniques to identify and quantitate the M-protein have been around for decades with little change. However, recent and drastic improvements in the treatment of MM are driving the development of more sensitive and specific testing methods. This chapter will discuss the different monoclonal gammopathies, the current technologies utilized for M-protein detection, and the techniques that are being developed to address the changing needs of patients, clinicians, and clinical laboratorians.

6.2 Monoclonal gammopathies

Monoclonal gammopathies are a group of diseases that range in severity. On one end of the spectrum, MGUS is an asymptomatic, premalignant monoclonal gammopathy that is present in approximately 3% of the white population over the age of 50 [1,2]. The incidence of MGUS increases with age, with a mean age at diagnosis of 70 years. The prevalence of MGUS is approximately 1.5-fold greater in men than in women and twofold greater in African Americans than in white Americans [2–5]. The condition is defined by the presence of an M-protein <3 g/dL, <10% clonal plasma cells in the bone marrow (BM), and the absence of symptoms related to the clonal proliferation of plasma cells. Because MGUS is asymptomatic, it is almost always detected incidentally during a laboratory workup for unrelated symptoms. However, once a patient is diagnosed with MGUS, it is important to monitor their condition closely due to the risk of progression to malignancy. While most MGUS patients will never develop cancer, nearly all cases of MM and other malignant monoclonal gammopathies arise from MGUS, and the risk of progression is approximately 1% per year. The risk of progression is increased in MGUS patients with M-proteins >1.5 g/dL, M-proteins with an IgM isotype, an abnormal serum free light chain (FLC) ratio, and/or immunoparesis.

Some patients with newly discovered M-proteins are found to be in an intermediate stage between MGUS and active malignancy and are considered to have SMM or, if the M-protein is of the IgM isotype, smoldering Waldenström's macroglobulinemia (SWM). These patients have an M-protein >3 g/dL and/or 10%–59% clonal plasma cells in their BM, but like MGUS patients, they lack symptoms associated with the proliferation of their clonal plasma cells. SMM/SWM carries a significantly higher likelihood of progression to active malignancy compared to MGUS. The risk of progression is approximately 10% per year for the first 5 years after diagnosis; however, the risk declines to 3% per year for the next 5 years and further decreases to 1% per year after that [6]. Patients with larger M-proteins and a higher percentage of clonal plasma cells are more likely to progress to malignancy.

MM is an incurable plasma cell neoplasm that accounts for approximately 2% of all cancers and 17% of hematologic cancers. The annual incidence rate of MM was 7 per 100,000 people between 2014 and 2018 and the annual death rate was 3.2 per 100,000 people between 2015 and 2019 [7]. The most recently reported 5-year relative survival rate was 55.6% [7]. MM primarily occurs in older adults, with most cases diagnosed in people between the ages of 65 and 74. Like MGUS, men are 1.5-fold more likely than women to develop MM and African Americans are about twice as likely to develop MM than white Americans. In addition, individuals with a first-degree relative who is diagnosed with MM have a fourfold higher risk of developing MM.

The diagnostic criteria for MM include the presence of 10% or more clonal plasma cells in the BM or a biopsy-proven plasmacytoma as well as symptoms and clinical findings associated with end-organ damage caused by the cancer. The hallmarks of end-organ damage that are characteristic of MM are often referred to using the acronym CRAB: **c**alcium elevation, **r**enal insufficiency, **a**nemia, and osteolytic **b**one disease. A normochromic, normocytic anemia is the most common presenting feature in MM patients and may be caused by multiple mechanisms related to the proliferation of the malignant plasma cells. Renal insufficiency in MM patients is most often caused by the formation of casts in the renal tubules composed of monoclonal FLCs. Hypercalcemia and bone disease are the result of bone tissue being destroyed by proliferating malignant plasma cells. Elevated calcium levels can contribute to the development of renal insufficiency and lytic bone lesions can cause bone pain and fractures. Even in the absence of findings consistent with CRAB, there are a few biomarkers that are associated with almost certain progression to end-organ damage and are considered diagnostic of MM, such as the presence of 60% or greater clonal plasma cells in the BM, an involved:uninvolved serum FLC ratio of 100 or greater, and/or an MRI showing more than 1 focal bone lesion. In addition to the symptoms associated with CRAB, immunoparesis is common in MM patients due to the excess of clonal plasma cells and circulating M-protein, making patients markedly more susceptible to infections.

While the presence of an M-protein is not required to make a diagnosis of MM, > 95% of MM patients have a measurable M-protein in serum and/or urine. In about half of MM patients, the M-protein is an IgG isotype. IgA and light chain only myelomas are the next most common, with each representing about 15%–20% of cases. IgD myeloma comprises approximately 2% of MM cases and IgE myeloma is exceedingly rare, with fewer than 70 total cases ever documented. Approximately 3% of MM patients do not have a detectable M-protein in serum or urine and have a normal serum FLC ratio and are thus considered to have nonsecretory myeloma. Finally, IgM MM is also quite rare, making up only 0.5% of cases. This is likely because IgM MGUS is more likely to progress to WM than to MM.

WM is an indolent B cell malignancy that is characterized by lymphoplasmacytic lymphoma (LPL) in the BM and an IgM M-protein in serum. With an annual incidence of approximately 3 per million people and only 1400 cases diagnosed in the United States each year, WM is considerably rarer than MM. It also occurs primarily in older adults, but unlike MM, WM is much more common in white Americans than in other races and is uncommon among African Americans. Individuals who have a first-degree relative with WM are 15- to 20-fold more likely to develop WM. The symptoms and clinical features of WM are generally distinct from MM and may be related to the infiltration of malignant B cells into tissues or to the unique characteristics of the IgM M-protein. For example, high concentrations of circulating IgM can cause serum hyperviscosity, which manifests as oronasal bleeding, visual disturbances, and other neurological symptoms. The circulating IgM may also act as an autoantibody, leading to symptoms such as neuropathy. Constitutional symptoms such as fatigue and weight loss are common, as are lymphadenopathy, hepatomegaly, and splenomegaly, which are not normally observed in MM. Excess IgM may be deposited in the kidney and GI tract, which can precipitate renal insufficiency and severe malabsorption, respectively. Notably, osteolytic bone lesions are very rare in WM patients. Patients with an IgM M-protein who present with bone pain and are found to have bone lesions are often diagnosed with IgM MM instead of WM.

AL amyloidosis is a plasma cell disorder in which amyloid fibrils composed of monoclonal FLCs are deposited in various organs and tissues in the body. AL amyloidosis may occur in isolation or arise in a patient who already has MM, WM, or non-Hodgkin lymphoma. It is a rare disorder, with an annual incidence of between 9 and 14 cases diagnosed per million people in the United States between 2007 and 2015. Like the other monoclonal gammopathies discussed here, AL amyloidosis occurs more commonly in men than in women and is primarily a disease of older adults, with a median age at diagnosis of 63–65 years. While nonspecific constitutional symptoms are common in patients with AL amyloidosis, other symptoms depend on the organs affected. In patients with kidney involvement, nephrotic syndrome is common; if the heart is involved, the patient may present with symptoms of heart failure. Similarly, amyloid deposits can cause neuropathy, liver dysfunction, purpura, or macroglossia due to light chain deposition in the associated tissues. Most AL amyloidosis patients have a measurable M-protein in serum and/or urine, and the M-protein is a FLC in approximately 50%. The lambda isotype is more prevalent than kappa in AL amyloidosis, accounting for approximately 70% of cases.

Monoclonal gammopathies are diagnosed using a wide range of data. A complete patient history is obtained and the patient is evaluated for the presence of symptoms such as bone pain, fatigue, unintentional weight loss, and neurological symptoms that could point to MM or another lymphocytic or plasmacytic disorder. BM aspiration and/or core biopsy is used to determine the percentage of clonal plasma cells present and to characterize their immunophenotype and cytogenetics, which is important for distinguishing between the different monoclonal gammopathies. Similarly, cross-sectional imaging studies aid in diagnosis and differential diagnosis of monoclonal gammopathies by revealing the presence or absence of osteolytic bone lesions.

The clinical laboratory also has a critical role to play in diagnosing and managing patients with monoclonal gammopathies. Complete blood counts and standard chemistry panels can be performed during initial evaluation of a patient and at regular intervals thereafter to assess kidney function and detect anemia, hypercalcemia, and immunoparesis. Importantly, laboratory tests are also used to measure, classify, and monitor the M-protein produced by the abnormal plasma cells. The information these tests provide is used not only for diagnosis, but also for risk stratification, predicting prognosis, monitoring patients' response to therapy, and detecting relapses.

6.3 Current technologies for M-protein detection

The main assays that are currently used for the routine detection of M-proteins include serum protein electrophoresis (SPEP), serum immunofixation electrophoresis (sIFE), immunosubtraction (ISUB), urine protein electrophoresis (UPEP), urine immunofixation (uIFE), and serum FLC quantitation. These are well-established assays that have been used for decades and are important components in the clinical management of patients with monoclonal gammopathies.

6.3.1 Serum protein electrophoresis

The purpose of SPEP is to detect and quantify the M-protein. In this technique, serum proteins are separated based on charge into six distinct fractions: albumin, alpha 1, alpha 2, beta 1, beta 2, and gamma regions (Fig. 6.1). Depending on the manufacturer, the beta 1 and beta 2 regions may be represented as a single region and therefore those SPEP traces contain only five fractions. The resulting pattern is visually inspected for the presence of an M-protein, which appears as a discrete "spike" above the normal background (Fig. 6.2). Immunoglobulins typically migrate in the gamma region but can be observed in other regions of the SPEP. For instance, IgA and IgM M-proteins often run in the beta region.

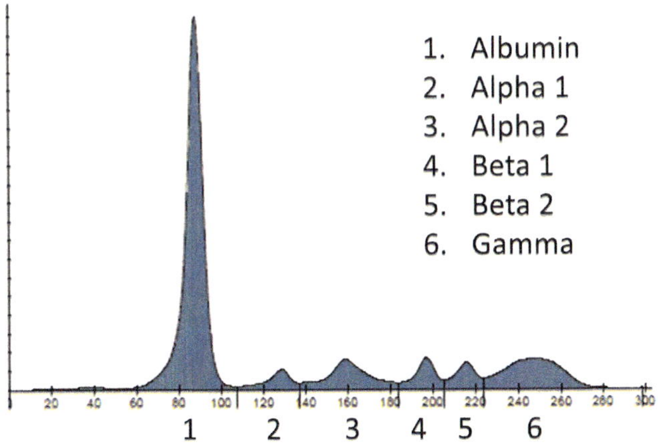

FIGURE 6.1

Serum protein electrophoresis (SPEP) of normal serum. This trace was generated with capillary electrophoresis using the Sebia Capillarys 3. The numbers (1–6) below the electropherogram indicate the different fractions of the SPEP.

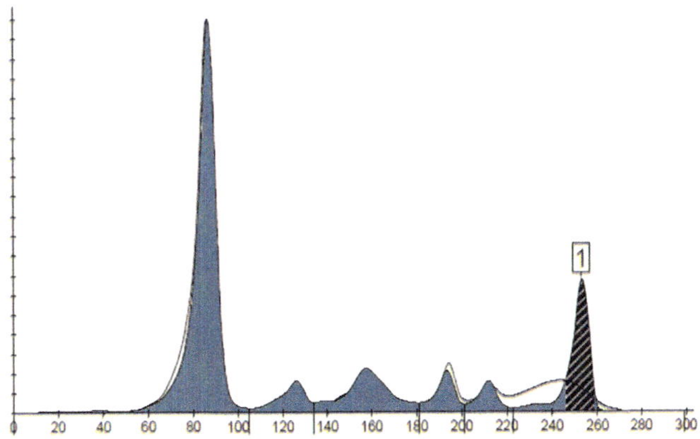

FIGURE 6.2

Abnormal serum protein electrophoresis. An M-spike (peak 1) is visible in the gamma region. The peak has been integrated for quantitation. The result from a normal control is overlayed (*black line*). This trace was generated with capillary electrophoresis using the Sebia Capillarys 3.

Most laboratories perform protein electrophoresis using agarose gel or capillary electrophoresis. With gel-based SPEP, proteins are separated using electrophoresis, the gel is stained with a protein-specific dye, and a densitometer is used to measure the intensity of staining. In capillary electrophoresis, proteins are detected directly based on their UV absorbance, typically at 200–215 nm, as they elute from the capillary. In both methods, the M-protein concentration is determined by relative quantification. From the densitometric scan (for gel-based SPEP) or the electropherogram (for capillary electrophoresis), the M-protein peak is integrated and the portion of total signal due to the M-protein is calculated. This percent is then multiplied by the total protein concentration of the sample to calculate the concentration of the M-protein.

Gel-based electrophoresis has been around longer and is used more commonly: 68.6% gel versus 31.4% capillary according to a 2016 College of American Pathologists (CAP) survey [8]. However, capillary electrophoresis offers a more automated method for M-protein testing and there is a trend, especially for larger laboratories, of adopting capillary-based methods.

6.3.2 Immunofixation and immunosubtraction

sIFE and ISUB are used to confirm that an immunoglobulin and not an interfering substance, is producing the spike observed on SPEP. This is important because some nonimmunoglobulin proteins and UV-absorbing substances can produce a false-positive result on SPEP. sIFE and ISUB offer better specificity than SPEP because they use immunoglobulin-specific reagents to confirm the identity of the SPEP spike. These techniques also are more sensitive than SPEP, with lower limits of detection about 0.01–0.03 g/dL [9], depending on the exact method used, the migration of the M-protein, and the polyclonal background of the sample.

In addition to confirming the presence of an M-protein, sIFE and ISUB are also used to determine the isotype of the M-protein. That is, they identify the type of light chains (κ or λ) and/or heavy chains (γ, α, μ, δ or ε) that comprise the immunoglobulin. This information, along with the migration pattern of the M-protein, helps distinguish the disease-related M-protein from the other monoclonal immunoglobulins that may be observed during the course of a patient's disease. Monoclonal proteins can be produced transiently in response to infection or after BM transplant, for example. In addition, monoclonal antibody therapeutics that are used for the treatment of monoclonal gammopathies can also be observed on these electrophoretic techniques [10]. In terms of patient management, it is most important to track the biomarker associated with the original plasma cell clone, and not these other monoclonal proteins.

In terms of the analytical steps to these methods, both sIFE and ISUB involve the electrophoretic separation of serum proteins and the use of antisera that is specific for γ-, α-, μ- heavy chains, or kappa or lambda light chains. Testing for IgD and IgE M-proteins is often performed as a second step since these M-proteins are rare. sIFE is a gel-based technique. Serum is applied to multiple lanes of an agarose gel and proteins are initially separated using electrophoresis. After separation, the reference lane is fixed with acid. In the other lanes, the patient serum is overlaid with antisera. If the antisera binds to antigen in the sample, an insoluble complex forms and precipitates, thus "fixing" it into the gel. Soluble proteins are washed away and the gel is stained with protein-specific dye. Some diffuse staining is observed in normal serum due to the presence of polyclonal immunoglobulins (Fig. 6.3A). But, an M-protein can be

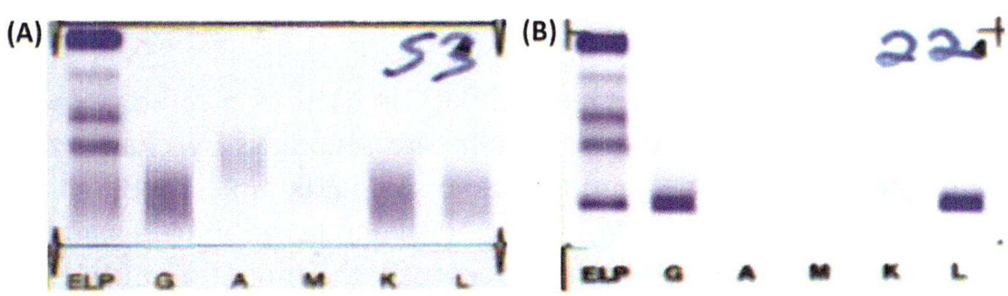

FIGURE 6.3

Immunofixation results from (A) normal serum corresponding to the SPEP shown in Fig. 6.1. Diffuse protein staining is observed due to the presence of polyclonal immunoglobulins, but no restricted band is visible. (B) Serum immunofixation result from a patient with multiple myeloma. Restricted bands are visible in the anti-IgG and anti-lambda lanes indicating the presence of an IgG lambda monoclonal protein. This result corresponds to the SPEP shown in Fig. 6.2 and confirms that the spike is due to a monoclonal protein.

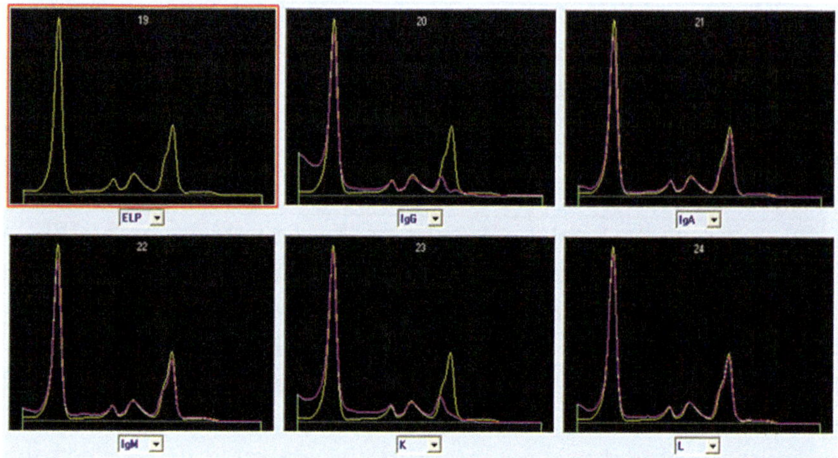

FIGURE 6.4

Immunosubtraction results from a patient with an IgG kappa M-protein. The reference trace (*yellow*) shows an M-spike in the beta 1 and beta 2 regions. The antisera-treated traces (*magenta*) show that the M-spike is subtracted by anti-IgG and anti-kappa reagent.

identified because it appears as a discrete band on immunofixation (Fig. 6.3B).

ISUB is a capillary electrophoresis-based method where six parallel capillaries are used. One capillary is used as a reference trace where serum is injected and proteins are separated by charge as in SPEP. In the other five capillaries, the sample is mixed with antisera specific to IgG, IgA, IgM, kappa, or lambda prior to separation. Because the antisera is negatively charged, the migration of the antisera–antigen complex is shifted compared to the reference trace. Disappearance of an abnormality in the antiserum-treated pattern compared to the reference trace indicates the presence of an M-protein and allows identification of its isotype (Fig. 6.4).

6.3.3 Urine protein electrophoresis and immunofixation

Protein electrophoresis and immunofixation can also be performed on urine. Urine testing is important for the detection of monoclonal

FLCs as these proteins are small enough to pass through the glomerulus and can be observed in the urine if produced in excess or if kidney damage prevents complete reabsorption. Urine testing offers better sensitivity compared to SPEP for the detection of monoclonal FLCs. This is largely due to the lower background signal compared to serum as there is little protein excreted into the urine when the kidneys are functioning properly. In addition, urine samples are often concentrated 50- to 100-fold prior to analysis in order to improve the limit of detection of these tests. Although urine is primarily used to detect monoclonal FLCs, intact monoclonal proteins can also be observed, especially when the kidney function becomes impaired and larger proteins are allowed to pass into the urine.

Similar to serum testing, UPEP and uIFE can be performed using gel or capillary electrophoresis. UPEP is used to detect and quantitate an M-protein in urine, and uIFE is used to confirm its presence and identify the isotype. Unlike SPEP, however, there is no "normal" protein pattern for healthy urine; typically, only a faint albumin band is observed. Thus there are no "normal" fractions, and the UPEP signal is divided into fractions numbered 1–5 (Fig. 6.5A and B). For uIFE testing, typically only FLC-specific antisera are used. Some uIFE kits offer a trivalent antiserum specific to IgG, IgA, and IgM to identify the presence of intact monoclonal immunoglobulins (Fig. 6.5C).

Detection and quantification of monoclonal proteins in urine is variable. This is related to the variable nature of the sample and to the variability of methods used for analysis. For instance, laboratories concentrate urine to various degrees, or sometimes not at all, prior to analysis. Twenty-four hour urine samples are recommended for UPEP and uIFE testing by the International Myeloma Working Group (IMWG) [11]. These samples are inconvenient to obtain, and as a result, urine testing is not performed as frequently as serum testing. As an alternative, some laboratories use urine from an early morning void [8]. In addition, several studies support the value of serum FLC measurement over urine studies for the detection and monitoring of monoclonal FLCs in many circumstances [12]. One important exception, however, is AL amyloidosis in which the combination of sIFE, serum FLC quantitation and

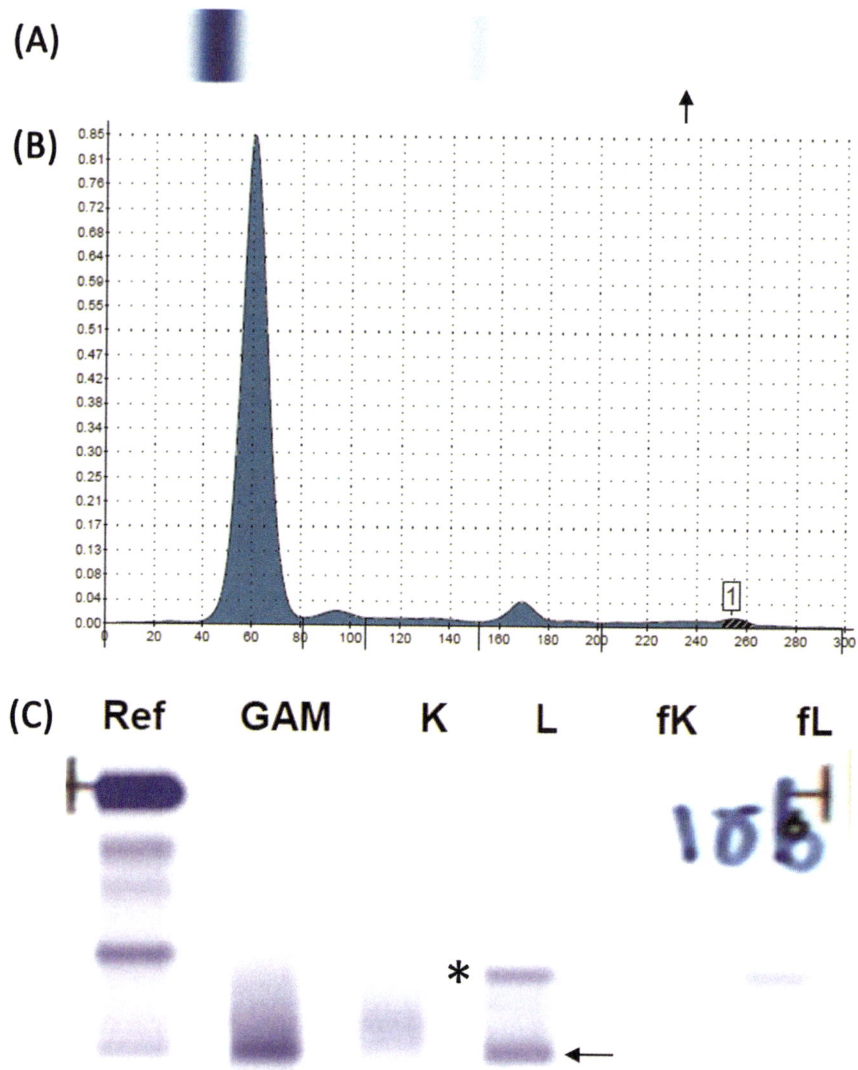

FIGURE 6.5

Urine protein electrophoresis and immunofixation results. (A) Image of the urine electrophoresis gel performed on a Sebia Hydrasys. The arrow indicates a monoclonal band. (B) Densitometer scan of the gel shown in (A). The band on the gel produced an M-spike in fraction 5 (peak 1). (C) uIFE result. In this method, the sample is applied to 6 lanes and electrophoresed. The reference lane is on the far left. The sample is treated with antisera against: IgG, IgA

(*Continued*)

uIFE provide the best sensitivity for detecting FLCs in this disease [13].

6.3.4 Free light chain quantitation

Since its introduction in 2001, FLC quantitation in serum has become an integral part of the diagnosis and monitoring of monoclonal gammopathies. Freelite (The Binding Site, Birmingham, United Kingdom) was the first commercially available assay and uses polyclonal antibody-coated latex particles that recognize only the exposed FLC epitopes which are hidden when the light chain is bound to the heavy chain. The reagent can be used with nephelometric or turbidimetric detection. The assay measures the concentration of free kappa and free lambda light chains and calculates the free kappa to free lambda ratio. It is important to note that this is not a direct measure of monoclonality. Instead, an abnormal FLC ratio indicates the presence of a monoclonal light chain.

For many years, the Freelite assay was the only immunoassay commercially available to measure FLC concentrations and all clinical utility studies were performed using this test. Currently, all of the international and national guidelines regarding serum FLC measurement are based on this assay. New serum FLC assays have been developed by other manufacturers using a variety of techniques (Table 6.1). Although these assays correlate to Freelite, the reference ranges and diagnostic cutoffs vary. Because there is no standardization for FLC testing, results between different assays should not be used interchangeably.

◀ and IgM heavy chains (lane 2), total kappa (lane 3), total lambda (lane 4), free kappa light chains (lane 5) and free lambda light chains (lane 6). The restricted bands in the "GAM" and total lambda lanes (noted with an arrow) indicates the presence of an intact monoclonal immunoglobulin. A free lambda light chain is also present (*) which produces monoclonal bands in both the total lambda and free lambda lanes.

Table 6.1 Free light chain assays.

Assay	Manufacturer	Methodology	Antibody reagents
Freelite	The Binding Site	Immunonephelometric or immunoturbidometric assay	Polyclonal antibodies (sheep): latex conjugates
N Latex FLC	Siemens	Immunonephelometric assay	Monoclonal antibodies: polystyrene-conjugates
SebiaFLC	Sebia	ELISA	Polyclonal antibodies (rabbit): latex-enhanced
Seralite	Abingdon	Lateral flow immunoassay	Monoclonal antibodies: gold-labeled
Diazyme Human Kappa and Lambda Free Light Chain	Diazyme	Immunoturbidometric	Polyclonal antibodies (rabbit and goat): latex-enhanced

6.3.5 Testing guidelines

Because many different tests are available, the process for identifying and monitoring patients with monoclonal gammopathies is complex. Currently, laboratories have adopted disparate strategies for testing and there is little standardization in which tests are used, how the tests are performed, and how results are interpreted [8]. Several national and international expert groups, such as the College Of American Pathologists (CAP) and the International Myeloma Working Group (IMWG), have developed evidence-based testing guidelines in an effort to standardize testing procedures [11,14]. One of the key messages from these guidelines is that individual tests should not be used in isolation for the detection of monoclonal gammopathies. Instead, a combination of tests is needed to provide the best sensitivity for M-protein detection. This is primarily due to the limited ability of SPEP or sIFE to detect monoclonal FLCs.

6.4 New and emerging technologies for M-protein detection

Although SPEP/sIFE, UPEP/uIFE, ISUB, and sFLC have served as the gold standard for M-protein detection for many years, drastic improvement in the treatment of monoclonal gammopathies has exposed some limitations of these techniques and new tests are being developed in response. In 2015 the FDA approved two monoclonal antibodies for the treatment of relapsed/refractory MM. Daratumumab, a fully human monoclonal antibody targeting CD38, was approved first, with elotuzumab, a humanized monoclonal antibody targeting SLAMF7, receiving approval just weeks later. Another anti-CD38 monoclonal antibody, isatuximab, was approved by the FDA in 2020, and several other drugs are being evaluated in clinical trials. The use of new monoclonal antibody therapies, especially in combination with other drugs, has significantly improved progression-free survival (PFS) and overall survival (OS) of MM patients.

However, the use of these new therapies presents two major challenges for current analytical laboratory techniques. Firstly, monoclonal antibody drugs can cause false-positive M-spikes on SPEP and sIFE (Fig. 6.6). This interference creates problems when evaluating a patient's response to treatment. For example, a complete response (CR) according to the IMWG's uniform response criteria requires the absence of the patient's disease-specific M-protein by sIFE. Since all the approved monoclonal antibodies are IgGκ immunoglobulins and the most common M-protein isotype is IgGκ, there is a high risk that the monoclonal antibody drug could be mistaken for the patient's M-protein, and the patient's response would be misclassified. Thus more specific tests are needed to distinguish between monoclonal antibody drugs and the disease-specific M-protein.

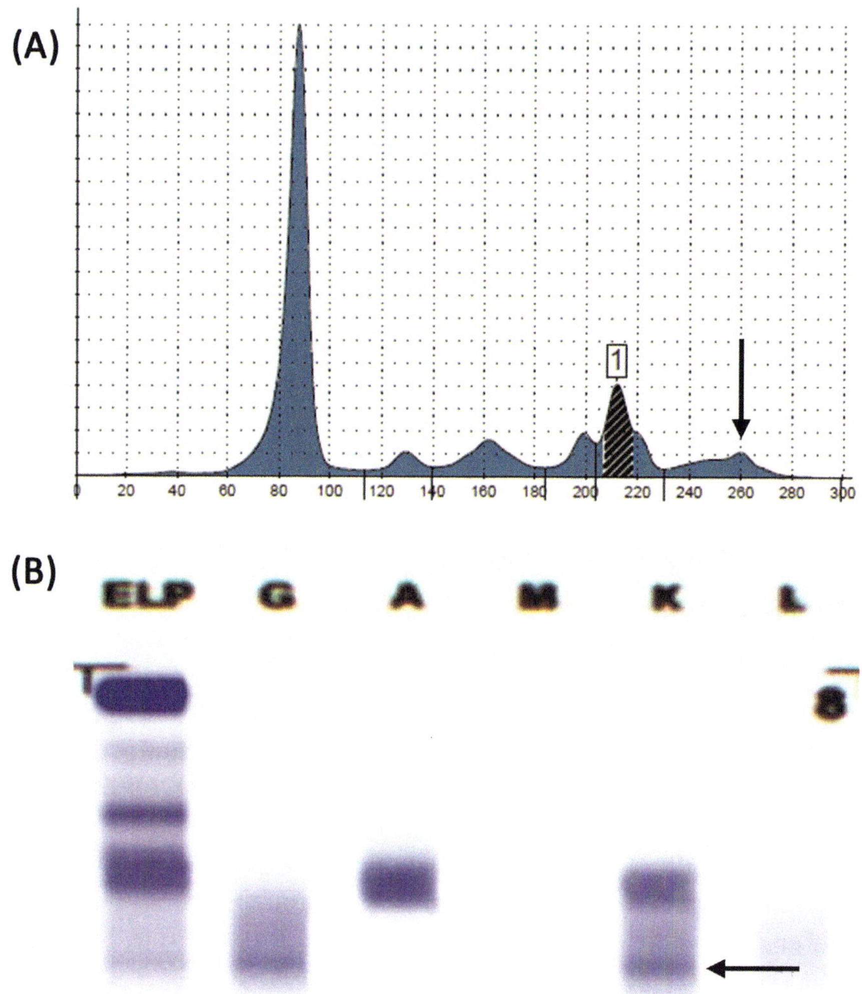

FIGURE 6.6

Monoclonal antibody drugs can cause a false-positive interference on SPEP and serum immunofixation electrophoresis (sIFE). (A) SPEP result showing an M-spike in beta 2 (peak 1) corresponding to the patient's original disease, and a peak in the gamma region (indicated with an arrow) due to daratumumab. (B) The corresponding immunofixation result showing the patient's M-protein, an IgA kappa, and an IgG kappa band due to daratumumab (arrow).

Secondly, more sensitive assays are needed as patients with MM and WM are surviving for substantially longer periods of time and reaching lower levels of disease. More than half of newly diagnosed MM patients now reach CR according to IMWG uniform response criteria. Unfortunately, most patients who achieve CR still relapse, suggesting the presence of persistent low levels of disease that are not detectable by conventional methods. Detection of minimal residual disease (MRD) has proven to have prognostic value, as MRD negativity is associated with improved PFS in MM patients. MRD is currently detected by multiparametric flow cytometry (MFC) or next-generation sequencing (NGS) on BM biopsy specimens. Both methods are highly sensitive, with limits of detection on the order of 1 myeloma cell per 10^5-10^6 nucleated cells. However, BM biopsies are invasive, painful, and expensive, making them poorly suited for serial measurements. In addition, patchy BM involvement by the clonal plasma cells can lead to a false negative biopsy result, and BM biopsies are unable to detect extramedullary disease. Because of the growing importance of measuring MRD and the drawbacks inherent in BM-based assays, there is a demand for highly sensitive assays that are based on peripheral blood (PB). PB collection is minimally invasive and cheaper than a BM biopsy. This makes serial measurements easier to obtain and allows clinicians to monitor a patient's disease more closely.

6.4.1 Shift assays

One strategy that has been developed to distinguish monoclonal antibody drugs from M-proteins is to use antidrug antibodies to shift their migration on immunofixation. The daratumumab-specific immunofixation electrophoresis reflex assay (DIRA) was first developed for patients receiving daratumumab during clinical trials [15]. DIRA was a laboratory-developed test, but the concept was later translated into a commercial assay. The Hydrashift 2/4 Daratumumab assay and the Hydrashift 2/4 Isatuximab assay were developed by Sebia and received FDA approval in 2018 and 2021, respectively. These assays are nearly identical to a standard Sebia sIFE assay, except that antiserum containing a murine monoclonal antiidiotypic antibody against daratumumab (or isatuximab) is applied to the IgG and κ lanes via a specific applicator. Similar to DIRA, the antidrug antiserum used in the Hydrashift assay shifts the migration of daratumumab or isatuximab into the alpha region during sIFE (Fig. 6.7). This removes the drug from the gamma region, allowing the interpreter to determine whether the patient's M-protein is present without interference. Although these assays have been very effective for accurately determining the presence of the patient's M-protein, they do not address

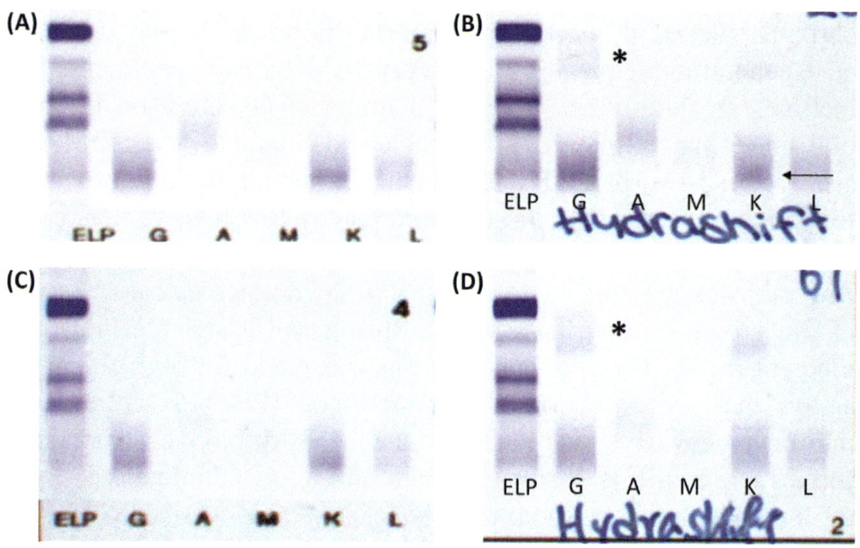

FIGURE 6.7

Hydrashift 2/4 Daratumumab removes monoclonal antibody drug interference. (A) Serum immunofixation and (B) hydrashift result for patient with IgG kappa multiple myeloma who is receiving daratumumab. One IgG kappa band in the gamma region is observed on the sIFE. In the hydrashift result, the IgG kappa band is visible in the gamma region (arrow) indicating that the patient's M-protein is still present. Daratumumab has been shifted away from the gamma region. The diffuse staining in the alpha region (*) in both the anti-IgG and anti-kappa lanes is due to the daratumumab/antidaratumumab complex. (C) Serum immunofixation and (D) hydrashift result for a second patient with IgG kappa multiple myeloma who is receiving daratumumab. In this instance, there is no IgG kappa band visible in the gamma region of the hydrashift. This confirms that the IgG kappa band on sIFE was due to daratumumab only.

the interference that daratumumab and isatuximab create on SPEP. The concentration of the M-protein will likely be overestimated if the drug comigrates with the patient's M-protein. In addition, these assays are specific for daratumumab and isatuximab only. As other drugs are approved and are used more widely, a more universal solution to the problem of therapeutic antibody interference will be needed.

6.4.2 Mass spectrometry

Another new strategy for M-protein detection is mass spectrometry (MS). Several methods have been developed which vary in terms of their analytical details (i.e., the mass spectrometers used, the portion of the M-protein that is followed and the sample preparation method that is employed). However, all are based on the concept that each M-protein has a specific amino acid sequence and therefore a specific molecular mass. This mass is constant over time and can be used as a marker to indicate the presence of clonal plasma cells. M-proteins can be detected by mass spectrometry in serum and urine. However, to date, most of the assay development and validation has focused on serum. Here, we will discuss the three most common methods currently being evaluated for M-protein detection.

The first two methods follow the intact light chain as a marker of the M-protein. In these methods, immunoglobulins are first purified from serum and disulfide bonds are reduced to separate the heavy chains from the light chains. The resulting sample is analyzed by mass spectrometry. Because the light chains ionize better than the heavy chains, they produce a larger signal. This signal is visually inspected for the presence of an M-protein, which appears as a sharp spike above the polyclonal background in the mass spectrum (Fig. 6.8). Originally described in 2014 by the Mayo Clinic, the monoclonal immunoglobulin rapid accurate mass measurement (miRAMM) method [16] uses liquid chromatography electrospray ionization quadrupole time-of-flight mass spectrometry (LC ESI-Q-TOF MS) for analysis.

The miRAMM method is highly sensitive with a limit of detection of about 0.005 g/dL. miRAMM also offers a universal solution for distinguishing monoclonal antibody drugs from a patient's M-

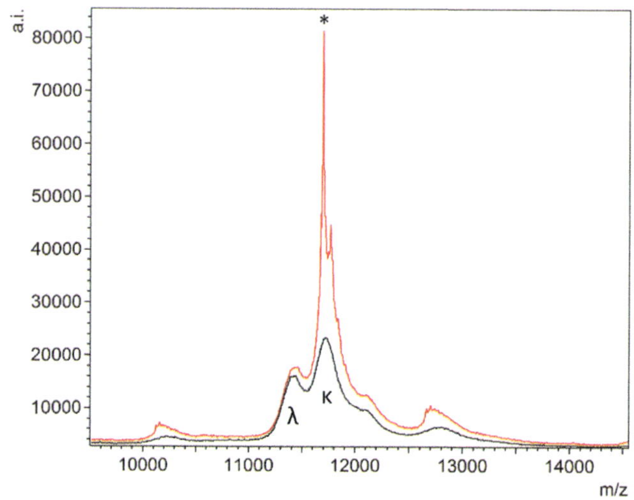

FIGURE 6.8

In intact light chain mass spectrometry methods, the M-protein appears as a spike above polyclonal background. These mass spectra were collected by matrix-assisted laser desorption/ionization time-of-flight mass spectrometry (MALDI-TOF MS). The signal of the doubly-charged light chains is shown for normal serum (black trace) and serum containing an M-protein (*red trace*). The normal polyclonal distribution of lambda and kappa light chains appear as smooth, Gaussian curves. When an M-protein is present, a spike (*) is visible in the mass spectrum.

protein because the proteins have different masses that appear as two distinct peaks in the mass spectrum [17]. However, the technique does present some challenges. The instrumentation is expensive and relatively complex, requiring highly trained individuals to develop tests and operate the mass spectrometer. The throughput of this method is limited due to the relatively long (~20 minutes) chromatography step. As such, this method is currently not appropriate for routine testing and is being evaluated for use as a blood based MRD test.

Not long after miRAMM was described, the Mayo Clinic developed a second mass spectrometry method, called MASS-FIX [18]. MASS-FIX also follows intact light chains as a marker of the

M-protein, except it employs a higher-throughput instrument called matrix-assisted laser desorption/ionization time-of-flight mass spectrometry (MALDI-TOF MS). This faster, higher-throughput analysis makes the method more amenable to routine testing in the clinical laboratory. In addition, MASS-FIX uses IgG-, IgA-, IgM-, κ- and λ-specific immunoaffinity reagents to isolate the immunoglobulins from serum and determine the isotype of the M-protein (Fig. 6.9). MASS-FIX is more analytically sensitive than sIFE [18]. It also provides better specificity than SPEP/sIFE for distinguishing monoclonal antibody drugs from the M-protein [19,20]. But the technique is less analytically sensitive compared to the miRAMM in part because of the lack of the chromatography step and the fact that MALDI-TOF MS offers lower mass resolution compared to Q-TOF MS.

Finally, the other common mass spectrometry approach is to follow a peptide that is specific for the M-protein, also known as a clonotypic peptide. Methods that use this approach first need to identify peptides that are specific to the M-protein. This requires knowledge of the M-protein sequence, which can be determined by *de novo* sequencing by mass spectrometry using a serum sample with abundant M-protein, or through DNA/RNA sequencing of the M-protein variable region using a BM sample with malignant plasma cells. Once the M-protein sequence is obtained, immunoglobulins are purified from serum and digested with proteases to produce peptides. The peptide mixture is analyzed by liquid chromatography high resolution mass spectrometry to select peptide(s) that are specific to the M-protein. Once identified, all subsequent analyses target this M-protein-specific peptide. Clonotypic peptide methods are more sensitive than MASS-FIX and miRAMM because M-protein detection is not limited by the polyclonal background. However, these methods are more labor-intensive because the workflow must be tailored to each patient. Although targeted peptide measurements can be completed in a matter of days, the initial sequencing of the patient's M-protein and identifying clonotypic peptides can take several weeks [21]. Similar to miRAMM, these methods may be more appropriate for MRD detection considering their high sensitivity but low throughput.

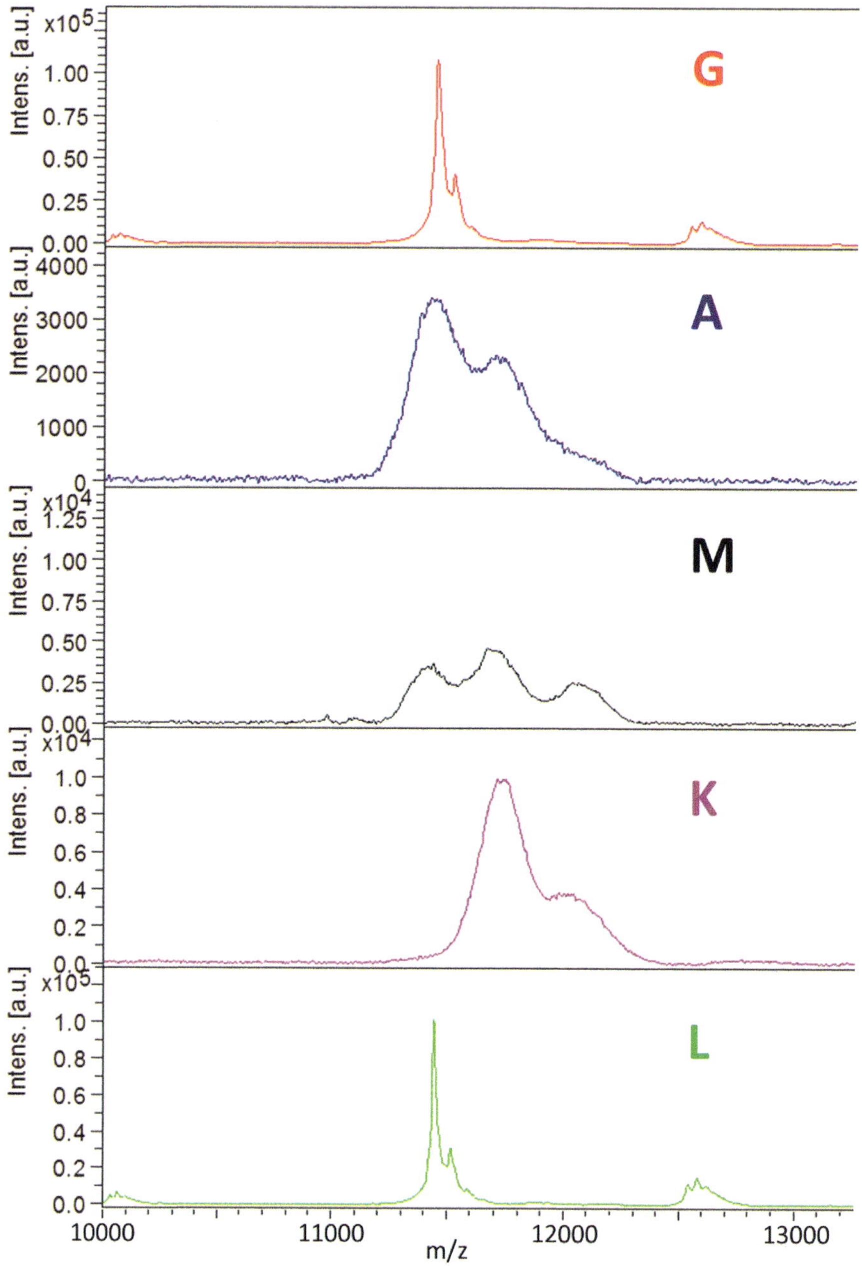

FIGURE 6.9

Determining M-protein isotype in MASS-FIX. In this method, a sample is purified with immunoaffinity reagents specific for IgG, IgA, IgM, kappa and lambda prior to analysis by MALDI-TOF MS. This sample contains an IgG lambda M-protein, so an M-spike is observed in the IgG- and the lambda-specific mass spectra.

6.4.3 Current status and potential role of mass spectrometry for M-protein detection

It is important to note that these mass spectrometry methods are relatively new. They are still largely used in research settings and are still being evaluated and optimized for clinical use. There is not yet a consensus about which approach is best. Studies have demonstrated that mass spectrometry offers several analytical advantages over SPEP and sIFE. The technique is more analytically sensitive, can help distinguish therapeutic monoclonal antibodies from M-proteins, and can detect the presence of posttranslational modifications (i.e., light chain glycosylation). Current work is now focused on understanding the clinical utility of these assays and which settings are most appropriate for use.

The use of mass spectrometry for detecting low levels of the M-protein may be the most impactful for patient care since few alternative PB-based methods are available. In the setting of MRD, several studies have shown that MS assays are able to match or exceed the sensitivity of MFC and NGS assays and are able to detect disease relapses earlier than SPEP and sIFE [22–27]. However, most of these studies examined relatively small cohorts of patients with relatively short follow up. More studies that compare test results to patient outcome will be critical for determining the clinical utility of mass spectrometry versus existing bone-marrow-based MRD assays. Thus far, the consensus from the available data is that PB MS assays complement BM-based MRD tests but will not replace them. Although BM-based assays have their drawbacks, they can provide information about BM reconstitution and the clonal evolution of a patient's disease that can't be obtained from a PB-based test. Similarly, although PB-based MS assays have several advantages, they are not without drawbacks. The long half-life of some immunoglobulins, especially IgA (7–14 days) and IgG (21–25 days), confounds M-protein measurement at low levels of disease. Because the M-protein may persist for weeks or even months after the death of malignant plasma cells, it can be difficult to distinguish between active production and delayed clearance of M-protein, although serial measurements may help mitigate this [23].

The advent of sensitive MS-based methods for M-protein detection also provides a potential opportunity for improved disease detection and monitoring in amyloidosis patients. AL amyloidosis is difficult to diagnose because the symptoms caused by organ damage are nonspecific and the offending plasma cell clone is often a small percentage ($<10\%$), meaning there is often no detectable M-protein. Therefore the enhanced analytical sensitivity of mass spectrometry may allow better detection of the low M-protein concentrations. In addition, mass spectrometry can also detect posttranslational modifications, such as light chain glycosylation. Studies using both MASS-FIX and miRAMM have found that light chain glycosylation is more common in AL amyloidosis compared to other monoclonal gammopathies [28]. Light chain glycosylation has also been shown to be present in patients with MGUS and to be a risk factor for progression to AL amyloidosis [29]. These findings suggest that MS-based techniques could lead to better identification of patients who are at high risk for progressing to AL amyloidosis and earlier diagnosis of AL amyloidosis, both of which have the potential to significantly improve patient outcomes. However, these findings still need to be validated in larger cohorts.

Finally, the clinical utility of mass spectrometry is also being evaluated for routine measurement of the M-protein at high levels of disease. Studies have shown that MASS-FIX provides comparable clinical sensitivity and specificity for detection of various monoclonal gammopathies compared to the standard panel of tests (SPEP, sIFE, FLC quantitation, UPEP, and uIFE) [28]. In 2018 the Mayo Clinic implemented MASS-FIX in routine clinical testing as a replacement for sIFE [30]. Following Mayo's experience with the assay, the IMWG endorsed the use of MASS-FIX as a replacement for sIFE in early 2021, but noted that CR rates obtained with different techniques should not be used interchangeably [31]. However, MASS-FIX has not yet been adopted on a larger scale for this purpose. This is largely because it is still a laboratory-developed test and requires a lot of resources to develop and validate. Efforts are underway to commercialize this assay and obtain FDA approval, which may lead to the more widespread adoption of this technique.

6.5 Summary Points

- The M-protein is one of the oldest tumor markers and continues to be an important part of diagnosing and managing patients with monoclonal gammopathies.
- Electrophoretic techniques such as protein electrophoresis and immunofixation have been used for decades to detect the M-protein, but improvements in the treatment of multiple myeloma over the last several years are driving many changes in diagnostic testing.
- Shift assays and assays leveraging mass spectrometry are being developed to address both the need to distinguish monoclonal antibody drugs from disease-related M-proteins and the need for greater sensitivity as patients live longer and reach lower levels of disease.

6.6 Future directions

Much work is still needed to determine how to best use these new techniques for M-protein testing. For mass spectrometry, current studies are focusing on the clinical utility of more sensitive M-protein detection at various stages of disease (i.e., diagnosis, routine monitoring, detecting early relapse, detecting MRD, etc.). In the future, it will also be important to understand the practical implications of these new tests, such as the cost, turn-around time, and ease of use. In the near future, we will likely see mass spectrometry assays limited to laboratories which have the resources and demand for these tests for clinical research purposes. More labs will likely adopt the technique once the test is commercially available and FDA approved.

In addition to better analytical performance, we will also likely see more automation and standardization of routine M-protein testing. Laboratories have seen a steady increase in the volume of M-protein tests in part because patients are surviving longer and are being followed more closely for relapse. Yet, many of the current techniques, particularly gel-based electrophoresis and immunofixation, are highly manual. In addition, many of the techniques are not standardized. Urine testing is incredibly variable, reporting of results is variable and FLC quantitation is not standardized. Efforts to address these issues are ongoing. Manufacturers are developing automation for sample handling, sample analysis, and result interpretation. In addition, national and international expert groups continue to provide guidelines for appropriate testing for monoclonal gammopathies. There are likely changes ahead, but the M-protein remains an important biomarker for monoclonal gammopathies.

References

[1] Kyle RA, Therneau TM, Rajkumar SV, Larson DR, Plevak MF, Offord JR, et al. Prevalence of monoclonal gammopathy of undetermined significance. N Engl J Med 2006;354:1362–9.

[2] Kyle RA, Vincent Rajkumar S. Monoclonal gammopathy of undetermined significance. Br J Haematol 2006;134:573–89.

[3] Singh J, Dudley AW, Kulig KA. Increased incidence of monoclonal gammopathy of undetermined significance in blacks and its age-related differences with whites on the basis of a study of 397 men and one woman in a hospital setting. J Lab Clin Med 1990;116:785–9.

[4] Cohen HJ, Crawford J, Rao MK, Pieper CF, Currie MS. Racial differences in the prevalence of monoclonal gammopathy in a community-based sample of the elderly. Am J Med 1998;104:439–44.

[5] Landgren O, Katzmann JA, Hsing AW, Pfeiffer RM, Kyle RA, Yeboah ED, et al. Prevalence of monoclonal gammopathy of undetermined significance among men in Ghana. Mayo Clin Proc 2007;82:1468–73.

[6] Kyle RA, Remstein ED, Therneau TM, Dispenzieri A, Kurtin PJ, Hodnefield JM, et al. Clinical course and prognosis of smoldering (asymptomatic) multiple myeloma. N Engl J Med 2007;356:2582–90.

[7] Myeloma–Cancer Stat Facts [Internet]. SEER, 2021. Available from: <https://seer.cancer.gov/statfacts/html/mulmy.html>; [accessed 21.12.2021].

[8] Genzen JR, Murray DL, Abel G, Meng QH, Baltaro RJ, Rhoads DD, et al. Screening and diagnosis of monoclonal gammopathies: an international survey of laboratory practice. Arch Pathol Lab Med 2018;142:507–15.

[9] Identification of monoclonal proteins by gel electrophoresis | Sebia | US [Internet]. Sebia, 2021. Available from: <https://www.sebia.com/en-us/tests/serum-and-urine-immunofixation/> [accessed 23.12.2021].

[10] Mills JR, Murray DL. Identification of friend or foe: the laboratory challenge of differentiating M-Proteins from monoclonal antibody therapies. J Appl Lab Med 2017;1:421–31.

[11] Kumar S, Paiva B, Anderson KC, Durie B, Landgren O, Moreau P, et al. International Myeloma Working Group consensus criteria for response and minimal residual disease assessment in multiple myeloma. Lancet Oncol 2016;17:e328–46.

[12] Dejoie T, Corre J, Caillon H, Hulin C, Perrot A, Caillot D, et al. Serum free light chains, not urine specimens, should be used to evaluate response in light-chain multiple myeloma. Blood 2016;128:2941–8.

[13] Palladini G, Russo P, Bosoni T, Verga L, Sarais G, Lavatelli F, et al. Identification of amyloidogenic light chains requires the combination of serum-free light chain assay with immunofixation of serum and urine. Clin Chem 2009;55:499–504.

[14] Keren DF, Bocsi G, Billman BL, Etzell J, Faix JD, Kumar S, et al. Laboratory detection and initial diagnosis of monoclonal gammopathies: guideline from the College of American Pathologists in collaboration with the American Association for Clinical Chemistry and the American Society for Clinical Pathology. Arch Pathol Lab Med 2021.

[15] McCudden C, Axel AE, Slaets D, Dejoie T, Clemens PL, Frans S, et al. Monitoring multiple myeloma patients treated with daratumumab: teasing out monoclonal antibody interference. Clin Chem Lab Med 2016;54:1095–104.

[16] Barnidge DR, Dasari S, Botz CM, Murray DH, Snyder MR, Katzmann JA, et al. Using mass spectrometry to monitor monoclonal immunoglobulins in patients with a monoclonal gammopathy. J Proteome Res 2014;13:1419–27.

[17] Mills JR, Kohlhagen MC, Willrich MAV, Kourelis T, Dispenzieri A, Murray DL. A universal solution for eliminating false positives in myeloma due to therapeutic monoclonal antibody interference. Blood 2018;132:670–2.

[18] Mills JR, Kohlhagen MC, Dasari S, Vanderboom PM, Kyle RA, Katzmann JA, et al. Comprehensive assessment of M-proteins using nanobody enrichment coupled to MALDI-TOF mass spectrometry. Clin Chem 2016;62:1334–44.

[19] Moore LM, Cho S, Thoren KL. MALDI-TOF mass spectrometry distinguishes daratumumab from M-proteins. Clin Chim Acta 2019;492:91–4.

[20] Kohlhagen MC, Mills JR, Willrich MAV, Dasari S, Dispenzieri A, Murray DL. Clearing drug interferences in myeloma treatment using mass spectrometry. Clin Biochem 2021;92:61–6.

[21] Langerhorst P, Noori S, Zajec M, De Rijke YB, Gloerich J, van Gool AJ, et al. Multiple myeloma minimal residual disease detection: targeted mass spectrometry in blood vs next-generation sequencing in bone marrow. Clin Chem 2021;67:1689–98.

[22] Derman BA, Stefka AT, Jiang K, McIver A, Kubicki T, Jasielec JK, et al. Measurable residual disease assessed by mass spectrometry in peripheral blood in multiple myeloma in a phase II trial of carfilzomib, lenalidomide, dexamethasone and autologous stem cell transplantation. Blood Cancer J 2021;11:1–4.

[23] Mills JR, Barnidge DR, Dispenzieri A, Murray DL. High sensitivity blood-based M-protein detection in sCR patients with multiple myeloma. Blood Cancer J 2017;7:e590 e590.

[24] Liyasova M, McDonald Z, Taylor P, Gorospe K, Xu X, Yao C, et al. A Personalized mass spectrometry–based assay to monitor m-protein in patients with multiple myeloma (EasyM). Clin Cancer Res 2021;27:5028–37 American Association for Cancer Research.

[25] Abeykoon JP, Murray DL, Murray I, Jevremovic D, Otteson GE, Dispenzieri A, et al. Implications of detecting serum monoclonal protein by MASS-fix following stem cell transplantation in multiple myeloma. Br J Haematol 2021;193:380–5.

[26] Santockyte R, Jin C, Pratt J, Ammar R, Desai K, Bolisetty M, et al. Sensitive multiple myeloma disease monitoring by mass spectrometry. Blood Cancer J 2021;11:78.

[27] Eveillard M, Rustad E, Roshal M, Zhang Y, Ciardiello A, Korde N, et al. Comparison of MALDI-TOF mass spectrometry analysis of peripheral blood and bone marrow-based flow cytometry for tracking measurable residual disease in patients with multiple myeloma. Br J Haematol 2020;189:904–7.

[28] Milani P, Murray DL, Barnidge DR, Kohlhagen MC, Mills JR, Merlini G, et al. The utility of MASS-FIX to detect and monitor monoclonal proteins in the clinic. Am J Hematol 2017;92:772–9.

[29] Dispenzieri A, Larson DR, Rajkumar SV, Kyle RA, Kumar SK, Kourelis T, et al. N-Glycosylation of monoclonal light chains on

routine MASS-FIX testing is a risk factor for MGUS progression. Leukemia. 2020;34:2749–53.

[30] Kohlhagen M, Dasari S, Willrich M, Hetrick M, Netzel B, Dispenzieri A, et al. Automation and validation of a MALDI-TOF MS (Mass-Fix) replacement of immunofixation electrophoresis in the clinical lab. Clin Chem Lab Med 2020;59:155–63.

[31] Murray DL, Puig N, Kristinsson S, Usmani SZ, Dispenzieri A, Bianchi G, et al. Mass spectrometry for the evaluation of monoclonal proteins in multiple myeloma and related disorders: an International Myeloma Working Group Mass Spectrometry Committee Report. Blood Cancer J 2021;11:1–6.

CHAPTER

7 Liquid biopsy as a cancer biomarker-potential, and challenges

Daniel C. Danila[1,2]

[1]*Department of Medicine, Memorial Sloan Kettering Cancer Center, New York, NY, United States*

[2]*Department of Medicine, Weill Cornell Medical College, New York, NY, United States*

Objectives

- Define liquid biomarkers.
- Understand the main aspects of analytical and clinical validation of a biomarker.
- List liquid-based biomarkers, potentials, and challenges.
- Discuss a strategy for guiding the decision-making process.
- Explain some of the challenges in developing and validating liquid biomarkers.

7.1 Introduction

Personalized cancer medicine requires the development of tumor-specific biomarkers to optimize selection of targeted therapies and to better assess response to therapy. Molecular profiling of liquid-based biomarkers, such as circulating tumor cells (CTCs) and circulating tumor DNA (ctDNA) in blood, can be used to assess the biologic determinants contributing to tumor growth. The liquid biopsy, a minimally invasive approach, has shown in the last decade prominent validity for diagnosis, prognosis, and for monitoring treatment efficacy in different cancer types. Throughout the years, several studies have explored the validity of different circulating analytes,

Cancer Biomarkers: Clinical Aspects and Laboratory Determination. DOI: https://doi.org/10.1016/B978-0-12-824302-2.00013-8

© 2022 Elsevier Inc. All rights reserved.

including CTCs, ctDNA, or cell-free DNA (cfDNA), cell-free RNAs (mRNAs, long noncoding RNAs, and microRNAs), extracellular vesicles, and proteins markers. However, CTCs and ctDNA remain the most widely studied circulating biomarkers so far.

Life expectancy and clinical benefit of patients with cancer vary across the disease spectrum, and the biological determinants of tumor sensitivity or resistance may change with exposure to therapy. Optimal treatment selection requires contemporaneous analysis of the tumor characteristics at the decision time. Tumor biopsies are invasive, costly, and hard to repeat. In addition, sampling of a single site may not be representative of all metastatic lesions [1]. Finding of tumor characteristics in liquid biopsies, from materials shed into fluid collections easily accessible from patients in clinic, holds the promise of determining prognosis, tumor sensitivity, and assessing the effects of treatment. Current efforts have shown that patients in whom tumor-specific markers are detected may have an inferior prognosis relative to those in whom these biomarkers are not detected and that the elimination or decrease of tumor-shedding following treatment is associated with improved clinical outcomes. Technological advances in the detection, isolation, capture, and characterization of tumor-specific markers in a phlebotomy sample obtained serially in routine clinical practice setting have enabled the evaluation of different biomarkers based on liquid biopsies.

7.2 Clinical utility of liquid biomarkers

A biomarker is a characteristic that is objectively measured and evaluated as an indicator of normal biological or pathogenic processes, or pharmacologic responses to therapeutic intervention [2]. Targeted approaches are useful for patients most likely to respond to the treatment as their tumors express the target. Another clear unmet need for patients with cancer is for reliable surrogate measures of the clinical benefit early into the treatment. Ideally, blood-based assays to profile patients in a "liquid biopsy" will provide biomarkers easy to use in clinic, to ultimately accelerate clinical trials and improve clinical care.

7.3 Clinical development and validation of liquid biomarkers

To fulfill its intended used, the development of a biomarker must follow a similar path to a drug approval, from defining the context of use to validation in clinical practice, embedded in prospective clinical studies [3,4].

7.3.1 Context of use

Represents the specific clinical settings for which the results of an assay will be used to inform a medical decision. As drugs are approved for indications, biomarkers are qualified for a specific context of use as detailed in the Food and Drug Administration (FDA) Critical Path [5,6]. Assays qualified for a specific context have to demonstrate specific analytical and clinical validation in other intended uses. The trial designs needed to demonstrate clinical validity for a new biomarker would also depend on existing assays that address the context of use.

The specific contexts of use where qualified biomarkers would influence medical decisions include:

7.3.1.1 Detection

The use of the biomarker to establish a **diagnosis**;

7.3.1.2 Prognosis

Measuring the probability of a specific clinical outcome, such as recurrence, progression, or **survival**;

7.3.1.3 Prediction

Identifying the chance of response to a specific therapy; **treatment resistance** biomarkers define biologic determinants of failure or progression, such as mutations or copy number alterations.

7.3.1.4 Response-indicator

Biomarkers demonstrate a pharmacologic or physiologic response from the treatment, which does not necessarily mean that the patient may benefit from a treatment;

7.3.1.5 Efficacy-response

Biomarkers are surrogates of how a patient feels or functions or ***how long he/she survives***, extrapolating the *clinical benefit*;

7.3.2 Analytical validation

Establishes the accuracy, and reproducibility of an assay in the range of conditions under which an assay is being proposed to be used. Clinical Laboratory Improvement Amendments (CLIA) regulatory requirements impose rigorous performance testing before use clinical testing, which can be considered in three steps: (1) preanalytical assessment of specimen selection, handling, processing, and storage parameters; (2) validation of analytical characteristics, establishing inter- and intraassay precision, linearity, range of detection, analyte recovery, and standardization. This step requires the development of a comprehensive quality control program; and (3) postanalytical parameters requiring data management. Refer to Chapter 1 for an extensive discussion on analytical validation.

7.3.3 Clinical qualification

Distinct from the analytical validation, is the evidentiary process of linking a biomarker with biological processes or clinical endpoints in the *context of an intended use* to inform a medical decision [5,6]. In this process, the analytically validated biomarker is tested for relevance in answering a specific clinical question and effectively add value to the existing assays in the clinical context. We refer the reader to Chapter 9 for additional discussion on this. As there are more biomarkers than questions and given the effort and cost to conduct these investigations, it is essential to develop metrics to determine which biomarkers that appear promising warrant prospective testing in large-scale phase III trials to establish qualification and clearance by FDA.

7.4 Circulating tumor cells

CTCs, first described in 1869 as cells identified in blood identical to those of the cancer itself [7], originate from the primary tumor or metastatic deposits after intravasating through the tumor vasculature by mechanisms that are not completely understood (Fig. 7.1, from Danila et al. [8]). Although most CTC die in the circulation, a proportion estimated to be 0.05% have the ability to spread, seed, and proliferate in distant sites to form secondary metastasis [9], or reestablish into the organ of origin to form new tumors [10], as demonstrated in short-term proliferation experiments [9,11]. First described by Paget in 1889 in his "seed and soil" hypothesis [12], the process of metastatic dissemination is driven by chemokines and integrins specific to the tumor type and by the adhesive interactions of the CTC to the vasculature and extracellular matrix components [10,13–15]. As an example, the β4 integrin signaling promotes tumor cell adhesion to endothelium through

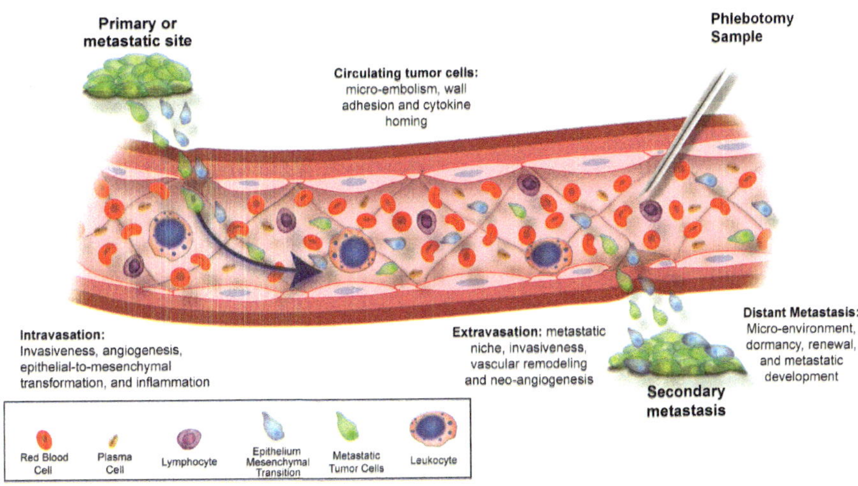

FIGURE 7.1

Liquid biopsy/circulating tumor cells (CTCs) in the process of metastatic progression. Tumor cells transition from the primary or metastatic site into circulation to establish secondary sites.

From Danila DC, et al. Circulating tumors cells as biomarkers: progress toward biomarker qualification. Cancer J 2011;17(6):438–50.

ErbB2-mediated VEGF secretion and has been involved in metastatic development [16]. Additionally, CTCs have the ability to secrete stem-cell–like factors to promote proliferation and self-renewal [17,18]. Microembolic clusters of CTC isolated using a variety of techniques are associated with a particularly poor prognosis, which may be related to the enhanced activation of survival pathways by cellular contact, and mechanical capture in the microvasculature, further enabling metastasis formation [19,20].

As a consequence of vascular spread and the microenvironment, tumor cells can be identified in the bone marrow as disseminated tumor cells (DTCs). These DTCs, which are in some cases detected at a higher frequency than CTC, are found concurrently in the blood, and they are associated with an increased risk of tumor recurrence when detected [21–26]. DTC can also serve as a reservoir organ of CTC for further dissemination [9,27,28]. At this time, testing for the presence of DTCs is not routinely performed in patients with solid malignancies because of the uncertainty of the information provided by the assay; however, they are performed at diagnosis and serially in patients with leukemias and lymphomas [29].

After seeding a potential metastatic site [30], tumor cells have to adapt to a hypoxic, nutrient deficient, and low pH environment [31] in order to remain in a viable yet dormant state for extended periods of time [22,32]. During disease remission, a dormant heterogeneous tumor cell population may undergo a dedifferentiation process known as epithelial–mesenchymal transition (EMT), adapt characteristics of stem cells through reprogramming or selectively enrich for tumor cells with stem-like features [33,34]. Eventually, dormant tumor cells are stimulated by intrinsic mechanisms or by stroma microenvironment to facilitate active proliferation into visible metastasis and to promote invasive features and spread hematogeneously as CTC again [34].

7.4.1 The morphology and characteristics of circulating tumor cells

The range of CTC phenotypes is representative of the heterogenic range of cells found in different metastatic sites, a source of tumor for profiling in an individual patient at the time when treatment decisions are made. As CTCs are shed into the peripheral circulation from both the primary tumor and from multiple heterogeneous metastases, they can represent the spectrum of the malignant phenotype including tumor initiating cells with stem or stem-cell–like properties, the full range of undifferentiated to differentiated phenotypes that lack tumor initiating capabilities, and cells that have undergone an EMT [35,36]. Therefore there is no single definition of a CTC, rather there are specific definitions that relate to a particular tumor cell type and profiled by physical characteristics such as size or shape, or by molecular markers.

Importantly, the EMT from an epithelial to a fibroblast-like cell morphology is accompanied by a decrease in expression of epithelial adhesion molecules (e.g., E-cadherin) and increased expression of mesenchymal molecules (e.g., vimentin, N-cadherin), leading to increased mobility and invasiveness [35,36], increasing important cancer stem cell properties [37,38]. EMT phenotypic change is not only important mechanistically, but also from the point of view of CTC detection and characterization, because many of the capture technologies utilize antibodies to epithelial cell adhesion molecule (EpCAM), which, although commonly expressed on the majority of primary tumor cells, may be down-regulated or absent on the subsets of CTCs that have undergone EMT [39,40]. Similarly, cells with stem or stem-cell–like properties may also lack EpCAM, or cytokeratin (CK) expression [41]. In addition, DTC in the bone marrow must undergo the reversal of EMT (called MET) to form a solid metastasis. It is, therefore, possible that cancer cells, with a high degree of plasticity to undergo both EMT and MET, are the "real" metastasis-initiator cells. Different degrees of EMT have been found to occur in solid tumors. Tumor cells with a partial EMT (or "intermediate phenotype" between epithelial and mesenchymal) have shown increased aggressiveness, and ability to survive in suspension as well as single cells [42].

7.4.1.1 Circulating tumor cells and cancer drug development

CTCs provide a "liquid biopsy" that can be obtained easily and repeatedly, and which may provide a more complete biologic profile of the patient's illness. The focus is on identifying and validating predictive biomarkers to inform treatment selection from banked specimens from patients obtained each time a change in therapy is required for disease progression. Biomarker results are then associated with clinical outcome in prospective clinical trials geared towards the specific context of use for such biomarkers. For example, Epic Sciences microscopy–based platform is a College of American Pathologists (CAP) accredited and CLIA certified technology that is analytically validated for the enumeration of CTCs [43] and detects a wide range of phenotypically distinct CTC subtypes [44–47]. Notably, each patient sample yields 10–12 slides for analysis. In patients with metastatic prostate cancer, nuclear expression of androgen receptor (AR) splice variant 7 was cleared to aid in selecting patients with metastatic prostate cancer least likely to respond to androgen receptor signaling inhibitor (ARSI). Further studies are needed to define the clinical utiliy of this biomarker [48].

7.4.1.2 Phenotypic heterogeneity of circulating tumor cells informs clinical decision-making

CTC phenotypes from individual patients vary significantly, as they are illustrative of the dynamic transformation of tumor cells from different metastatic sites, undergoing possible further morphological changes from lack of cell-adhesion signaling while circulating in blood. To increase the proportion of patients for whom predictive biomarkers are detected, quantified digital pathology features of individual CTCs were collected from metastatic castration resistant prostate cancer (mCRPC) patients [49]. Heterogeneity of phenotypically distinct cell types in individual patient samples was quantified using the Shannon index and associated with overall survival (OS) in the specimens collected prior to initiation of second or later lines of therapy. Low CTC phenotypic heterogeneity was associated with better OS in patients treated with ARSI, whereas high heterogeneity was associated with better OS in patients treated with taxane chemotherapy. Overall, the results show that quantifying CTC phenotypic heterogeneity can help inform the choice between ARSI and taxanes in mCRPC patients.

7.4.2 Detection methods

CTCs are rare events in the peripheral blood of patients with a variety of metastatic carcinomas and are currently estimated to account for 1 cell in a billion nucleated cells. In contrast to invasive procedures like biopsies, a blood test for CTC or bone marrow aspirate and biopsy for DTC are safe and can be performed with minimal patient discomfort on a repeated basis at low cost.

Different assays are needed for different contexts of use in the decision the test result will be used to inform, be it prognosis, prediction, pharmacodynamic, or a measure of patient benefit (Fig. 7.2; [50]). As examples, these assays may enable a determination of the risk of relapse, risk of metastasis development, and guide the decision on adjuvant therapy based on posttreatment CTC enumeration.

The definition of a CTC varies with the assay used to isolate, capture, visualize and/or enumerate cells. Different assays detect different subsets of cells and different biologic determinants in respective subsets. Therefore, an individual assay must be evaluated critically from the point of view of which subsets of cells are being captured and analyzed, the performance characteristics of the assay that is used for what type of measurement, and how it is reported [20,40,51,52].

CellSearch (Menarini Silicon Biosystems) immunomagnetic isolation technique uses antibodies to EpCAM conjugated to magnetic beads for the immunomagnetic capture of EpCAM-expressing epithelial cells. The EpCAM expressing cells are automatically displayed visually and scored, followed by adjudication by a trained technologist as a CTC based on strict morphologic and immunohistochemical staining characteristics. A captured cell is classed as a CTC if it expresses CK, displays a nucleus when stained with 4′,6-diamidino-2-phenylindole (DAPI), and is not a white blood cell as determined by a lack of expression of CD-45 [53]. The results are reported as the number of cells meeting the morphologic and immunohistochemical definition per 7.5 mL of blood. This has been approved by the FDA as an aid in the monitoring of patients with metastatic breast [54–57], colorectal [58,59], or prostate cancer [60–62]. Importantly, the number of cells meeting the criteria for a CTC is only a small proportion of the EpCAM positive events displayed [63]. In this assay, EpCAM positive cell fragments, cells undergoing apoptosis, necrotic and anucleated cells are not counted towards the CTC number.

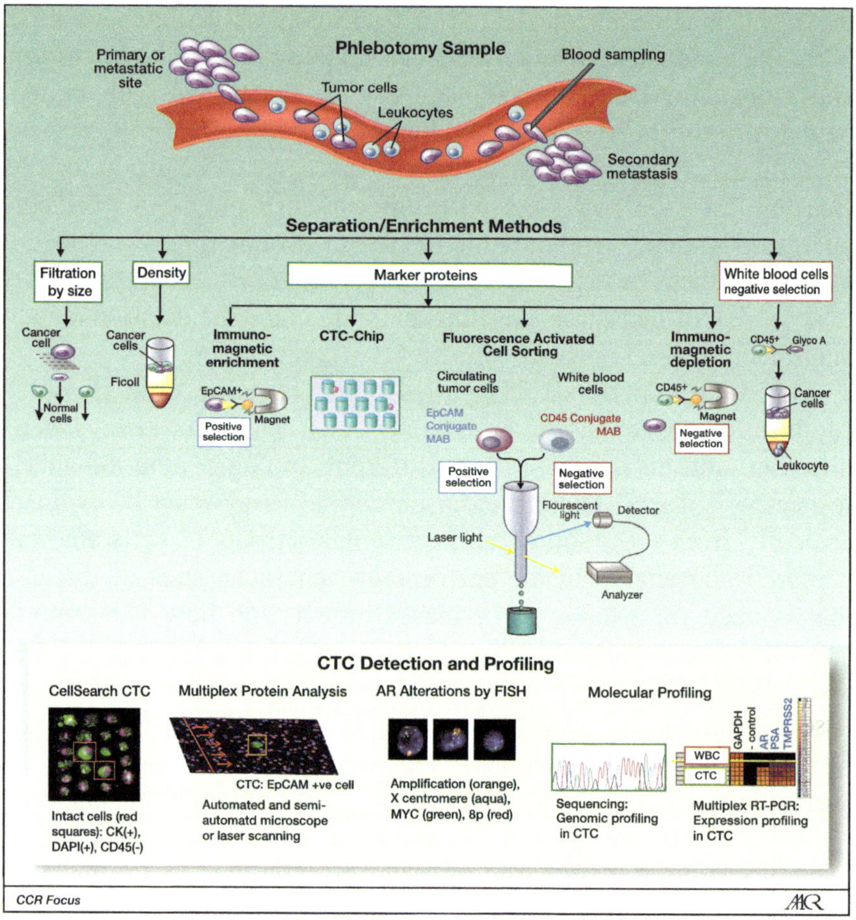

FIGURE 7.2

Circulating tumor cells (CTCs) sampled by phlebotomy at the time when treatment is being considered. The definition of a CTC varies with the assay used to isolate, capture, visualize and/or enumerate cells.

From Danila DC, Fleisher M, Scher HI. Circulating tumor cells as biomarkers in prostate cancer. Clin Cancer Res 2011;17(12):3903–12.

For patients with metastatic prostate cancer, CTCs isolated by this method have the biologic determinants unique to this malignancy including, expressing prostate-specific antigen (PSA), alpha-methylacyl-CoA racemase, and prostate-specific genomic abnormalities such as AR gene copy number amplifications, phosphatase,tensin homolog deletions, and TMPRSS:ETV fusion products [64,65]. Underappreciated is that the proportion of cells visible in the chamber meeting the strict criteria defining a CTC represents only 1%–10% of the cells actually present. The sensitivity of the test can be increased if all $EpCAM^{+}CK^{+}CD45^{-}$ events are counted, a measurement prognostic for survival [63]. This finding, however, has not been analytically or clinically validated, and is not currently reported routinely.

As the assay employs a positive selection based on immunomagnetic enrichment for tumor cells expressing EpCAM, it is not useful for the isolation of cells from tumors that do not express EpCAM, such as malignant melanoma and renal cell carcinoma. For these tumors, a different "hook" is necessary to capture tumor cells from circulation. To capture these cells require non-EpCAM-based methods such as negative depletion strategies that remove leukocytes, which is followed by specific staining of the remaining mononuclear cell fraction with antibodies to markers specific to the tumor under study [17,18]. In an exploratory study in patients with melanoma, the isolation of cells based on the expression of a melanoma-associated chondroitin sulfate proteoglycan allowed the recognition of circulating melanoma cells based on morphology [66].

Flow cytometry methods are used to enrich for more homogeneous and highly purified tumor cell populations separated based on multiple cell surface and intracellular markers. The technique is useful for the evaluation of expression profiles in CTC [67–69], such as the isolation and capture of $EpCAM^{+}$, $CD45^{-}$ cells from a mononucleated layer separated through Ficoll-Hypaque gradient from blood. DAPI is used as a vital stain to exclude permeable and apoptotic cells. The cells isolated by FACS were shown by multiplex RT-PCR to express tumor-specific mRNAs, indicating that these $EpCAM^{+}$ events are *bona fide* CTCs [50]. Other techniques use photoacoustic properties of tumors cells, such as melanoma, to enrich CTC after stimulation by laser light by flow cytometry for further purification and isolation [70]. By isolating CTC with higher purity, these approaches have the potential to enable in-depth molecular profiling in CTC in a larger cohort of patients, enabling the study of gene signature patterns in the heterogeneous population of CTC. Whole-genome amplification from the cellular events isolated by sorting also enables mutational analyses and studies of copy number alterations at the level of single cells, as prognostic and predictive biomarkers [71]. Flow cytometry or fluorescence in situ hybridization–based methods to identify sarcoma-specific fusion genes are currently being tested [72–74].

Microfluidic capture platforms allow sensitive and selective detection of CTC for enumeration and further molecular profiling at genomic, transcriptional, and translational levels in CTC isolated from patients with lung, breast, and prostate cancer [11,20,75]. Reports using this technology have shown CTC isolation and capture rates at points in an illness where CellSearch does not detect cells, such as localized prostate cancer [52]. Configurations designed to create turbulence in the capture platform maximize gentle cell contact with the surface of the chip while minimizing cell trauma. These platforms allow the isolation and capture of viable cells harboring mutations that predict tumor sensitivity to targeted therapies and clinical outcome, such as the epidermal growth factor receptor T790M activating mutation in nonsmall-cell lung cancer (NSCLC) [76,77] or the KRAS mutations in colorectal cancer (CRC) [78]. Additional designs optimize flow to increase contact with capture antibody laden surfaces, where prostate-specific membrane antigen–based immunoselection allows enrichment of EpCAM-negative tumor cells [79]. Other technologies allow the isolation of CTC based on an inertial focusing-enhanced microfluidic chips or ferrofluidic separation in strategies that are either dependent or independent of tumor membrane epitopes on CTC from patients in epithelial and nonepithelial cancers such as lung, prostate, pancreas, breast, and melanoma [80,81].

Functional assays allow detection of CTC based on tumor-specific metabolic capabilities. EPISPOT (EPithelial ImmunoSPOT) assays detect CTC based on the release of full-length CK19 by carcinoma cells of various origins [18]. In patients with breast cancer, detecting CK-19 releasing cells in blood by EPISPOT predicted worse overall survival, in addition to the CellSearch assay [82]. Tumor antigens mucin 1 or PSA were detected in the majority of patients with metastatic breast and prostate cancer, respectively, whereas the test results were negative in healthy controls or in patients with benign prostatic hyperplasia [83]. Collagen adhesion matrix assays allow enriching CTC based on their ability to ingest and invade into the connective tissue-like material. This method could allow for a functional separation of more aggressive, invasive types of CTC from patients with prostate or breast cancers [84]. Dielectrophoretic-array technologies displace blood cells when they pass over microelectrode array configurations designed to separate cells as a function of cell size and electrostatic charges. Cells isolated this way can be characterized by microscopy or could be analyzed for predictive molecular biomarkers based on relative fluorescence parameters combined with morphology [85–87].

Filtration assays capture CTC based on larger size and inflexible biophysical proprieties relative to hematopoietic cells when passed through a porous barrier. Several different pore sizes and shapes have been tested. Advantages of these systems are the ability to perform CTC analysis, including immunocytochemistry, directly on membrane, as well as the agnostic enrichment of CTC, indifferent of EpCAM expression that allows collection of EMT, CTC with stem-cell–like features and circulating cancer-associated macrophage-like cells [88–92]. These approaches miss small cells, however, in the process of recovering CTC, cancer-associated macrophage-like cell populations have been noted to associate with CTC and provide highly sensitive biomarkers for solid tumors. These tumor-associated giant cells are capable of binding to CTC and facilitate their transport and mobilization through the peripheral circulation of cancer patients [92,93].

Genomic, transcriptional, or translational-based assays have been long utilized to detect low levels of cancer-specific markers in the mononuclear cell fraction of the blood that is presumed to be from CTC. Messenger RNA for PSA has been tested in peripheral blood of patients with prostatic cancer [94,95], as well as detecting carcinoembryonic antigen (CEA), cytokeratin (CK), and CD133 in patients with CRC [96]. Additional markers are under investigation for detection of CTC from patients with breast, prostate, lung, melanoma, and head and neck cancers [97–104]. An

RT-PCR method targeting EWS-FLI-1 or EWS-ERG transcripts was studied in patients with Ewing tumors to detect evidence of occult tumor cells [105]. Additional methods isolate CTC from blood with immunomagnetic methods to preenrich for increased detection of tumor-specific mRNA to breast, prostate, EMT, or stem-cell–like cells [106]. In some cases, the tumor-specific mRNA detected in the mononuclear cell fraction represents phagocytosed tumor cells in macrophages and does not represent intact tumor cells. To increase detection sensitivity and tumor-specific molecular profiling of CTC from the blood sample, selection techniques have been used to enrich for CTC (MagSweeper, AdnaTest, IsoFlux) or deplete for white blood cells before in depth molecular profiling [81,107–109].

Whole blood methods to characterize CTC do not enrich for a specific epitope marker of physical size-specific capture methods, instead all nucleated blood cells are analyzed by immunofluorescence and cytometry for tumor-specific markers by high-throughput multiparametric imaging systems [93]. Such technologies have allowed the identification and characterization of tumor cells in relation to their primary tumors in patients with breast, colorectal, and lung cancers [93,110].

Prognosis and response-indicator biomarkers of CTC enumeration by CellSearch were tested in a series of clinical trials of similar design in patients with metastatic breast [54–57,111] prostate [61,62], and colorectal [58,59] cancer about to start a new line of therapy. CTC measured at baseline and with subsequent follow-up visits were used to define patient groups with a favorable and unfavorable prognosis based on disease-specific cutoffs (four or fewer CTC per 7.5 mL of blood for breast and prostate cancer, and two or fewer for CRC for the favorable group). In this context, CTC enumeration was prognostic of median progression-free survival (PFS) and overall survival (Table 7.1, [9]). Posttreatment CTC enumeration remains prognostic at 4, 8, or 12 weeks after initiating treatments in this context [58,59,61,62,112]. There studies resulted in the FDA-clearance of the CellSearch system to measure CTC as a marker of prognosis in the monitoring of patients with metastatic breast, prostate, and CRC, and as a response-indicator biomarker by

Table 7.1 Context of use for baseline CTC counts: a prognosis biomarker for overall survival for patients with metastatic disease.

Baseline CTC counts in patients with metastatic cancer	Threshold for unfavorable CTC	Number of patients (%) with unfavorable baseline counts	Overall survival based on CTC counts. Months (95% CI)		Cox hazard ratio; log-rank *P*	References
			Favorable	Unfavorable		
Prostate cancer	≥ 5	219 (43%)	21.7 (21.3– > 26)	11.5 (9.3–13.7)	3.3; $P < .0001$	[61,62]
Breast cancer		177 (49%)	21.9 (20.1– > 25)	10.9 (6.4–15.1)	2.45 (1.64–3.65); $P < .0001$	[57]
Colorectal cancer	≥ 3	413 (26%)	18.5 (15.5–21.2)	9.4 (7.5–11.6)	2.45 (1.77–3.39); $P < .0001$	[58,59]

Reproduced from Danila DC, et al. Circulating tumors cells as biomarkers: progress toward biomarker qualification. Cancer J 2011;17(6):438–50.

serial testing for CTC in conjunction with other clinical methods, as described in the 510(K) document [113].

When compared head-to-head in patients with metastatic mCRPC starting a new line of chemotherapy, posttreatment CTC number was a stronger prognostic factor for survival than a 50% decline in PSA [61,62]. The discriminatory power of baseline CTC counts on survival was analyzed in a multivariate proportional hazards model with the concordance probability estimate [114]. Baseline CTC counts (analyzed with the dichotomous definition of unfavorable vs favorable) had the strongest association with survival time [hazard ratio (HR), 1.58; 95% confidence interval (95% CI), 1.41–1.77], compared to PSA (HR, 1.26; 95% CI, 1.10–1.45). Additional baseline factors associated with decreased overall survival included high lactate dehydrogenase (LDH) levels (HR, 6.44; 95% CI, 4.24–9.79) and low hemoglobin (HR, 0.72; 95% CI, 0.64–0.81).

In a separate analysis, CTC counts were related to survival time as a continuous function using nonparametric smoothing analysis [115], without a threshold effect [60]. This argues against the use of discrete CTC cutoff values in decision-making, a finding confirmed in multiple studies [62].

As a response indicator of treatment outcome, CTC conversion is more associated with survival than is a decline in PSA, as measured at 4, 8, and 12 weeks [61,62]. The most prognostic factors for survival were baseline LDH levels and the fold change in CTC count from baseline to the follow-up time of measurement [62]. These findings suggest that monitoring CTC number as early as 4 weeks posttreatment may add clinical data that assist in treatment decisions.

Molecular characteristics of CTC by PCR-based assays have been tested as biomarkers of prognosis and response indicators. The frequency of detection of PSA mRNA in peripheral blood of patients with prostatic cancer correlated with tumor stage [94,95]. Detecting additional prostate cancer enriched genes in whole blood collected in PAXgene tubes by a sensitive and analytically validated PCR assay was demonstrated to have predictive value for overall survival in patients with metastatic castration-resistant prostate cancer, complimentary with CellSearch CTC enumeration [104]. Detecting CEA, CK 19, CK20, and/or CD133 mRNA by RT-PCR demonstrated a significant prognostic factor for overall survival OS (HR, 3.84; 95% CI, 2.41–6.22; $P < .001$) and disease-free survival DFS (HR, 3.02; 95% CI, 1.83–5.00; $P < .001$) especially in patients with Dukes' stage B and C CRC [96]. Serial monitoring by RT-PCR of a multimarker panel for CTC during biochemotherapy may be useful for predicting therapeutic efficacy and disease outcome in patients with metastatic melanoma, although additional studies are necessary [116].

CTC as biomarkers predictive of tumor sensitivity have the potential to provide a snapshot of the molecular makeup of an individual patient's tumor to profile for determinants that predict for sensitivity or resistance to treatment at the time of treatment decision. Such a "liquid biopsy" will improve the understating of the tumor biology that can change during the course of disease. To predict the likelihood a patient will benefit from a targeted therapy requires a demonstration that the "target" is present in the tumor when treatment is being considered. Currently, the molecular profiles of multiple metastatic sites in a patient at time of treatment decision is limited, given the lack of representative tumor samples in routine clinical practice setting in a continuously evolving disease [117].

In prostate cancer, recent reports evaluated alterations in the AR at the protein, RNA and DNA level [65,118], IGF-1R expression [119], and TMPRSS2-ETV fusions [65,107], which have been postulated to increase the likelihood of response to a therapy targeting the specific signaling pathway. AR amplification, although infrequent in primary/diagnostic tumor specimens, is detected in upwards of 50% castration-resistant lesions [120,121]. AR genomic amplification and copy number gain have been documented in CTC from patients with CRPC with frequencies similar to that reported for late-stage tumors [65,118].

Detection of a kinase mutation predictive of tumor sensitivity in CTCs has potential for clinical application in selecting patients with metastatic colorectal or lung cancer most likely to benefit from epidermal growth factor receptor (EGFR)-targeted therapies [77,122,123]. Combining a microfluidic enrichment system and mutation assay for KRAS has been proposed to improve sensitivity to track longitudinally of such oncogene mutational changes with high success rates [78]. The prognostic value of HER2/neu and IGF-1R expression in CTC is being currently studied in the context of clinical trials with novel therapies targeting these markers, as predictive of tumor response and optimizing pharmacodynamic measurements in dose finding studies [64,118,119,124].

Association of AR splice variant 7 (AR-V7) on CTCs as a treatment specific biomarker in CRPC. Multiple therapies for patients with mCRPC are available leaving the clinician in a quandary as to which intervention to use at which timepoint. AR-V7 results in a truncated receptor, which leads to ligand independent constitutive activation that is not inhibited by antiandrogen therapies, including abiraterone or enzalutamide. Quantification and localization of protein expression is used to measure markers such as AR and its splice variants, which led to the analytical and clinical validation of Epic's AR-V7 assay [125,126]. In collaboration with Epic Sciences, a cross-sectional cohort study in 161 men with progressive mCRPC undergoing a change in treatment was tested using an AR-V7 epitope specific antibody to the coded protein. A nuclear localized AR-V7 was a highly specific predictor of resistance to ARSIs. In this study, survival times were prolonged when patients with CTCs that expressed the AR-V7 protein were treated with taxane. These findings have been proposed for prospective validation in larger phase III trials [49].

Prospective validation in PROPHECY trial (ClinicalTrials.gov identifier: NCT02269982), a multicenter, prospective blinded study of men with high-risk mCRPC starting abiraterone acetate or enzalutamide treatment, had as the primary objective the validation of baseline CTC AR-V7 as prognostic biomarkers [127]. OS and PSA responses were secondary endpoints in this study of 118 men with mCRPC starting a new treatment. AR-V7 detection by both Johns Hopkins and Epic AR-V7 assays was independently associated with shorter PFS [HR, 1.9 (95% CI, 1.1–3.3; $P = .032$) and HR, 2.4 (95% CI, 1.1–5.1; $P = .020$), respectively] and OS [HR, 4.2 (95% CI, 2.1–8.5) and 3.5 (95% CI, 1.6–8.1), respectively] after adjusting for CTC number and clinical prognostic factors. Men with AR-V7–positive mCRPC had fewer confirmed PSA responses (0%–11%) or soft tissue responses (0%–6%). The observed percentage agreement between the two

AR-V7 assays was 82% [127]. Detection of AR-V7 in CTCs by two blood-based assays is independently associated with shorter PFS and OS with abiraterone or enzalutamide, and such men with mCRPC should be offered alternative treatments.

7.5 Circulating tumor DNA

DNA fragments released into the blood stream by tumor cells have been considered representative of cancer biology, ideal for biomarker development. ctDNA is short fragments of DNA, averaging 160–180 bp, with short half-life [128]. The amount and quality of ctDNA, released mostly by dying cancer cells and possibly through active secretion, influences the total cfDNA present in plasma of individual subjects. The total quantity of circulating DNA has been considered as a function of tumor burden and aggressiveness, yet, high cfDNA levels have been related to other pathological injuries such as smoking, exercise infections/inflammation, or mechanical trauma.

First described by Mandel in 1948 [129], cfDNA concentration, ranging between 1 and 10 ng/mL in plasma of healthy individuals [130], is often increased in the serum of cancer patients [131,132]. In addition to quantifying ctDNA shedding into the blood stream, mutation and methylation status in tumor-derived DNA has been studied as noninvasive biomarkers of tumor sensitivity to target therapies, treatment monitoring, and detection of residual disease. Clonal evolution of ctDNA may identify secondary resistance mechanisms before clinical progression is observed. Deep genomic sequencing studies in cfDNA have proposed ctDNA as an accessible source for molecular profiling at the time of treatment decision, bypassing the need for a painful and invasive biopsy.

Methods of detection of multiple proof-of-principle, lab-based studies, have been published [133]. There have been few assays that have been analytically validated and demonstrate clinical utility that are approved by FDA for clinical management. Novel assays used for cfDNA detection and analysis have reported different ranges of ctDNA. Detecting ctDNA in plasma samples from patients can be focused on known point mutations or limited sequence alterations (targeted ctDNA analysis) versus more genome-wide analysis by next-generation sequencing (NGS) [134] (Fig. 7.3). The utility of technology platforms for ctDNA analysis is summarized in Table 7.2.

Targeted ctDNA analysis involves targeted DNA sequencing techniques, such as digital PCR (dPCR), beads–emulsion–amplification–magnetics (BEAMing) technology, the safe-sequencing system (Safe-SeqS), cancer personalized profiling by deep sequencing (CAPP-Seq), and tagged-amplicon deep sequencing (TAmSeq). Based on their focus and very sensitive techniques, these methods are used in biomarkers of detection and diagnosis of minimal residual disease (MRD) [160].

Detection of KRAS mutations in patients with CRC predicts worse overall prognosis and assists in treatment selection. Sensitive and quantitative methods of detection of mutant allele fractions in plasma of patients with CRC were proposed as a prognostic biomarker at different stages of diagnosis and were studied as a method to detect MRD after surgery or chemotherapy [128,161–162].

Current treatments for patients with NSCLC are based on tumor characteristics at diagnosis of metastatic disease [163,164]. Optimized molecular testing in liquid biopsies, with reflex testing in

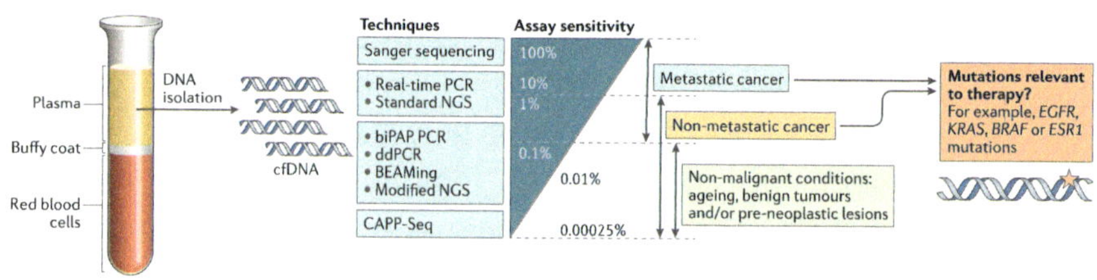

FIGURE 7.3

Methods of detecting circulating tumor DNA (ctDNA). BEAMing, beads–emulsion–amplification–magnetics; biPAP, bidirectional pyrophosphorolysis-activated polymerization; CAPP-Seq, cancer personalized profiling by deep sequencing; ddPCR, droplet digital PCR; NGS, next-generation sequencing. *From Pantel K at al. Liquid biopsy and minimal residual disease - latest advances and implications for cure. Nat Rev Clin Oncol. 2019 Jul;16(7):409-424.*

Table 7.2 Comparison and utility of technology platforms for ctDNA analysis [128,135,139,179].

Scale of analysis	Example technologies [139,140,179]	Loci interrogated	Indicative limit of detection (mutant allele fraction or concentration)	Clinical utility
Single-locus or multiplexed assays	PCR-based: • Digital PCR [135,136,137,138] • BEAMing [139,140] • Intplex [141]	• 1–10 loci • Both ctDNA and cfDNA (IntPlex)	Varies by method, optimal implementations can reach sensitivity of 0.001%–0.01% or individual mutant copies/mL	• Detection and quantification of recurrent hotspot mutations • Monitoring for recurrent resistance mutations • Rapid turnaround time
	Enrichment for mutant alleles: • COLD-PCR [142] • SCODA [143,144] • NaME-PrO [145]	• 10–100 loci		
	Allele-specific or ARMS-PCR kits for companion diagnostics (CDx): • Cobas EGFR [146] • Therascreen EGFR [147]	Cobas EGFR: 7 mutation assays covering multiple variants Therascreen EGFR: 3 mutation assays covering multiple variants	Limit of detection (≥95% sensitivity): Cobas EGFR: 25–100 copies/mL Therascreen EGFR: median 1.42% (range 0.05%–12.47% for different variants)	• Approved for in vitro diagnostic use: • Cobas EGFR: FDA approved • Therascreen EGFR: CE marked

(*Continued*)

Table 7.2 Comparison and utility of technology platforms for ctDNA analysis [128,135,139,179]. *Continued*

Scale of analysis	Example technologies [139,140,179]	Loci interrogated	Indicative limit of detection (mutant allele fraction or concentration)	Clinical utility
Targeted sequencing approaches	Amplicon-based: • TAm-Seq [148] • Enhanced TAm-Seq [149] • Safe-SeqS [150] Hybrid capture-based: • Exome sequencing [151] • CAPP-Seq [152] • Digital Sequencing [153–155]	10 kb to 50 Mb	<0.01%–0.5% for purpose-built panels [14] 1% for off-the-shelf multiplexed panels 5% for exome sequencing	• Profiling gene panels Monitoring for *de novo* resistance mutations. Monitoring clonal evolution in response to therapy. • Sensitivity for disease burden can be increased by testing multiple loci in parallel

Genome-wide	WGS: • Plasma-Seq [156] • PARE [157] Amplicon-based: • FAST-SeqS [158] • mFAST-SeqS [159]	3.2 gigabases (whole genome) 21.6 unique kilobases of LINE-1	5%–10%	• Identification of structural variants • Stratification of patient samples based on disease burden • Detecting the presence of chromosomal aberrations

ARMS, *amplification-refractory mutation system;* BEAMing, *Beads, emulsion, amplification, and magnetics;* CAPP-Seq, *cancer personalized profiling by deep sequencing;* COLD-PCR, *coamplification at lower denaturation temperature PCR;* FAST-SeqS, *fast aneuploidy screening test-sequencing system;* LINE-1, *long interspersed nucleotide element-1;* mFAST-SeqS, *modified fast aneuploidy screening test-sequencing system;* NaME-PrO, *nuclease-assisted minor-allele enrichment with probe-overlap;* PARE, *personalized analysis of rearranged ends;* SCODA, *synchronous coefficient of drag alteration;* TAm-Seq, *tagged amplicon deep sequencing.*

From Wan JCM, et al. Liquid biopsies come of age: towards implementation of circulating tumour DNA. Nat Rev Cancer. 2017 Apr;17(4):223-238.

tissue-based assays, allows to identify *EGFR* mutations in serum or plasma ctDNA of lung cancer patients [165]. Decreased mutated EGFR allele frequency with treatment has been proposed as a biomarker of response. Additional genomic rearrangements, such as echinoderm microtubule-associated protein-like 4-ALK (*EML4-ALK*) fusions, ROS proto-oncogene 1 (*ROS1*) rearrangements, and B-Raf proto-oncogene (*BRAF*) mutations, resulting in aberrant signaling, have FDA-approved targeted therapeutic options, for which informative ctDNA-based assays have clear clinical utility [163].

Patients with metastatic breast cancer have higher levels of cfDNA and ctDNA, and plasma concentrations of cancer antigen 15–3 in ctDNA have been proposed as biomarkers of treatment response [166]. Detecting mutations in PIK3CA and ESR1 by ctDNA-based assays have been studied in clinical trials with CDK4/CDK6 inhibitors [167].

NGS allows studying ctDNA in genome-wide analysis in a less targeted approach. Advances in NGS-based technologies have allowed analysis of multiple genes in cfDNA in selected tumor gene panels of tagged amplicons to analyze ctDNA at each stage of patient management. Whole-exome and whole-genome sequencing in cfDNA is limited by low abundance of tumor-derived DNA fragments, and different methods to increase sensitivity for ctDNA detection in plasma have been shown successful in patients with melanoma, lung, renal, glioma, and breast cancer [168].

With more targeted therapeutic options available, comprehensive genomic profiling can help guide treatment strategies and help predict patient benefit across multiple cancer indications.

Selected sequencing panels have been FDA cleared and are currently available in clinical practice. FoundationOne Liquid CDx© obtained FDA clearance as a companion diagnostic in clinical use to test approximately 300 genes of interest in oncology in cfDNA. In their label, a reflex analysis using an FDA-cleared tumor tissue test is recommended to confirm a negative ctDNA mutation status [FoundationOne Liquid CDx – P190032 | FDA]. FDA-cleared Guardant360 CDx cfDNA test [Guardant360 CDx – P200010 | FDA] demonstrated excellent concordance with tissue-based tests in a prospective study of patients with NSCLC for FDA-approved targets (EGFR, ALK, ROS1, BRAF).

Utilizing cfDNA in one study, in addition to tissue, increased detection by 48%, from 60 to 89 patients, including those with negative, not assessed, or insufficient tissue results [169].

The MSK-ACCESS clinical test was designed and analytically validated at Memorial Sloan Kettering Cancer Center, New York, NY, to detect somatic genetic alterations (single nucleotide variants, indels, copy number alterations, and structural variants) in cfDNA specimens based on hybridization capture and deep sequencing, and it is now clinically approved for clinical use by the New York State Department of Health [170]. Copy number analysis in whole-genome sequencing may detect tumor-derived chromosomal aberrations, including deletions and amplifications, gene rearrangements, and indirectly DNA damage repair defective tumors.

Liquid biopsies for immunotherapy in oncology may allow the selection of patients most likely to benefit clinically. Immunotherapy is a novel pillar of cancer care, from the neoadjuvant and adjuvant settings to the metastatic stage of many cancers. Not every patient will respond to such therapies, and liquid-based assays have been proposed as predictive biomarkers. The presence of programmed cell death 1 (PD1) in tumor cells inhibit the immune mediated effects through conjunction with PD1 ligand 1 (PDL1). Therapeutic antibodies that block the PD1–PDL1 pathway have demonstrated antitumor activity and are currently used in multiple cancer types. PDL1 expression on CTCs have been proposed as biomarkers for stratification and monitoring of cancer patients undergoing immune checkpoint blockade [171,172]. As mismatch repair deficiency and hypermutations predict response of solid tumors to PD-1 blockade [173,174], ctDNA-based assays detect high tumor mutational burden status based on hybrid-capture sequencing of panels of genes [175,176].

A comprehensive list of studies suggesting a clinical use for liquid biopsies is found in Martins et al. [133].

The limited numbers of CTC or small amounts of ctDNA, in a high normal blood cells or cfDNA, respectively, represent a challenge in the development of clinically valuable tests. Additionally, underperforming assays with limited analytical validation have reduced the applicability of liquid-based biomarkers in large scale clinical trials or practice, A summary of the challenges and possible

Table 7.3 Challenges and possible solutions to the use of various liquid biopsy components.

Analyte	Challenges unique to analyte	Possible solutions
CTCs	Low abundance	Sampling of larger volumes
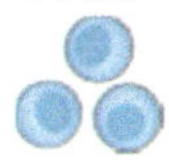	Special equipment required for detection and/or isolation	Development of enrichment technologies, which enable subsequent molecular analysis and high throughput at low cost
	High degree of heterogeneity	Marker-independent enrichment
	No high-throughput procedures available	Development of appropriate devices for high-throughput procedures
	No universal marker available; mostly enrichment of epithelial cells	Enrichment based on multiple CTC properties
ctDNA	High degree of fragmentation	Sampling of larger volumes
	Low overall amount of cfDNA	Further improvement of analytical sensitivity
	Often very low levels of ctDNA (< 0.1%)	Detection of multiple alterations
	Highly variable tumor fractions even in advanced stages	Better understanding of ctDNA and cfDNA biology
EVs 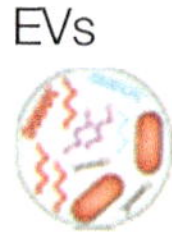	Hundreds of thousands of EVs are released every day by many cell types	Enrichment of tumor-specific EVs
	Unknown composition of EV content	Better understanding of EV content
	Isolation of tumor-specific EVs is challenging	Identification of tumor-specific EV marker
cfRNA	Intrinsic instability of mRNA	Use of stabilizing agents
	RNAs may be tightly linked to EVs	Standardization of preanalytics
miRNA	May be tightly linked to EVs	Guidelines for data normalization
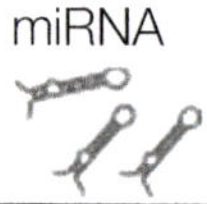	Data normalization is challenging	Use of multiple markers for normalization

cfDNA, *Circulating cell-free DNA;* cfRNA, *circulating cell-free RNA;* CTC, *circulating tumor cell;* ctDNA, *circulating tumor DNA;* EVs, *extracellular vesicles;* miRNA, *microRNA.*

From Heitzer E, et al. Current and future perspectives of liquid biopsies in genomics-driven oncology. Nat Rev Genet 2019;20(2):71–88. PMID: 30410101.

solutions to the use of various liquid biopsy components is presented in Table 7.3, from Heitzer et al. [134].

7.6 Conclusion

Tissue-based analysis, the current standard of care, is often limited by lack of contemporaneous samples for treatment decision-making. Furthermore, obtaining a biopsy can be painful as well as being costly. Recent work in several tumor types has shown that patients in whom liquid biomarkers are detected with a variety of different assays have an inferior prognosis relative to those in whom they are not detected. The decrease or elimination of the biomarker with treatment was associated with improved clinical outcomes, possible representing a surrogate clinical endpoint. Liquid biopsies may also detect disease recurrence prior to symptomatic disease or other imaging or laboratory markers. Additionally, biologic characterization in a liquid biopsy of a patient's tumor may be used to guide treatment selection, enabling personalized medicine, and superior outcomes. There are many more potential contexts where blood-based assays are studied. The analytical validation of these assays requires a rigorous process across independent cohorts of patients. Prioritizing biomarker development for a medical decision in a specific context of use should follow drug-development strategies, beginning with signal-finding short trials before large, costly validation trials are performed.

7.7 Summary points and future directions

- Liquid biopsy based biomarkers allow characterization of a patient's tumor at the time of treatment consideration.
- Developing a biomarker for a medical decision in a specific context of use requires stringent, lengthy, and costly analytical and clinical validation steps. This validation should follow drug-development strategies.
- Most promising liquid biomarkers are based on the detection CTC or ctDNA to predict prognosis, monitor treatment response

for clinical benefit, and profile an individual tumor biological characteristics at time of treatment decision. Limited amounts of circulating tumor materials, in a high normal background, in addition to underperforming assays, represent challenges in the development of clinically valuable tests.
- Liquid-based biomarkers hold the promise of noninvasive tools to improve clinical decision-making process.

Further readings

Liquid biopsies come of age: towards implementation of circulating tumor DNA [132].

Liquid biopsies: applications for cancer diagnosis and monitoring [133].

Current and future perspectives of liquid biopsies in genomics-driven oncology [134].

Liquid Biopsy: From Discovery to Clinical Application [178].

Liquid Biopsy-Based Biomarkers of Treatment Response and Resistance [179].

References

[1] Shah RB, et al. Androgen-independent prostate cancer is a heterogeneous group of diseases: lessons from a rapid autopsy program. Cancer Res 2004;64(24):9209–16.

[2] Atkinson A, et al. Biomarkers and surrogate endpoints: preferred definitions and conceptual framework. Clin Pharmacol Ther 2001;69 (3):89–95.

[3] Danila DC, et al. Clinical validity of detecting circulating tumor cells by adnatest assay compared with direct detection of tumor mRNA in stabilized whole blood, as a biomarker predicting overall survival for metastatic castration-resistant prostate cancer patients. Cancer J 2016;22(5):315–20.

[4] Danila Danila D, Scher HI, Fleisher M. Circulating tumor cells as an analytical tool in the management of patients with cancer. Manual of Molecular and Clinical Laboratory Immunology. 8th ed. Wiley; 2016.

[5] FDA/NCI/CMS Oncology Biomarker Qualification Initiative. Memorandum of understanding between the FDA, NCI, and CMS for

the FDA/NCI/CMS Oncology Biomarker Qualification Initiative. Document MOU 225-06-8001. January 2006. Last updated April 30, 2009. Available from: http://www.fda.gov/AboutFDA/Partnerships Collaborations/MemorandaofUnderstandingMOUs/DomesticMOUs/ ucm115681.htm [accessed March 16, 2011].

[6] Altar CA. The biomarkers consortium: on the critical path of drug discovery. Clin Pharmacol Ther 2008;83(2):361–4.

[7] Ashworth T. A case of cancer in which cells similar to those in the tumours were seen in the blood after death. Aus Med J 1869;14:146–9.

[8] Danila DC, et al. Circulating tumors cells as biomarkers: progress toward biomarker qualification. Cancer J 2011;17(6):438–50.

[9] Muller V, et al. Circulating tumor cells in breast cancer: correlation to bone marrow micrometastases, heterogeneous response to systemic therapy and low proliferative activity. Clin Cancer Res 2005;11(10):3678–85.

[10] Kim MY, et al. Tumor self-seeding by circulating cancer cells. Cell 2009;139(7):1315–26.

[11] Stott SL, et al. Isolation of circulating tumor cells using a microvortex-generating herringbone-chip. Proc Natl Acad Sci U S A 2010;107 (43):18392–7.

[12] Paget S. The distribution of secondary growths in cancer of the breast. Lancet 1889;1:571–3.

[13] Fidler IJ. The pathogenesis of cancer metastasis: the 'seed and soil' hypothesis revisited. Nat Rev Cancer 2003;3(6):453–8.

[14] Gupta GP, Massague J. Cancer metastasis: building a framework. Cell 2006;127(4):679–95.

[15] Ruoslahti E, Giancotti FG. Integrins and tumor cell dissemination. Cancer Cell 1989;1(4):119–26.

[16] Fan J, et al. Integrin beta4 signaling promotes mammary tumor cell adhesion to brain microvascular endothelium by inducing ErbB2-mediated secretion of VEGF. Ann Biomed Eng 2011;39(8): 2223–41.

[17] Alix-Panabieres C, et al. Detection and characterization of putative metastatic precursor cells in cancer patients. Clin Chem 2007;53(3):537–9.

[18] Alix-Panabieres C, et al. Full-length cytokeratin-19 is released by human tumor cells: a potential role in metastatic progression of breast cancer. Breast Cancer Res 2009;11(3):R39.

[19] Krebs MG, et al. Evaluation and prognostic significance of circulating tumor cells in patients with non-small-cell lung cancer. J Clin Oncol 2011;29(12):1556–63.

[20] Nagrath S, et al. Isolation of rare circulating tumour cells in cancer patients by microchip technology. Nature 2007;450(7173):1235–9.
[21] Riethdorf S, Wikman H, Pantel K. Review: Biological relevance of disseminated tumor cells in cancer patients. Int J Cancer 2008;123(9):1991–2006.
[22] Pantel K, Alix-Panabieres C, Riethdorf S. Cancer micrometastases. Nat Rev Clin Oncol 2009;6(6):339–51.
[23] Schindlbeck C, et al. Comparison of circulating tumor cells (CTC) in peripheral blood and disseminated tumor cells in the bone marrow (DTC-BM) of breast cancer patients. J Cancer Res Clin Oncol 2013;139(6):1055–62.
[24] Braun S, et al. A pooled analysis of bone marrow micrometastasis in breast cancer. N Engl J Med 2005;353(8):793–802.
[25] Weckermann D, et al. Perioperative activation of disseminated tumor cells in bone marrow of patients with prostate cancer. J Clin Oncol 2009;27(10):1549–56.
[26] Kollermann J, et al. Prognostic significance of disseminated tumor cells in the bone marrow of prostate cancer patients treated with neoadjuvant hormone treatment. J Clin Oncol 2008;26(30):4928–33.
[27] Pierga JY, et al. Clinical significance of immunocytochemical detection of tumor cells using digital microscopy in peripheral blood and bone marrow of breast cancer patients. Clin Cancer Res 2004;10(4):1392–400.
[28] Bidard FC, et al. Disseminated tumor cells of breast cancer patients: a strong prognostic factor for distant and local relapse. Clin Cancer Res 2008;14(11):3306–11.
[29] Scher HI, Pantel K. Bone marrow aspiration for disseminated tumor cell detection: a must-have test or is the jury still out? J Clin Oncol 2009;27(10):1531–3.
[30] Husemann Y, et al. Systemic spread is an early step in breast cancer. Cancer Cell 2008;13(1):58–68.
[31] Bartkowiak K, et al. Two-dimensional differential gel electrophoresis of a cell line derived from a breast cancer micrometastasis revealed a stem/ progenitor cell protein profile. J Proteome Res 2009;8(4):2004–14.
[32] Uhr JW, Pantel K. Controversies in clinical cancer dormancy. Proc Natl Acad Sci U S A 2011;108(30):12396–400.
[33] Lin WC, Rajbhandari N, Wagner KU. Cancer cell dormancy in novel mouse models for reversible pancreatic cancer: a lingering challenge in the development of targeted therapies. Cancer Res 2014;74(8):2138–43.

[34] Borovski T, et al. Cancer stem cell niche: the place to be. Cancer Res 2011;71(3):634–9.
[35] Thiery JP, Sleeman JP. Complex networks orchestrate epithelial-mesenchymal transitions. Nat Rev Mol Cell Biol 2006;7(2):131–42.
[36] Kalluri R, Weinberg RA. The basics of epithelial-mesenchymal transition. J Clin Invest 2009;119(6):1420–8.
[37] Mani SA, et al. The epithelial-mesenchymal transition generates cells with properties of stem cells. Cell 2008;133(4):704–15.
[38] Polyak K, Weinberg RA. Transitions between epithelial and mesenchymal states: acquisition of malignant and stem cell traits. Nat Rev Cancer 2009;9(4):265–73.
[39] Thurm H, et al. Rare expression of epithelial cell adhesion molecule on residual micrometastatic breast cancer cells after adjuvant chemotherapy. Clin Cancer Res 2003;9(7):2598–604.
[40] Deng G, et al. A new system for enrichment and detection of circulating tumor cells in the peripheral blood of patients with metastatic breast cancer. Breast Cancer Res Treat 2007;106:S24.
[41] Mikolajczyk SD, et al. Detection of EpCAM-negative and cytokeratin-negative circulating tumor cells in peripheral blood. J Oncol 2011;2011:252361.
[42] Huang RY, et al. An EMT spectrum defines an anoikis-resistant and spheroidogenic intermediate mesenchymal state that is sensitive to e-cadherin restoration by a src-kinase inhibitor, saracatinib (AZD0530). Cell Death Dis 2013;4:e915.
[43] Werner SL, et al. Analytical validation and capabilities of the Epic CTC platform: enrichment-free circulating tumour cell detection and characterization. J Circ Biomark 2015;4:3.
[44] Beltran H, et al. The initial detection and partial characterization of circulating tumor cells in neuroendocrine prostate cancer. Clin Cancer Res 2016;22(6):1510–19.
[45] Punnoose EA, et al. PTEN loss in circulating tumour cells correlates with PTEN loss in fresh tumour tissue from castration-resistant prostate cancer patients. Br J Cancer 2015;113(8):1225–33.
[46] Anantharaman A, et al. Programmed death-ligand 1 (PD-L1) characterization of circulating tumor cells (CTCs) and white blood cells (WBCs) in muscle invasive and metastatic bladder cancer patients. J Clin Oncol 2016;34.
[47] McDaniel AS, et al. Phenotypic diversity of circulating tumour cells in patients with metastatic castration-resistant prostate cancer. BJU Int 2017;120(5B):E30–44.

[48] Scher HI, et al. Association of AR-V7 on circulating tumor cells as a treatment-specific biomarker with outcomes and survival in castration-resistant prostate cancer. JAMA Oncol 2016;2(11):1441–9.
[49] Scher HI, et al. Phenotypic heterogeneity of circulating tumor cells informs clinical decisions between AR signaling inhibitors and taxanes in metastatic prostate cancer. Cancer Res 2017;77(20):5687–98.
[50] Danila DC, Fleisher M, Scher HI. Circulating tumor cells as biomarkers in prostate cancer. Clin Cancer Res 2011;17(12):3903–12.
[51] Yu M, et al. Circulating tumor cells: approaches to isolation and characterization. J Cell Biol 2011;192(3):373–82.
[52] Stott SL, et al. Isolation and characterization of circulating tumor cells from patients with localized and metastatic prostate cancer. Sci Transl Med 2010;2(25):25ra23.
[53] Allard WJ, et al. Tumor cells circulate in the peripheral blood of all major carcinomas but not in healthy subjects or patients with nonmalignant diseases. Clin Cancer Res 2004;10(20):6897–904.
[54] Budd GT, et al. Circulating tumor cells vs imaging—predicting overall survival in metastatic breast cancer. Clin Cancer Res 2006;12(21):6403–9.
[55] Cristofanilli M, et al. Circulating tumor cells, disease progression, and survival in metastatic breast cancer. N Engl J Med 2004;351(8):781–91.
[56] Cristofanilli M, et al. Circulating tumor cells: a novel prognostic factor for newly diagnosed metastatic breast cancer. J Clin Oncol 2005;23(7):1420–30.
[57] Hayes DF, et al. Circulating tumor cells at each follow-up time point during therapy of metastatic breast cancer patients predict progression-free and overall survival. Clin Cancer Res 2006;12(14 Pt 1):4218–24.
[58] Cohen SJ, et al. Relationship of circulating tumor cells to tumor response, progression-free survival, and overall survival in patients with metastatic colorectal cancer. J Clin Oncol 2008;26(19):3213–21.
[59] Cohen SJ, et al. Prognostic significance of circulating tumor cells in patients with metastatic colorectal cancer. Ann Oncol 2009;20(7):1223–9.
[60] Danila DC, et al. Circulating tumor cell number and prognosis in progressive castration-resistant prostate cancer. Clin Cancer Res 2007;13(23):7053–8.
[61] de Bono JS, et al. Circulating tumor cells predict survival benefit from treatment in metastatic castration-resistant prostate cancer. Clin Cancer Res 2008;14(19):6302–9.

[62] Scher HI, et al. Circulating tumour cells as prognostic markers in progressive, castration-resistant prostate cancer: a reanalysis of IMMC38 trial data. Lancet Oncol 2009;10(3):233–9.
[63] Coumans FA, et al. All circulating EpCAM + CK + CD45− objects predict overall survival in castration-resistant prostate cancer. Ann Oncol 2010;21(9):1851–7.
[64] Shaffer DR, et al. Circulating tumor cell analysis in patients with progressive castration-resistant prostate cancer. Clin Cancer Res 2007;13(7):2023–9.
[65] Attard G, et al. Characterization of ERG, AR and PTEN gene status in circulating tumor cells from patients with castration-resistant prostate cancer. Cancer Res 2009;69(7):2912–18.
[66] Ulmer A, et al. Immunomagnetic enrichment, genomic characterization, and prognostic impact of circulating melanoma cells. Clin Cancer Res 2004;10(2):531–7.
[67] Racila E, et al. Detection and characterization of carcinoma cells in the blood. Proc Natl Acad Sci U S A 1998;95:4589–94.
[68] Scholtens TM, et al. CellTracks TDI: an image cytometer for cell characterization. Cytometry A 2011;79:203–13.
[69] Bocsi J, et al. Scanning fluorescent microscopy analysis is applicable for absolute and relative cell frequency determinations. Cytometry A 2004;61(1):1–8.
[70] Weight RM, Viator JA. Detection of circulating tumor cells by photoacoustic flowmetry. Methods Mol Biol 2014;1102:655–63.
[71] Navin N, et al. Tumour evolution inferred by single-cell sequencing. Nature 2011;472(7341):90–4.
[72] Yaniv I, et al. Tumor cells are present in stem cell harvests of Ewings sarcoma patients and their persistence following transplantation is associated with relapse. Pediatr Blood Cancer 2004;42 (5):404–9.
[73] Wong IH, Chan AT, Johnson PJ. Quantitative analysis of circulating tumor cells in peripheral blood of osteosarcoma patients using osteoblast-specific messenger RNA markers: a pilot study. Clin Cancer Res 2000;6(6):2183–8.
[74] Dubois SG, et al. Flow cytometric detection of Ewing sarcoma cells in peripheral blood and bone marrow. Pediatr Blood Cancer 2010;54 (1):13–18.
[75] Helzer KT, et al. Circulating tumor cells are transcriptionally similar to the primary tumor in a murine prostate model. Cancer Res 2009;69(19):7860–6.

[76] Giordano A, et al. Circulating tumor cells in immunohistochemical subtypes of metastatic breast cancer: lack of prediction in HER2-positive disease treated with targeted therapy. Ann Oncol 2012;23(5):1144–50.

[77] Maheswaran S, et al. Detection of mutations in EGFR in circulating lung-cancer cells. N Engl J Med 2008;359(4):366–77.

[78] Harb W, et al. Mutational analysis of circulating tumor cells using a novel microfluidic collection device and qPCR assay. Transl Oncol 2013;6(5):528–38.

[79] Gleghorn JP, et al. Capture of circulating tumor cells from whole blood of prostate cancer patients using geometrically enhanced differential immunocapture (GEDI) and a prostate-specific antibody. Lab Chip 2010;10(1):27–9.

[80] Ozkumur E, et al. Inertial focusing for tumor antigen-dependent and -independent sorting of rare circulating tumor cells. Sci Transl Med 2013;5(179):179ra47.

[81] Winer-Jones JP, et al. Circulating tumor cells: clinically relevant molecular access based on a novel CTC flow cell. PLoS One 2014;9(1):e86717.

[82] Ramirez JM, et al. Prognostic relevance of viable circulating tumor cells detected by EPISPOT in metastatic breast cancer patients. Clin Chem 2014;60(1):214–21.

[83] Alix-Panabieres C, et al. Detection of circulating prostate-specific antigen-secreting cells in prostate cancer patients. Clin Chem 2005;51(8):1538–41.

[84] Paris PL, et al. Functional phenotyping and genotyping of circulating tumor cells from patients with castration resistant prostate cancer. Cancer Lett 2009;277(2):164–73.

[85] Shim S, et al. Dynamic physical properties of dissociated tumor cells revealed by dielectrophoretic field-flow fractionation. Integr Biol (Camb) 2011;3(8):850–62.

[86] Gel M, et al. Dielectrophoretic cell trapping and parallel one-to-one fusion based on field constriction created by a micro-orifice array. Biomicrofluidics 2010;4(2).

[87] Borgatti M, et al. Dielectrophoresis-based 'Lab-on-a-chip' devices for programmable binding of microspheres to target cells. Int J Oncol 2005;27(6):1559–66.

[88] Lecharpentier A, et al. Detection of circulating tumour cells with a hybrid (epithelial/mesenchymal) phenotype in patients with metastatic non-small cell lung cancer. Br J Cancer 2011;105(9):1338–41.

[89] Lin HK, et al. Portable filter-based microdevice for detection and characterization of circulating tumor cells. Clin Cancer Res 2010;16 (20):5011–18.

[90] Zheng S, et al. 3D microfilter device for viable circulating tumor cell (CTC) enrichment from blood. Biomed Microdevices 2011;13 (1):203–13.

[91] Farace F, et al. A direct comparison of CellSearch and ISET for circulating tumour-cell detection in patients with metastatic carcinomas. Br J Cancer 2011;105(6):847–53.

[92] Adams DL, et al. Circulating giant macrophages as a potential biomarker of solid tumors. Proc Natl Acad Sci U S A 2014;111 (9):3514–19.

[93] Marrinucci D, et al. Fluid biopsy in patients with metastatic prostate, pancreatic and breast cancers. Phys Biol 2012;9(1):016003.

[94] Ghossein RA, et al. Detection of circulating tumor cells in patients with localized and metastatic prostatic carcinoma: clinical implications. J Clin Oncol 1995;13(5):1195–200.

[95] Helo P, et al. Circulating prostate tumor cells detected by reverse transcription-PCR in men with localized or castration-refractory prostate cancer: concordance with CellSearch assay and association with bone metastases and with survival. Clin Chem 2009;55 (4):765–73.

[96] Iinuma H, et al. Clinical significance of circulating tumor cells, including cancer stem-like cells, in peripheral blood for recurrence and prognosis in patients with Dukes' stage B and C colorectal cancer. J Clin Oncol 2011;29(12):1547–55.

[97] Xi L, et al. Optimal markers for real-time quantitative reverse transcription PCR detection of circulating tumor cells from melanoma, breast, colon, esophageal, head and neck, and lung cancers. Clin Chem 2007;53(7):1206–15.

[98] Reinholz MM, et al. Cytokeratin-19 and mammaglobin gene expression in circulating tumor cells from metastatic breast cancer patients enrolled in North Central Cancer Treatment Group (NCCTG) trials, N0234/336/436/437. Clin Cancer Res 2011;17 (22):7183–93.

[99] Sieuwerts AM, et al. mRNA and microRNA expression profiles in circulating tumor cells and primary tumors of metastatic breast cancer patients. Clin Cancer Res 2011;17(11):3600–18.

[100] Zhao S, et al. The prognostic role of circulating tumor cells (CTCs) detected by RT-PCR in breast cancer: a meta-analysis of published literature. Breast Cancer Res Treat 2011;130(3):809–16.
[101] Yoon SO, et al. TTF-1 mRNA-positive circulating tumor cells in the peripheral blood predict poor prognosis in surgically resected non-small cell lung cancer patients. Lung Cancer 2011;71(2):209–16.
[102] Ignatiadis M, et al. Prognostic value of the molecular detection of circulating tumor cells using a multimarker reverse transcription-PCR assay for cytokeratin 19, mammaglobin A, and HER2 in early breast cancer. Clin Cancer Res 2008;14(9):2593–600.
[103] Nakagawa T, et al. Detection of circulating tumor cells in early-stage breast cancer metastasis to axillary lymph nodes. Clin Cancer Res 2007;13(14):4105–10.
[104] Danila DC, et al. Analytic and clinical validation of a prostate cancer-enhanced messenger RNA detection assay in whole blood as a prognostic biomarker for survival. Eur Urol 2014;65(6):1191–7.
[105] Schleiermacher G, et al. Increased risk of systemic relapses associated with bone marrow micrometastasis and circulating tumor cells in localized ewing tumor. J Clin Oncol 2003;21(1):85–91.
[106] Aktas B, et al. Evaluation and correlation of risk recurrence in early breast cancer assessed by Oncotype DX, clinicopathological markers and tumor cell dissemination in the blood and bone marrow. Mol Clin Oncol 2013;1(6):1049–54.
[107] Danila DC, et al. TMPRSS2-ERG status in circulating tumor cells as a predictive biomarker of sensitivity in castration-resistant prostate cancer patients treated with abiraterone acetate. Eur Urol 2011;60(5):897–904.
[108] Andreopoulou E, et al. Comparison of assay methods for detection of circulating tumor cells in metastatic breast cancer: AdnaGen AdnaTest BreastCancer Select/Detect vs Veridex CellSearch system. Int J Cancer 2012;130(7):1590–7.
[109] Berruti A, Pia A, Terzolo M. Abiraterone and increased survival in metastatic prostate cancer. N Engl J Med 2011;365(8):766 Author reply 767–768.
[110] Wendel M, et al. Fluid biopsy for circulating tumor cell identification in patients with early-and late-stage non-small cell lung cancer: a glimpse into lung cancer biology. Phys Biol 2012;9(1):016005.
[111] Botteri E, et al. Modeling the relationship between circulating tumour cells number and prognosis of metastatic breast cancer. Breast Cancer Res Treat 2010;122(1):211–17.

[112] Scher HI, et al. Antitumour activity of MDV3100 in castration-resistant prostate cancer: a phase 1–2 study. Lancet 2010;375 (9724):1437–46.

[113] Veridex. 510k for CellSearch Veridex; 2008. Available from: http://www.accessdata.fda.gov/cdrh_docs/pdf7/K073338.pdf.

[114] Gonen M, Heller G. Concordance probability and discriminative power of proportional hazards regression. Biometrika 2005;92:965–70.

[115] Gentlemen R, Crowley J. Graphical methods for censored data. J Am Stat Assoc 1991;86:678–83.

[116] Koyanagi K, et al. Serial monitoring of circulating tumor cells predicts outcome of induction biochemotherapy plus maintenance biotherapy for metastatic melanoma. Clin Cancer Res 2010;16 (8):2402–8.

[117] McShane LM, et al. Reporting recommendations for tumor marker prognostic studies (REMARK). J Natl Cancer Inst 2005;97 (16):1180–4.

[118] Leversha MA, et al. Fluorescence in situ hybridization analysis of circulating tumor cells in metastatic prostate cancer. Clin Cancer Res 2009;15(6):2091–7.

[119] de Bono JS, et al. Potential applications for circulating tumor cells expressing the insulin-like growth factor-I receptor. Clin Cancer Res 2007;13(12):3611–16.

[120] Holzbeierlein J, et al. Gene expression analysis of human prostate carcinoma during hormonal therapy identifies androgen-responsive genes and mechanisms of therapy resistance. Am J Pathol 2004;164 (1):217–27.

[121] Scher HI, Kelly WK. The flutamide withdrawal syndrome: its impact on clinical trials in hormone-refractory prostatic cancer. J Clin Oncol 1993;11:1566–72.

[122] Yang MJ, et al. Enhancing detection of circulating tumor cells with activating KRAS oncogene in patients with colorectal cancer by weighted chemiluminescent membrane array method. Ann Surg Oncol 2010;17(2):624–33.

[123] Punnoose EA, et al. Molecular biomarker analyses using circulating tumor cells. PLoS One 2010;5(9):e12517.

[124] Karp DD, et al. Phase II study of the anti-insulin-like growth factor type 1 receptor antibody CP-751,871 in combination with paclitaxel and carboplatin in previously untreated, locally advanced, or metastatic non-small-cell lung cancer. J Clin Oncol 2009;27 (15):2516–22.

[125] Scher HI, et al. Association of AR-V7 on circulating tumor cells as a treatment-specific biomarker with outcomes and survival in castration-resistant prostate cancer. JAMA Oncol 2016;2(11):1441–9.

[126] Scher HI, et al. Nuclear-specific AR-V7 protein localization is necessary to guide treatment selection in metastatic castration-resistant prostate cancer. Eur Urol 2017;71(6):874–82.

[127] Armstrong AJ, et al. Prospective multicenter validation of androgen receptor splice variant 7 and hormone therapy resistance in high-risk castration-resistant prostate cancer: The PROPHECY Study. J Clin Oncol 2019;37(13):1120–9.

[128] Diehl F, et al. Circulating mutant DNA to assess tumor dynamics. Nat Med 2008;14(9):985–90.

[129] Mandel P, Metais P. Nuclear acids in human blood plasma. C R Seances Soc Biol Fil 1948;142(3–4):241–3.

[130] Mouliere F, et al. Multi-marker analysis of circulating cell-free DNA toward personalized medicine for colorectal cancer. Mol Oncol 2014;8(5):927–41.

[131] Leon SA, et al. Free DNA in the serum of cancer patients and the effect of therapy. Cancer Res 1977;37(3):646–50.

[132] Wan JCM, et al. Liquid biopsies come of age: towards implementation of circulating tumour DNA. Nat Rev Cancer 2017;17 (4):223–38.

[133] Martins I, et al. Liquid biopsies: applications for cancer diagnosis and monitoring. Genes (Basel) 2021;12(3).

[134] Heitzer E, et al. Current and future perspectives of liquid biopsies in genomics-driven oncology. Nat Rev Genet 2019;20(2):71–88.

[135] Vogelstein B. & Kinzler K.W. Digital PCR. Proc. Natl Acad. Sci. *USA* 96, 9236–9241 (1999).

[136] Taly V, et al. Multiplex picodroplet digital PCR to detect *KRAS* mutations in circulating DNA from the plasma of colorectal cancer patients. Clin Chem 2013;59:1722–31.

[137] Yung TKF, et al. Single-molecule detection of epidermal growth factor receptor mutations in plasma by microfluidics digital PCR in non-small cell lung cancer patients. Clin Cancer Res 2009;15:2076–84.

[138] Holm M, et al. Detection of KRAS mutations in liquid biopsies from metastatic colorectal cancer patients using droplet digital PCR, Idylla, and next generation sequencing. PLoS One. 2020 Nov 25;15(11):e0239819. doi: 10.1371/journal.pone.0239819. PMID: 33237900; PMCID: PMC7688175.

[139] Dressman D, Yan H, Traverso G, Kinzler KW, Vogelstein B. Transforming single DNA molecules into fluorescent magnetic particles for detection and enumeration of genetic variations. Proc. Natl Acad. Sci. USA 2003;100:8817–22.

[140] Diehl F, et al. Detection and quantification of mutations in the plasma of patients with colorectal tumors. Proc. Natl Acad. Sci. USA 2005;102:16368–73.

[141] Kloten V, et al. Liquid biopsy in colon cancer: comparison of different circulating DNA extraction systems following absolute quantification of *KRAS* mutations using Intplex allele-specific PCR. Oncotarget. 2017 Sep 21;8(49):86253-86263.

[142] Li J, et al. Replacing PCR with COLD-PCR enriches variant DNA sequences and redefines the sensitivity of genetic testing. Nat Med 2008;14:579–84.

[143] Marziali A, Pel J, Bizzotto D, Whitehead LA. Novel electrophoresis mechanism based on synchronous alternating drag perturbation. Electrophoresis 2005;26:82–90.

[144] Thompson JD, Shibahara G, Rajan S, Pel J, Marziali A. Winnowing DNA for rare sequences: highly specific sequence and methylation based enrichment. PLoS ONE 2012;7e31597.

[145] Song C, et al. Elimination of unaltered DNA in mixed clinical samples via nuclease-assisted minor-allele enrichment. Nucleic Acids Res 2016;44:1–11.

[146] US Food & Drug Administration. Premarket approval P150044 — Cobas EGFR MUTATION TEST V2. *FDA*Available from: http://www.accessdata.fda.gov/scripts/cdrh/cfdocs/cfpma/pma.cfm?id = P150044.

[147] https://www.accessdata.fda.gov/cdrh_docs/pdf12/p120022c.pdf.

[148] Forshew T, et al. Noninvasive identification and monitoring of cancer mutations by targeted deep sequencing of plasma DNA. Sci Transl Med 2012;4:136ra68.

[149] Gale D, et al. Analytical performance and validation of an enhanced TAm-Seq circulating tumor DNA sequencing assay. Inivata 2016;. Available from: http://inivata.com/wp-content/uploads/2016/06/AACRposter2016.pdf.

[150] Kinde I, Wu J, Papadopoulos N, Kinzler KW, Vogelstein B. Detection and quantification of rare mutations with massively parallel sequencing. Proc. Natl Acad. Sci. USA 2011;108:9530–5.

[151] Murtaza M, et al. Non-invasive analysis of acquired resistance to cancer therapy by sequencing of plasma DNA. Nature 2013;497:108–12.

[152] Newman AM, et al. An ultrasensitive method for quantitating circulating tumor DNA with broad patient coverage. Nat. Med. 2014;20:548–54.
[153] Lanman RB, et al. Analytical and clinical validation of a digital sequencing panel for quantitative, highly accurate evaluation of cell-free circulating tumor DNA. PLoS ONE 2015;10e0140712.
[154] Schwaederle MC, et al. Detection rate of actionable mutations in diverse cancers using a biopsy-free (blood) circulating tumor DNA assay. Oncotarget 2015;33:11004.
[155] Abdueva D. et al. Detection, frequency and actionability of recurrent copy number gains detected by non-invasive liquid biopsy of 3,942 lung and breast cancer samples. J Clin Oncol 2016; 34: abstr. 11541.
[156] Heitzer E, et al. Tumor-associated copy number changes in the circulation of patients with prostate cancer identified through whole-genome sequencing. Genome Med 2013;5:30.
[157] Leary RJ, et al. Development of personalized tumor biomarkers using massively parallel sequencing. Sci. Transl Med. 2010;2:20ra14.
[158] Kinde I, Papadopoulos N, Kinzler KW, Vogelstein B. FAST-SeqS: a simple and efficient method for the detection of aneuploidy by massively parallel sequencing. PLoS ONE 2012;7e41162.
[159] Belic J, et al. Rapid identification of plasma DNA samples with increased ctDNA levels by a modified FAST-SeqS approach. Clin. Chem. 2015;61:838–49.
[160] Pantel K, Alix-Panabieres C. Liquid biopsy and minimal residual disease – latest advances and implications for cure. Nat Rev Clin Oncol 2019;16(7):409–24.
[161] Dressman D, et al. Transforming single DNA molecules into fluorescent magnetic particles for detection and enumeration of genetic variations. Proc Natl Acad Sci U S A 2003;100(15):8817–22.
[162] Khatua S, et al. Overexpression of the EGFR/FKBP12/HIF-2alpha pathway identified in childhood astrocytomas by angiogenesis gene profiling. Cancer Res 2003;63(8):1865–70.
[163] Gregg JP, Li T, Yoneda KY. Molecular testing strategies in non-small cell lung cancer: optimizing the diagnostic journey. Transl Lung Cancer Res 2019;8(3):286–301.
[164] Li N, et al. Perioperative circulating tumor DNA as a potential prognostic marker for operable stage I to IIIA non-small cell lung cancer. Cancer 2022;128(4):708–18.

[165] Yung TK, et al. Single-molecule detection of epidermal growth factor receptor mutations in plasma by microfluidics digital PCR in non-small cell lung cancer patients. Clin Cancer Res 2009;15 (6):2076–84.

[166] Dawson SJ, et al. Analysis of circulating tumor DNA to monitor metastatic breast cancer. N Engl J Med 2013;368(13):1199–209.

[167] Tolaney SM, et al. Clinical significance of PIK3CA and ESR1 mutations in circulating tumor DNA: analysis from the MONARCH 2 study of abemaciclib plus fulvestrant. Clin Cancer Res 2022. Available from: http://doi.org/10.1158/1078-0432.CCR-21-3276.

[168] Wan JCM, et al. ctDNA monitoring using patient-specific sequencing and integration of variant reads. Sci Transl Med 2020;12(548).

[169] Leighl NB, et al. Clinical utility of comprehensive cell-free DNA analysis to identify genomic biomarkers in patients with newly diagnosed metastatic non-small cell lung cancer. Clin Cancer Res 2019;25(15):4691–700.

[170] Rose Brannon A, et al. Enhanced specificity of clinical high-sensitivity tumor mutation profiling in cell-free DNA via paired normal sequencing using MSK-ACCESS. Nat Commun 2021;12 (1):3770.

[171] Mazel M, et al. Frequent expression of PD-L1 on circulating breast cancer cells. Mol Oncol 2015;9(9):1773–82.

[172] Asante DB, et al. Multi-marker immunofluorescent staining and PD-L1 detection on circulating tumour cells from ovarian cancer patients. Cancers (Basel) 2021;13(24).

[173] Graham LS, Pritchard CC, Schweizer MT. Hypermutation, mismatch repair deficiency, and defining predictors of response to checkpoint blockade. Clin Cancer Res 2021;27(24):6662–5.

[174] Le DT, et al. Mismatch repair deficiency predicts response of solid tumors to PD-1 blockade. Science 2017;357(6349):409–13.

[175] Khagi Y, et al. Hypermutated circulating tumor DNA: correlation with response to checkpoint inhibitor-based immunotherapy. Clin Cancer Res 2017;23(19):5729–36.

[176] Murtaza M, et al. Non-invasive analysis of acquired resistance to cancer therapy by sequencing of plasma DNA. Nature 2013;497 (7447):108–12.

[177] Alix-Panabières C, Pantel K. Liquid Biopsy: From Discovery to Clinical Application. Cancer Discov. 2021 Apr;11(4):858-873. doi: 10.1158/2159-8290.CD-20-1311. PMID: 33811121.

[178] Kilgour E et al. Liquid Biopsy-Based Biomarkers of Treatment Response and Resistance. Cancer Cell. 2020 Apr 13;37(4):485-495.

[179] O'Leary B et al. Comparison of BEAMing and Droplet Digital PCR for Circulating Tumor DNA Analysis. Clin Chem. 2019 Nov;65(11):1405-1413. Available from: http://doi.org/10.1373/clinchem.2019.305805. Epub 2019 Sep 24. PMID: 31551314.

CHAPTER

8 Chimeric antigen receptor T cells and management of toxicities: implications of biomarkers

Kitsada Wudhikarn[1,2], Ana Alarcón Tomás[1,3], Kazunori Murata[4] and Miguel-Angel Perales[1,5]

[1]*Adult Bone Marrow Transplantation, Department of Medicine, Memorial Sloan Kettering Cancer Center, New York, NY, United States*

[2]*Division of Hematology and Research Unit in Translational Hematology, Department of Medicine, Chulalongkorn University, Bangkok, Thailand*

[3]*PhD Program in Signals Integration and Modulation in Biomedicine, Cell Therapy and Translational Medicine, University of Murcia, Murcia, Spain*

[4]*Department of Pathology and Laboratory Medicine, Memorial Sloan Kettering Cancer Center, New York, NY, United States*

[5]*Department of Medicine, Weil Cornell College of Medicine, New York, NY, United States*

Objectives

- Describe the current use and limitations of chimeric antigen receptor T-cell therapy in B lymphoid malignancy.
- Discuss the available evidence and ongoing studies of biomarkers in patients who receive chimeric antigen receptor T cells.
- Highlight the implications and limitations of biomarkers in routine clinical practice for patients who receive chimeric antigen receptor T cells.
- Propose the future approach in research and implementation of biomarkers to the care of CAR T-cell recipients in the routine oncology practice.

Cancer Biomarkers: Clinical Aspects and Laboratory Determination. DOI: https://doi.org/10.1016/B978-0-12-824302-2.00012-6
© 2022 Elsevier Inc. All rights reserved.

8.1 Introduction

Immunotherapy has become the major pillar of oncology treatment, beginning with the introduction of hematopoietic stem cell transplantation in the 1950s and followed by advances in cellular biology and bioengineering that enabled the development of adoptive cellular therapy over the past two decades. This approach involves manipulation of natural immune effector cells to enhance their efficacy to target cancer cells. Chimeric antigen receptor (CAR) T cells are a recent, novel platform of cell-based immunotherapy [1]. By exploiting viral-based gene transduction or other gene-editing techniques, immune cells are engineered to express the modified receptor (i.e., CAR or engineered T-cell receptors) with explicit affinity to tumor-associated antigens. CD19-targeted CAR T cells have revolutionized the treatment landscape of relapsed/refractory (RR) B lymphoid hematologic malignancies, enabling unprecedented response rates, and durable remissions.

As of 2021, there are five Food and Drug Administration (FDA)-approved CAR T-cell products licensed for RR hematologic malignancies: axicabtagene ciloleucel (axi-cel) for RR aggressive diffuse large B-cell lymphoma (DLBCL) and follicular lymphoma (FL); tisagenlecleucel (tisa-cel) for RR precursor B-cell acute lymphoblastic leukemia (B-ALL) in pediatric/young adult (age <26 years old) patients and for RR aggressive large B-cell lymphoma (LBCL); lisocabtagene maraleucel (liso-cel) for RR aggressive LBCL; brexucabtagene autoleucel (brexu-cel) for RR mantle cell lymphoma (MCL) and RR B-ALL; and idecabtagene vicleucel (ide-cel) for RR multiple myeloma. Ide-cel is the first approved B-cell maturation antigen–targeted CAR T-cell product, whereas all others target cell surface protein CD19 on B lymphocytes [2–9]. In addition to these commercially available CD19 CAR T cells, several CAR T-cell products with different targets are under investigation in other hematologic malignancies, such as multiple myeloma, chronic lymphocytic leukemia (CLL), Hodgkin lymphoma, and T-cell lymphoma.

Despite their impressive antitumor activity, CD19 CAR T cells have unique immune-mediated toxicities, including cytokine release syndrome (CRS) and immune effector cell-associated neurotoxicity syndrome (ICANS) [10]. Moreover, a substantial proportion of patients relapses or fails to respond after receiving CAR T-cell therapy. The efficacy and safety of CAR T-cell therapy could be improved with better ability to predict toxicities and responses by understanding the complex array of patient-, tumor-, and immune-associated factors involved. In this chapter, we will outline results of CD19 CAR T-cell therapy in patients with B lymphoid hematologic malignancies, implications of biomarkers in patients treated with CAR T cells, and future directions.

8.2 Construct and manufacturing process of CD19 chimeric antigen receptor T cells

Manufacturing of autologous CAR T cells involves a multistep process including leukapheresis, lymphocyte activation and selection, CAR gene transduction, ex vivo T-cell expansion, and product cryopreservation [11]. Manufacturing requires substantial infrastructure investment and multidisciplinary collaboration and generally takes 3–4 weeks from cell collection to product availability.

The typical CAR construct has three major components: (1) extracellular single-chain variable fragment derived receptor, (2) transmembrane hinge and linker region, and (3) intracellular costimulatory and CD3 stimulatory domain. All five FDA-approved and most experimental CAR T cells are second-generation models with a single costimulatory domain, usually CD28 or 4–1BB (Fig. 8.1). These costimulatory molecules are critical for in vivo CAR T-cell expansion and persistence. CAR gene transduction is achieved via various gene editing techniques, most commonly the viral-vector approach. The various CD19-targeted CAR T-cell approaches are described below according to therapeutic indication.

8.2.1 Relapsed/refractory large B-cell lymphoma

CD19 CAR T-cell therapy was approved in RR aggressive DLBCL based on the results of two pivotal studies, the ZUMA-1 [4] and JULIET trials [5]. ZUMA-1, a Phase 2 study, explored the efficacy of axi-cel, a retroviral-based CD19-CD28 CAR T-cell product. The

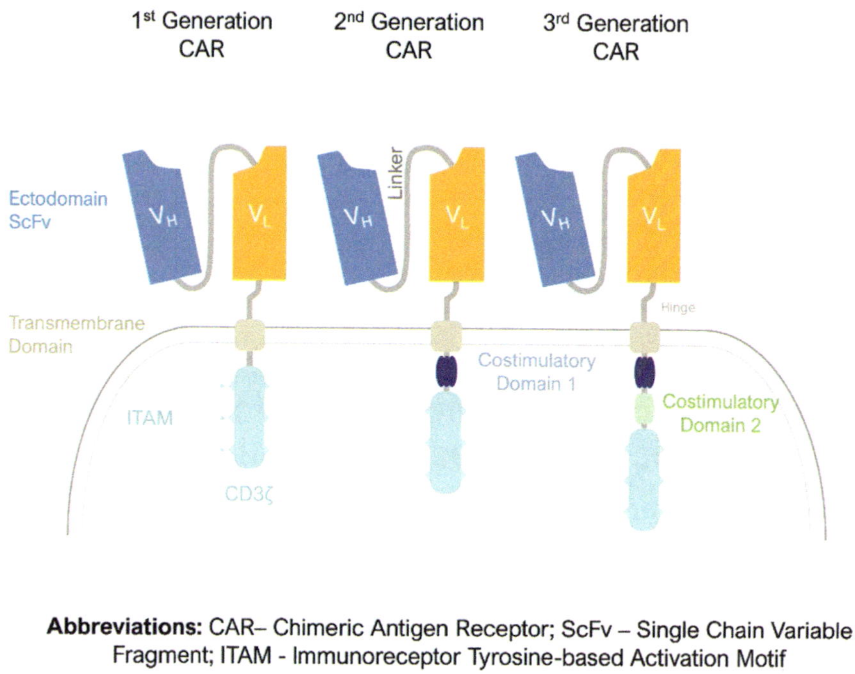

FIGURE 8.1

The structure and component of chimeric antigen receptor. *CAR*, chimeric antigen receptor; *ScFv*, single-chain variable fragment; V_H, heavy-chain variable region; V_L, light-chain variable region; *ITAM*, immunoreceptor tyrosine-based activation motif.

objective response rate was 82%, complete response (CR) rate 54%, and median duration of remission was 11.1 months among 101 evaluable patients. The incidence of CRS and ICANS was 93% and 64%, respectively. JULIET, a Phase 2 study, investigated the efficacy and safety of tisa-cel, a lentiviral-based CD19–4-1BB CAR T-cell product. The best overall response rate (ORR) of tisa-cel from the JULIET study was 52%, with a CR rate of 40%, and median duration of remission not reached among 93 evaluable patients. CRS and ICANS were observed in 58% and 21% of evaluable patients, respectively. Long-term follow-up of both trials showed sustained durable remission, indicating potential cure after CAR T-cell therapy [12,13]. Liso-cel was approved for patients with RR LBCL in early 2021 based on the result of the large Phase 1b

TRANSCEND NHL-001 trial showing its excellent safety and efficacy [14]. There are three major clinical trials comparing treatment outcomes between various CAR T-cell products and autologous stem cell transplantation in the front-line salvage setting recently completed patient accrual. The preliminary results of some of these studies (ZUMA-7 and TRANSFORM) demonstrated superior event-free survival in patients with RR LBCL treated with CD19 CAR T cells compared to those who underwent autologous stem cell transplantation [15,16].

8.2.2 Relapsed/refractory B-cell acute lymphoblastic leukemia

Several trials have investigated the safety and efficacy of CD19 CAR T cells in patients with RR B-ALL [17–19]. Tisa-cel was the first FDA-approved product, approved only for patients younger than 26 years old based on the results of the ELIANA trial, a Phase 1/2a study [17]. The overall remission rate at 3 months post-treatment was 81%, in children and young adults, with 1-year event-free and overall survival (OS) rates of 50% and 76%, respectively. The incidences of CRS and ICANS were 77% and 40%, respectively. Despite the similarly excellent response rates in all adult RR B-ALL, approval of CD19 CAR T cells in this population was initially hindered by the higher incidences of relapse and treatment-related toxicities. Brexu-cel was subsequently approved for adult RR B-ALL based on the results of Phase 1/2 ZUMA-3 study, which demonstrated high CR rate and acceptable CAR T cell-associated toxicities (24% grade ≥ 3 CRS and 25% grade ≥ 3 ICANS) [9,20]. This improved safety profile may be due to improvements in supportive management and better understanding of clinical course during CAR T-cell therapy.

8.2.3 Relapsed/refractory mantle cell lymphoma

In 2020, brexu-cel became the first CAR T-cell product approved for patients with RR MCL who have received or are not eligible for autologous stem cell transplant. Brexu-cel has the same construct as axi-cel but during the manufacturing process adds the step of removing circulating tumor cells from autologous T cells [21]. In the ZUMA-2 study, 68 of 74 enrolled patients received brexu-cel, and in the intention-to-treat analysis the study reported ORR of 85%, CR rate of 59%, and progression-free survival (PFS) and OS rates of 61% and 83%, respectively, at 12 months [8]. The incidence of severe (grade $\geq$3) CRS and ICANS was 15% and 31%, respectively, with no CAR T cell-associated mortality.

8.2.4 Relapsed/refractory follicular lymphoma

Following the success of CD19 CAR T cells in aggressive B-cell non-Hodgkin lymphoma (B-NHL) and B-ALL, several studies also explored their safety and efficacy in indolent lymphoma. Recently, the ZUMA-5 study demonstrated high and durable responses (ORR 94%, CR 80%, and 12-month PFS 74%) in patients with RR FL [7], which led to the approval of axi-cel for the treatment of RR FL after two or more lines of treatment.

8.2.5 Relapsed/refractory chronic lymphocytic leukemia

Several clinical trials have explored the efficacy of CD19 CAR T cells in heavily treated RR CLL, showing objective response rates of 21%–38% [22–24]. However, the responses were short-lived and relapse rates were substantial.

8.3 Immune-mediated toxicities of CD19 chimeric antigen receptor T-cell therapy

Treatment with CD19 CAR T cells is known to result in multiorgan systemic toxicities. The most common are CRS and ICANS; other adverse effects include B-cell aplasia, cytopenia, and hypogammaglobulinemia. The risk of immune-mediated toxicities after CAR T-cell therapy is determined by several factors. Patients with B-ALL appear to develop more severe toxicity than patients with lymphoma [4–6,17]. Baseline disease burden is associated with a higher incidence of CRS and ICANS [25]. In addition, therapy-related factors such as the intensity of conditioning therapy, higher CAR T-cell dose, and CAR design (e.g., CD28 over 4–1BB) may lead to increased CAR T-cell expansion and higher toxicity. Nearly all patients treated with CD19 CAR T cells develop some level of CRS. In the registration trials, up to one-third of patients with B-ALL developed severe CRS, and up to 50% of patients developed ICANS, although the real-world rates of CRS and ICANS may be lower [26–28]. Despite some overlapping features, CRS and ICANS have distinctive pathophysiology and different clinical courses.

8.3.1 Cytokine release syndrome

CRS is the most frequent side effect of CD19 CAR T-cell therapy, involving many immune system components including cellular and cytokine milieu. Preclinical data show that interleukin 6 (IL-6) is one of the primary drivers of CRS [29], activating downstream immunologic signaling cascades, while other critical cytokines involved include tumor necrotic factor-alpha (TNF-α), IL-2, and interferon-alpha (IFN-α). In murine models, IL-1 and macrophages have also been shown to play critical roles in CRS [30].

CRS has a wide range of symptoms, including fever, arrhythmia, hypoxia, hypotension, multiorgan dysfunction, and macrophage activation syndrome. The reported incidence of CRS in CD19 CAR T cells ranges from 50% to 90%, varying by disease- and CAR T-associated factors, as well as the varied definitions and grading systems used to assess CRS [4–6,17,18,31,32]. Pennisi et al. investigated discordant CRS grading, showing that the University of Pennsylvania grading system overestimated the severity of CRS compared to other criteria [33]. Thus comparisons of CRS rates among publications should account for the grading system used. In 2019 the American Society of Transplant and Cellular Therapy (ASTCT) released a consensus statement on the definition and grading system for CRS (Table 8.1) and ICANS (Tables 8.2 and 8.3) [31].

CRS management is dictated by its extent and severity. Treatment generally focuses on providing aggressive supportive care (i.e., oxygen

Table 8.1 Diagnostic criteria and severity grading of cytokine release syndrome stratified by the American Society of Transplant and Cellular Therapy consensus criteria.

Grading system	Grade 1	Grade 2	Grade 3	Grade 4
Temperature[a]	Temperature $\geq 38°C$ without hypotension or hypoxia	Temperature $\geq 38°C$	Temperature $\geq 38°C$	Temperature $\geq 38°C$
Hypotension[a]	None	Not requiring vasopressors	Requiring a vasopressor with or without vasopressin	Requiring >1 vasopressor excluding vasopressin
Hypoxia[a]	None	Requiring low flow canula[b] or blow-by	Requiring high-flow oxygen	Requiring positive pressure ventilatory support

CRS grade is determined by the more severe event.
[a]*Fever, hypotension, or hypoxia are not attributable to any other cause.*
[b]*Low-flow nasal cannula is defined as oxygen delivered at <6 L/min.*
Reproduced from Lee DW, Santomasso BD, Locke FL, et al. ASTCT consensus grading for cytokine release syndrome and neurologic toxicity associated with immune effector cells. Biol Blood Marrow Transpl 2019;25(4):625–38. https://doi.org/10.1016/j.bbmt.2018.12.758.

Table 8.2 Diagnostic criteria and severity grading of immune effector cell-associated neurotoxicity syndrome (ICANS) by the American Society of Transplant and Cellular Therapy consensus criteria.

Neurotoxicity domain	Grade 1	Grade 2	Grade 3	Grade 4
ICE score	7–9	3–6	0–2	0 (patient is unarousable and unable to perform ICE)
Depressed level of consciousness	Awakens spontaneously	Awakens to voice	Awakens only to tactile stimulus	Patient is unarousable or requires vigorous or repetitive tactile stimuli to arouse. Stupor or coma
Seizure	N/A	N/A	Any clinical seizure focal or generalized that resolves rapidly or nonconvulsive seizures on EEG that resolve with intervention	Life-threatening prolonged seizure (>5 min); or repetitive clinical or electrical seizures without return to baseline n between
Motor findings	N/A	N/A	No	Deep focal motor weakness such as hemiparesis or paraparesis
Elevated ICP/ cerebral edema	N/A	N/A	Focal/local edema on neuroimaging	Diffuse cerebral edema on neuroimaging; decerebrate or decorticate posturing or cranial nerve VI palsy; or papilledema or Cushing's triad

ICANS grade is determined by the most severe event not attributable to any other causes. EEG, Electroencephalography; *ICE,* immune effector cell encephalopathy; *ICP,* intracranial pressure; *N/A,* not applicable.

Reproduced from Lee DW, Santomasso BD, Locke FL, et al. ASTCT consensus grading for cytokine release syndrome and neurologic toxicity associated with immune effector cells. Biol Blood Marrow Transpl 2019;25(4):625–38. https://doi.org/10.1016/j.bbmt.2018.12.758.

Table 8.3 Encephalopathy assessment tools for grading of immune effector cell-associated neurotoxicity syndrome.

ICE score
Orientation: orientation to year, month, city, hospital: 4 points
Naming: ability to name 3 objects (e.g., point to clock, pen, button): 3 points
Following commands: ability to follow simple commands (e.g., "Show me 2 fingers" or "Close your eyes and stick out your tongue"): 1 point
Writing: ability to write a standard sentence (e.g., "Our national bird is the bald eagle"): 1 point
Attention: ability to count backwards from 100 by 10: 1 point

ICE, immune effector cell encephalopathy.
Reproduced from Lee DW, Santomasso BD, Locke FL, et al. ASTCT consensus grading for cytokine release syndrome and neurologic toxicity associated with immune effector cells. Biol Blood Marrow Transpl 2019;25(4):625–38. https://doi.org/10.1016/j.bbmt.2018.12.758.

supplement, intravenous fluids, antipyretic medications), inotropic drugs, ventilation support, IL-6/IL-6R antagonists, and systemic corticosteroids [34,35].

8.3.2 Immune effector cell-associated neurotoxicity syndrome

The pathophysiology of ICANS, previously known as CAR-related encephalopathy syndrome (CRES), is complicated and not fully understood. Similar to CRS, ICANS can include wide-ranging neurological findings, such as altered mental status, confusion, aphasia, Parkinsonism, as well as focal neurological deficit, seizure, or cerebral edema in its severe form. Most criteria for ICANS diagnosis and severity assessment require comprehensive neurological examination. The ASTCT classification is the most recent and commonly used of several diagnostic criteria [31,36]. Treatment of ICANS is mainly supportive; in severe cases, patients may require airway protection, systemic corticosteroid and antiseizure therapies, and neurocritical care management [34,35,37].

8.3.3 Other chimeric antigen receptor T cell-associated toxicities

Other important CAR T cell-associated toxicities include prolonged cytopenia, macrophage activation syndrome/hemophagocytic lymphohistiocytosis (MAS/HLH), B-cell aplasia, and hypogammaglobulinemia. Prolonged cytopenia has several mechanisms and has clinical sequelae such as infection. Coagulation abnormalities have also been reported, including asymptomatic laboratory derangement to clinically significant hemorrhage or thrombosis. MAS/HLH, a rare hematologic consequence of overwhelming immune responsiveness, has been increasingly observed and can lead to life-threatening multi-organ dysfunction. The pathophysiology and mechanisms of MAS/HLH in patients who received CAR T-cell therapy overlap with those of other secondary HLH but are not well defined, warranting further studies to enable prediction and early intervention [36,38].

8.4 Clinical application of biomarkers in CD19 chimeric antigen receptor T-cell therapy

Despite excellent antitumor effects of CAR T cells, with responses seen across many patient subsets and clinical presentations, a better understanding of why many patients fail to respond or eventually relapse after CAR T-cell therapy is needed. Biomarkers are useful for predicting toxicity (both short term and long term), response, and relapse; for risk assessment (safety); and for identifying therapeutic targets (drug development) (Fig. 8.2) [39,40].

This section focuses on biologic or radiologic biomarkers as predictors of toxicity (particularly CRS and ICANS) and outcomes after CAR T-cell therapy. Laboratory biomarker measures vary from basic assays to complex high-dimensional cellular genomic sequencing requiring bioinformatic analyses [40]. CAR T-cell biomarkers have complex inputs including characteristics of the underlying malignancy, patient characteristics, prior treatment exposure, the construct of CAR T cells, among other factors, which can confound their predictive power (Fig. 8.3). Combinatorial algorithms including cytokine levels and clinical parameters may be useful in predicting toxicity associated with CAR T-cell therapy.

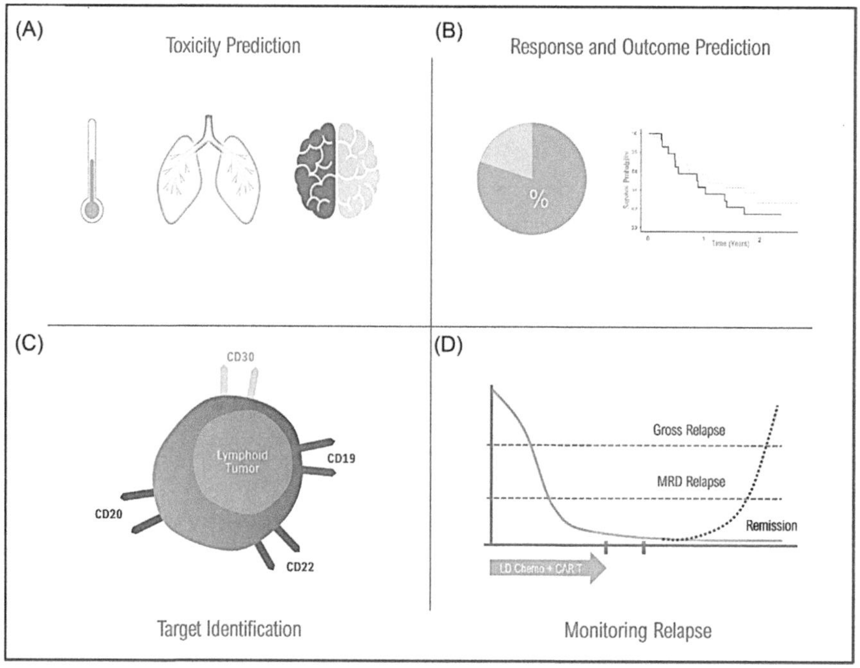

FIGURE 8.2

The potential implication of biomarkers in chimeric antigen receptor T-cell therapy. (A) Prediction of CAR T cell–associated toxicities that can identify patients who are at risk of developing treatment-related complications to allocate proper prophylactic or preemptive strategies. (B) Prediction of response and outcomes after CAR T-cell therapy that could distinguish favorable-risk patients from unfavorable-risk patients. (C) Target identification of novel tumor-associated/specific antigens for CAR T cell. (D) Response monitoring for residual diseases and surveillance for relapse after CAR T-cell therapy.

8.5 Predictive biomarkers of severe chimeric antigen receptor and immune effector cell-associated neurotoxicity syndrome

8.5.1 Chimeric antigen receptor T-cell dose

Higher cell dose and conditioning chemotherapy containing fludarabine have been associated with severity of CRS and ICANS

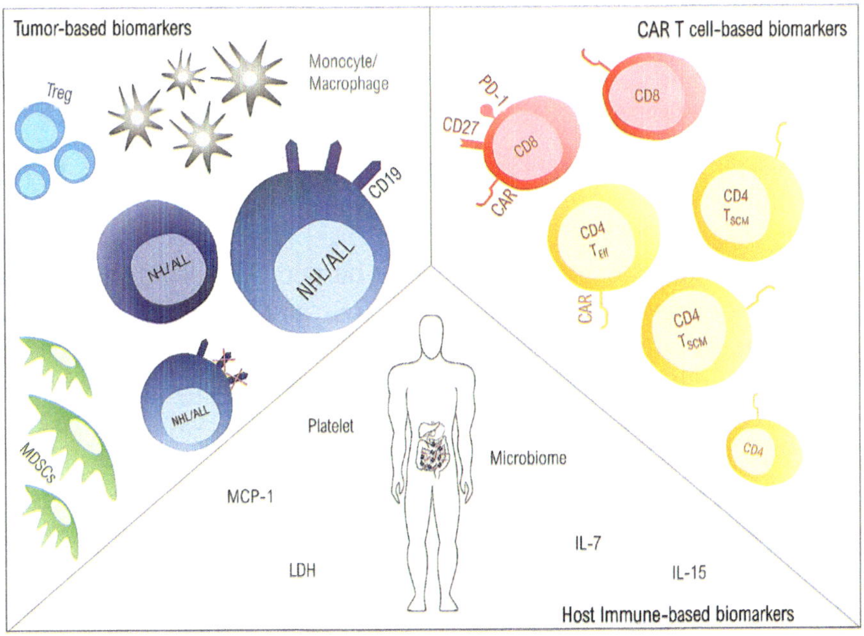

FIGURE 8.3

Potential biomarkers for predicting toxicities, efficacy, and outcomes after chimeric antigen receptor T-cell therapy. *ALL*, acute lymphoblastic leukemia; *CAR*, chimeric antigen receptor; *IL*, interleukin; *LDH*, lactate dehydrogenase; *MCP-1*, monocyte chemoattractant protein-1; *MDSCs*, myeloid-dervied suppressor cells.

[41–46]. However, the effect of CAR T-cell dose on toxicities of CAR T-cell therapy is influenced by several other factors, such as tumor burden and the component and phenotype of CAR T cells. Therefore the association between infused CAR T-cell dose and severity of toxicity highlights the need for dose-escalation design when testing a new CAR T-cell product.

8.5.2 Tumor burden

Early in vivo expansion of CAR T-cells is driven by CAR binding to the tumor and normal cells. Tumor burden is associated with the peak expansion of CAR T cells in vivo and is among the most robust pretreatment predictors of CRS [17,29,41,47–49]. Among patients with B-ALL, >5% blasts in the bone marrow are associated with higher CRS risk [48]; in B-NHL, patients with higher tumor burden are more likely to experience severe CRS [29].

Disease burden is also associated with ICANS [19,43,49]. Leukemic blasts >50% were associated with severe neurotoxicity in the trial of JCAR015 for B-ALL [43]. Lactate dehydrogenase (LDH), a surrogate marker of tumor burden and proliferation, has also been implicated in severe CRS and ICANS [30,42,50]. Karschnia et al. reported elevated (≥400 U/L) levels of LDH at baseline in 69% of patients with severe neurotoxicity [50]. Overall, tumor burden should always be considered when evaluating risk for severe toxicities.

8.5.3 Inflammatory cytokines

Several cytokines have been associated with severe CRS and/or ICANS, including effector cytokines such as IFN-γ, IL-2, IL-6, granulocyte-macrophage colony-stimulating factor (GM-CSF), those secreted by monocytes and/or macrophages: IL-1, IL-6, IL-8, IL-10, IL-12, TNF-α, IFN-γ, monocyte chemotactic protein (MCP)-1, and macrophage inflammatory protein (MIP) [29,44–46,51]. Deng et al. confirmed the role of monocytes and macrophages in severe ICANS, using a single-cell transcriptome analysis to identify cells with monocyte-like transcriptional signature [52].

The quantitation of cytokines in the clinical setting may be problematic for some institutions, as there are currently no FDA-approved assays available for the analysis of cytokines. There are, however, many immunoassays available on the market for research use. All assays currently available are immmunoassays, ranging from enzyme linked immunosorbent assays (ELISA) kits to multiplex bead-based Luminex assays. These immunoassays have varying degrees of precision. Many of the multiplex assays are geared more toward discovery applications rather than for clinical use due to their high imprecision [53].

Another aspect to keep in mind is that as of writing, there is no standardization between currently available assays for cytokine quantitation such as is available for other biomarkers for clinical use, for example, glucose or creatinine. This makes compliance with clinical regulatory requirements, including meeting proficiency requirements, difficult. In addition, reference ranges, as well as other clinical decision limits, must be determined for each analyte, by the performing laboratory. An additional option for institutions wishing to implement cytokine testing for clinical use is testing by reference laboratories. Testing for many of the cytokines mentioned in this chapter is available from reference laboratories such as Quest Diagnostics, ARUP Laboratories, Mayo Clinic Labs, and Labcorp. Physicians wishing to set up cytokine quantitation for clinical use are strongly encouraged to work with the medical directors of the clinical laboratories at their respective institutions to select the appropriate testing option that meets their needs. However, currently there is no uniform level of these inflammatory biomarkers to define patients at the highest risk of developing severe toxicity.

IL-6 has been identified as a crucial mediator of CRS, and tocilizumab, a monoclonal antibody that blocks signaling through the IL-6 receptor, is often used in CRS management. Several studies have associated elevated pretreatment IL-6 levels with the risk of CRS and ICANS development [4,5,29,42,47,51]. Davila et al. identified elevated IL-6 levels (≥40 pg/mL) to be associated with severe toxicity and death in patients treated with axi-cel therapy [47].

Together with IL-6, levels of IL-1β and its natural antagonist IL-1 receptor agonist (IL-1Ra), as well as GM-CSF, tend to peak earlier than other proinflammatory cytokines, suggesting that these cytokines might have a role in initiating the inflammatory cascade [54]. Norelli et al. prevented both CRS and neurotoxicity in humanized mice by administering the IL-1R antagonist anakinra [55]. Strati et al. published their experience with anakinra as a potential steroid-sparing strategy for treating CAR T cell-associated toxicities, mainly ICANS [56]. Although overall outcomes were poor when anakinra was used as a last resource, ongoing trials (ClinicalTrials. gov NCT04148430, NCT04205838, NCT03430011, NCT04432506, and NCT04359784) using anakinra prophylactically or as an early treatment for CRS and ICANS are actively enrolling patients. The results of these studies should provide us with better insights on treatment of immune-mediated adverse events using a prophylactic or preemptive approach.

Levels of serum cytokines including IL-6, IL-10, IFN-γ, TNF-α, and angiopoietin-2 (ANG-2) continuously increase in ICANS. While the mechanisms underlying the development of ICANS remain unclear, these high concentrations of inflammatory cytokines appear to directly affect vascular endothelial cells, leading to increased epithelial blood–brain barrier permeability and dysfunction. Several studies have suggested that patients with severe ICANS exhibit a higher concentration of cytokines in their cerebrospinal fluid but did not reveal specific cytokine patterns for neurotoxicity [43,46]. In a study of axi-cel, ANG-2 levels and the ANG-2:ANG-1 ratio were higher even before lymphodepleting chemotherapy in patients who subsequently developed grade ≥ 3 ICANS [57]. The standard treatment for CRS, tocilizumab, does not appear to improve ICANS, for which the current standard of care is symptomatic management and corticosteroids. In fact, tocilizumab, a receptor antagonist, may result in higher levels of IL-6 in the CSF and may lead to increased ICANS when administered in early stages.

Despite substantial differences in the signaling domain structure of CAR T cells, the dynamics of cytokines in each CAR T cell product among trials share an overlapping pattern reflecting a common fate and similar kinetics of CAR T cells after infusion.

8.5.4 Serum inflammatory markers including C-reactive protein and ferritin

General inflammatory biomarkers including C-reactive protein (CRP) and ferritin have been associated with the development of CRS and ICANS [29,43,46,47,50,57]. However, their predictive power is limited and thresholds for significant elevation in these serum inflammatory markers remain unclear.

Teachey et al. demonstrated that patients with grade 4 or 5 CRS had higher peak levels of CRP and ferritin versus those with grade ≤ 3 CRS, but CRP and ferritin levels did not improve CRS prediction in the first 3 days post-CAR T-cell infusion [29]. Davila et al. proposed that CRP of ≥ 20 mg/dL can be used as a simple indicator of severe CRS when the availability of other cytokines is limited [47]. Hu et al. also reported that CRP, serum ferritin, and D-dimer levels are associated with severe CRS, with reduced levels of each indicating a favorable response to tocilizumab or corticosteroids [57].

One study found higher CRP levels in patients who experienced high-grade ICANS [43], whereas in two different studies CRP levels peaked in all patients before the development of neurologic symptoms and did not discriminate between patients with mild compared with severe symptoms [46,50]. Karschnia et al. reported that peak concentrations for ferritin 2 days after onset of neurotoxicity were higher in patients with high-grade versus low-grade neurotoxicity [50].

8.5.5 Preexisting endothelial activation

Patients with preexisting endothelial activation before infusion of CAR T cells are at higher risk of developing severe CRS and/or neurotoxicity [19,42,43,49]. Severe thrombocytopenia before lymphodepletion chemotherapy is associated with a higher risk of developing CRS [42], while low platelet count ($\leq$50,000/μL) is correlated with severe neurotoxicity [43]. Additionally, platelet count at the time of CAR T-cell infusion has been shown to be significantly lower in patients who then developed severe neurotoxicity versus those with mild symptoms. Thus patients with severe thrombocytopenia might be prone to endothelial activation because platelets are one of the few sources of the endothelial stabilizing cytokine, ANG-1 [49].

Some authors suggest that examining endothelial activation biomarkers such as von Willebrand factor and ANG-2 before and after CAR T-cell infusion be recommended for clinical management during CAR T-cell therapy.

8.6 Predictive biomarkers for hematologic toxicities

There is limited understanding of the frequency or severity of cytopenia after CAR T-cell therapy, as well as of hematopoietic recovery and its underlying mechanisms. The development of early and late cytopenias is correlated with CRS and ICANS severity and lymphodepleting chemotherapy regimen [58]. However, there is no single definitive biomarker or patient characteristic that can precisely predict these complications prior to CAR T-cell infusion.

A clinical trial of CD19 CAR T cells in $CD19^+$ B-cell malignancies showed that bone marrow tumor burden, the number of prior therapies, and CRS occurrence were associated with longer hematologic recovery, and demonstrated that absolute neutrophil count, hematocrit, and platelet nadirs were lower in patients with more severe CRS [42]. The time to hematologic recovery was longer than expected in most patients with grade 4 CRS and delayed in patients with grades 1–3 CRS versus patients without CRS. Jain et al. validated these findings, showing in a study with different B malignancies and CAR T products that grade ≥ 3 CRS or ICANS was associated with the absence of complete count recovery at 1 month [59]. Fried et al. found late hematologic toxicity was more common in patients with high-grade CRS and patients treated after a recent stem cell transplantation [58].

Rejeski et al. developed and validated the CAR-HEMATOTOX model using a multicenter database to predict the likelihood of prolonged/profound hematologic toxicity [60]. The score incorporates baseline hemoglobin, absolute neutrophil count, platelet count, CRP, and ferritin levels before lymphodepletion chemotherapy. Patients with high scores (>2) had a higher risk of developing prolonged and severe cytopenia, especially neutropenia.

Further research is needed to characterize the course and pathophysiology of cytopenias from population-based registries and long-term follow-up of clinical trials.

8.7 Predictive biomarkers for hemophagocytic lymphohistiocytosis and macrophage activation syndrome

HLH is a rare but severe complication of CAR T-cell therapy with a reported incidence of 1%–4% [29,36,61–63]. The incidence is likely underreported owing to diagnostic uncertainty, and the mechanism is not well understood. Post-CAR T-cell secondary HLH may represent the severe form of CRS, which has a similar cytokine pattern to hemophagocytic syndrome. Teachey et al. described a similar pattern of elevated cytokines in patients with HLH compared with severe CRS: IFNγ, IL-10, IL-6, IL-8, IP-10, MCP-1, MIP-1β, and IL-2Rα, which are expected to be elevated in HLH, were also elevated in patients with severe CRS (grade >3) compared with grades 0–3 CRS [29]. Recently, Lichtenstein et al. described HLH-like syndrome manifested as a CRS variant in patients with RR B lymphoid malignancy receiving CD22 CAR T cells. The authors observed preinfusion natural killer (NK)-cell lymphopenia, higher bone marrow T/NK-cell ratio, and postinfusion CD8 T-cell expansion in patients who developed HLH [64].

Severe CRS and HLH/MAS likely exist on a spectrum of systemic hyperinflammatory disorders, yet classical criteria for HLH are unhelpful in the CAR T-cell context because patients receiving CAR T cells often exhibit all the diagnostic criteria for HLH, including high fever, multiorgan dysfunction, CNS disturbances, high serum levels of ferritin, lactate dehydrogenase, soluble CD25, cytokines (such as IFNγ and IL-6), and hypofibrinogenemia. Neelapu et al. have proposed a diagnostic criterion for CAR T-associated HLH/MAS, defined as peak ferritin $>$10,000 mg/mL with at least two of the following additional toxicities: elevated hepatic transaminases or bilirubin grade ≥ 3, creatinine grade ≥ 3, pulmonary edema grade ≥ 3, and/or evidence of HLH on bone marrow biopsy [36].

8.8 Predictive models for chimeric antigen receptor and immune effector cell-associated neurotoxicity syndrome

Optimal predictive biomarkers for CRS and ICANS have not been established. The majority of biomarkers described above are suboptimal because evaluation is not performed early enough to predict severe CRS, which can occur 24–72 hours after CAR T-cell infusion. In addition, these nonspecific biomarkers lack sensitivity and specificity to predict CRS or ICANS. However, several models with combination biomarkers have been developed, providing useful clinical tools in predicting severe toxicity.

Teachey et al. reported a model that accurately predicts severe CRS in patients with RR B-ALL based on serial levels of various inflammatory cytokines after CAR T-cell infusion. In pediatric patients, the predictive model for severe CRS, including IFN-γ, IL-13, and MIP-1α concentrations after CAR T-cell infusion, had exceptionally high accuracy, with sensitivity of 100% and specificity of 96% [29]. Some baseline patient characteristics were also proposed as predictive biomarkers for severe neurotoxicity, including baseline blood platelet level $<60{,}000/\mu L$ or mean corpuscular hemoglobin concentration $>33.2\%$ and morphologic disease ($>5\%$ blasts), which had 95% sensitivity and 70% specificity.

Pennisi et al. proposed two new versions of the EASIX (Endothelial Activation and Stress Index) score [65]: modified EASIX (m-EASIX), in which CRP (mg/dL) replaces creatinine as a readily available surrogate for IL-6; and simplified EASIX (s-EASIX), which excludes creatinine. EASIX, m-EASIX, and s-EASIX scores were associated with severe CRS prior to CAR T-cell infusion and early postinfusion, preceding the onset of severe symptoms. The three formulas also had a moderate discriminatory ability for severe ICANS [66]. Similarly, Greenbaum et al. proposed EASIX combined with ferritin and CRP. These scores, EASIX-F (for CRS) and EASIX-FC (for ICANS) measured before lymphodepletion, were associated with the incidence and severity of CRS and ICANS in patients with LBCL treated with axi-cel [67].

Hay et al. investigated classification-tree modeling to identify patients at risk of developing severe CRS and ICANS that includes testing serum MCP-1 in patients with fever $\geq 38.9°C$ within 36 hours of infusion [42]. By combining IL-6 levels ≥ 16 pg/mL in the first 36 hours with MCP-1 levels and fever, this revised algorithm was highly predictive of grade ≥ 4 neurotoxicity, with sensitivity of 100% and specificity of 94% [46].

Rubin et al. reported a model to predict the occurrence of ICANS in patients with LBCL treated with axi-cel, weighting various parameters: age at CAR T-cell therapy, histology subtypes, maximal temperature, maximum CRP, maximum ferritin levels, minimum white blood cell count, CRS maximum severity, CRS onset day from infusion, and number of tocilizumab doses [68]. Patients with score >6 of a total 14 possible points had higher chance of developing ICANS (accuracy 77%, sensitivity 82%, specificity 70%). It is not yet clear if these predictive models will be generalizable as confirmation and validation are required in further studies. Therefore modeling using a combination of clinical parameters may be more widely used in practice.

8.9 Biomarkers of chimeric antigen receptor T-cell response, outcome, and treatment failure

The ability to identify or predict response or failure after CAR T-cell therapy would enable better selection of patients and consideration of consolidative interventions post-CAR T-cell therapy in high-risk patients. These studies are challenging because the mechanisms of relapse after CAR T-cell therapy are multifactorial, encompassing CAR T-cell kinetic failure, underlying host immune dysfunction, immune–tumor interactions, and inherited/acquired intrinsic tumor biology. Biomarkers of efficacy include short-term endpoints such as response rate and long-term outcomes such as survival which are often overlapping.

8.9.1 Chimeric antigen receptor T cell and host-immune-based biomarkers

8.9.1.1 Inflammatory cytokines

After CAR T-cell infusion, the engagement between the engineered lymphocytes and tumor cells results in CAR T-cell activation. The resulting robust immunologic response, including supraphysiologic surge of many inflammatory cytokines and immune cell activation, is primarily responsible for tumor cell killing. Cytokine levels during the acute phase of CAR T-cell therapy can predict response and survival outcomes.

Elevated serum IL-15 is associated with objective response in patients with NHL treated with CAR T cells [69], and elevated baseline IL-15 prior to lymphodepleting chemotherapy is associated with CAR T-cell persistence and response across all B-cell lymphoid malignancy [70]. Higher levels of MCP-1 before lymphodepletion chemotherapy are associated with improved response to CD19 CAR T-cell therapy in aggressive NHL; higher serum MCP-1 on the day of CAR T-cell infusion is also associated with superior PFS, as is a higher IL-7 serum peak [71]. Data from a study of tisa-cel for patients with RR CLL suggested that superior inflammatory cytokine production (CCL20, IL-21, IL-22, IL-17, and IL-6) and release of CAR T cells are associated with CR [72]. Several nonspecific inflammatory parameters and acute-phase reactant proteins—such as CRP, platelet count, and serum ferritin—have been reported as potential biomarkers to predict response in patients with different hematology malignancies treated with CD19 CAR T cells [71,73,74].

8.9.1.2 T-cell phenotypes and functional attributes

T-cell functional attributes and durability influence the efficacy of CAR T-cell therapy. The repertoires, phenotypes, and proportions of helper/cytotoxic T cells can affect the replication potential and durability of CAR T cells. The memory-like and naive phenotypes of T cells and $CD4^+$ T-helper cells preserve higher proliferative ability and are less susceptible to activation-induced exhaustion; therefore, the premanufactured apheresis or final CAR T product which contains higher proportion of these T cell subset was shown to be associated with better response and outcomes after CAR T cells [52,72,75–78]. In addition, T-cell exhaustion and activation-induced T-cell apoptosis have been well described after chronic T-cell stimulation. The expression of T-cell exhaustion markers on the surface of CAR T cells can be a predictor of response failure and shorter remission after CAR T-cell therapy. Besides T-cell phenotypes, the composition of CAR T cells may also determine the efficacy of CAR T-cells. The positive impact of $CD4^+/CD8^+$ compartment on CAR T-cell efficacy has been documented in preclinical models, and investigation of CAR T-cell products with predefined $CD4^+/CD8^+$ ratio in various hematologic malignancies is ongoing [14,45,79–81].

Owing to the complexity of T-cell phenotypic and functional subsets, the polyfunctional strength index (PSI) of CAR T product using a high-content single-cell multiplex cytokine analysis has been explored to determine CAR T-cell response in RR NHL [82]. Higher PSI of preinfusion CAR T-cell product, especially in the $CD4^+$ subset, is associated with clinical response [83].

8.9.1.3 T-cell kinetics

Data from several groups indicate that in vivo CAR T-cell expansion and persistence are measured by either maximum peak CAR T-cell level (C_{max}) or the area under the curve during the first 28 days (AUC_{0-28d}). There have been some evidence indicating that higher C_{max} and AUC_{0-28d} were associated with response rate and remission duration after CAR T-cell therapy; however, the available data are still conflicting among studies attributed to the intricate interaction between CAR T-construct and primary malignancy [4,5,17]. Furthermore, B-cell recovery after CAR T-cell therapy is correlated with and can be used as a simple indicator to indirectly monitor CAR T-cell persistence, which may predict subsequent relapse of malignancy [23].

8.9.1.4 Host intestinal microbiome

The human intestinal microbiome has been correlated with response and survival after various immune-based therapies [84]. Smith et al. reported that enrichment of intestinal *Oscillospiraceae*, *Ruminococcaceae*, and *Lachnospiraceae* species was associated with complete remission of DLBCL and B-ALL after patients received CD19 CAR T-cells [85]. However, these findings require further investigation and validation in larger studies.

8.9.2 Tumor-based biomarkers

8.9.2.1 Tumor and tumor microenvironment immunophenotypes

The tumor microenvironment (TME) and its interaction with the host immune system are highly complex. Aberrant immune checkpoint protein expression on the tumor cell surface and immunosuppressive TME influence response to CAR T-cell therapy in solid tumors and multiple myeloma [86,87], but this effect is less well defined in other lymphoid hematologic malignancies. Generally, the expression of immune inhibitory markers in bulky disease and the presence of immune suppressive cells such as myeloid-derived suppressor and dendritic cells can compromise the efficacy of CAR T cells and negate their cytotoxic function, respectively. The JULIET study did not support the association between PD-L1 expression in pretreatment tumor specimens and outcome after tisa-cel therapy [5]. In contrast, the TRANSCEND-NHL001 study of liso-cel in patients with NHL demonstrated that the abundance of lymphoma-infiltrating lymphocytes expressing PD-1 and lower number of intratumoral $CD163^+$ monocytes correlated with longer duration of remission after CAR T-cell therapy [88]. Thus the role of TME as a predictive biomarker for post-CAR-cell T response and outcome requires further exploration.

8.9.2.2 Mutational attributes

Although most data indicate no difference in survival among patients harboring various genetic or molecular aberrations, these derangements have been implicated in response in some preclinical data. Mancikova et al. demonstrated that TP53 aberration of CLL cells could compromise CAR T-cell engraftment in a xenograft model [89], although survival outcomes were not affected. Shouval et al. also showed the importance of TP53 alteration on the outcomes of DLBCL patients treated with CD19 CAR T cells [90].

8.9.2.3 Disease status/burden and kinetics of response

In B-ALL, different parameters of pretreatment disease burden, including baseline peripheral white blood cell count ($\geq$30,000/μL), bone marrow blast percentage ($<$5%), and minimal residual disease negativity, have been correlated with response rate and survival outcomes after CD19 CAR T-cell therapy [18,91]. Baseline LDH, explored as a predictor of outcomes, correlated with post-CAR T-cell survival for both ALL and LBCL [5,71,73].

Pre- and on-treatment tumor burden as measured by radiographic parameters such as total metabolic tumor volume (TMTV) and maximal tumor diameter, as well as sequencing, have been studied for associations with therapy response or failure. In a small series, Shah et al. reported a durable >2-year remission in three of seven patients treated with CD19 CAR T cells. Patients had no residual TMTV on PET-CT at 1 month after CAR T-cell therapy [92]. Dean et al. demonstrated baseline preinfusion TMTV below a median of 147.5 mL as a predictive marker for improved outcomes after axi-cel in RR DLBCL, including superior PFS and OS [93]. Recently, Vercellino et al. showed that high TMTV (>80 mL) before CAR T-cell therapy was associated with higher risk of relapse within 1 month posttreatment. Additionally, Sesques et al. suggested that dynamic changes of TMTV and total lesion glycolysis might have superior correlation with survival outcomes post-CAR T-cell therapy [94].

Maximal tumor diameter predicts response in DLBCL patients treated with CD19 CAR T cells. The ZUMA-1 study and real-world outcome reports revealed that patients with bulky disease at the higher quartile (determined by the sum of perpendicular diameters of target lesions) were associated with shorter duration of remission [26,95,96]. Circulating tumor DNA sequencing after CAR T-cell infusion likewise correlated with clinical response; in particular, change by day 7 was associated with remission [52].

Several groups have developed predictive tools incorporating different parameters and statistical models to assess response and failure after CAR T-cell therapy. Vercellino et al. proposed a model including extranodal involvement (≥2 sites), baseline CRP, and TMTV at the time of treatment to identify patients with DLBCL at higher risk of progression following CD19 CAR T-cell therapy [74]. Conventional prognostic score (International Prognostic Index) has been correlated with inferior outcomes in patients with DLBCL after CD19 CAR T-cell therapy in some reports [97], but these findings require further validation in larger cohorts.

8.10 Summary points and future directions

- CAR T-cell therapy represents a novel individualized therapy which has been integrated into the current cancer treatment paradigm.

- Learning how to customize the delivery of CAR T-cell therapy to patients according to their profiles is extremely important to maximize treatment efficacy and improve patients' experience.
- Due to the complex interactions among underlying malignancy, CAR T cells, and host immunity, a combination of biomarkers and risk-assessment tools are required to predict toxicity and outcomes.
- Advances in cellular biology and next-generation sequencing, including omics analyses, and more recently novel techniques such as single-cell analysis, are the mainstay approaches leading to the discovery of predictive biomarkers that can be used to risk-stratify patients or newer tumor-associated molecules that could serve as targets for immune-based therapy.
- Multidisciplinary collaboration, comprehensive laboratory infrastructure, and simplified biomarker assays are required to enable the integration of biomarkers into routine oncology practices.
- Biomarkers will be a critical tool to accommodate the risk-stratified and patient-centered approach to deliver CAR T-cell therapy to appropriate patients.

Disclosures

K.W. and A.A.T. have no conflict of interest to disclose. K.M. receives research funding from Abbott Laboratories. M.A.P. reports honoraria from AbbVie, Astellas, Bristol Myers Squibb, Celgene, Equilium, Incyte, Karyopharm, Kite/Gilead, Merck, Miltenyi Biotec, MorphoSys, Novartis, Nektar Therapeutics, Omeros, OrcaBio, Takeda, and VectivBio AG, Vor Biopharma. He serves on Data Safety Monitoring Boards for Cidara Therapeutics, Medigene, Sellas Life Sciences, and Servier, and the scientific advisory board of NexImmune. He has ownership interests in NexImmune and Omeros. He has received research support for clinical trials from Incyte, Kite/Gilead, Miltenyi Biotec, and Novartis. He serves in a volunteer capacity as a member of the Board of Directors of the American Society for Transplantation and Cellular Therapy (ASTCT) and Be The Match (National Marrow Donor Program, NMDP), as well as on the Center for International Blood and Marrow Transplant Registry (CIBMTR) Cellular Immunotherapy Data Resource (CIDR) Executive Committee.

Acknowledgment

Hannah Rice, BA, ELS provided editorial support.

References

[1] June CH, O'Connor RS, Kawalekar OU, Ghassemi S, Milone MC. CAR T cell immunotherapy for human cancer. Science 2018;359 (6382):1361–5. Available from: https://doi.org/10.1126/science.aar6711.

[2] Mullard A. FDA approves first BCMA-targeted CAR-T cell therapy. Nat Rev Drug Discov 2021;20(5):332. Available from: https://doi.org/10.1038/d41573-021-00063-1.

[3] Munshi NC, Anderson Jr. LD, Shah N, et al. Idecabtagene vicleucel in relapsed and refractory multiple myeloma. N Engl J Med 2021;384 (8):705–16. Available from: https://doi.org/10.1056/NEJMoa2024850.

[4] Neelapu SS, Locke FL, Bartlett NL, et al. Axicabtagene ciloleucel CAR T-cell therapy in refractory large B-cell lymphoma. N Engl J Med 2017;377(26):2531–44. Available from: https://doi.org/10.1056/NEJMoa1707447.

[5] Schuster SJ, Bishop MR, Tam CS, et al. Tisagenlecleucel in adult relapsed or refractory diffuse large B-cell lymphoma. N Engl J Med 2019;380 (1):45–56. Available from: https://doi.org/10.1056/NEJMoa1804980.

[6] Abramson JS, Palomba ML, Gordon LI, et al. Lisocabtagene maraleucel for patients with relapsed or refractory large B-cell lymphomas (TRANSCEND NHL 001): a multicentre seamless design study. Lancet. 2020;396(10254):839–52. Available from: https://doi.org/10.1016/S0140-6736(20)31366-0.

[7] Jacobson C, Chavez JC, Sehgal AR, et al. Primary analysis of Zuma-5: a phase 2 study of axicabtagene ciloleucel (axi-cel) in patients with relapsed/refractory (R/R) indolent non-Hodgkin lymphoma (iNHL). Blood. 2020;136(Supplement 1):40–1. Available from: https://doi.org/10.1182/blood-2020-136834.

[8] Wang M, Munoz J, Goy A, et al. KTE-X19 CAR T-cell therapy in relapsed or refractory mantle-cell lymphoma. N Engl J Med 2020;382

(14):1331–42. Available from: https://doi.org/10.1056/NEJMoa1914347.
[9] Shah BD, Ghobadi A, Oluwole OO, et al. KTE-X19 for relapsed or refractory adult B-cell acute lymphoblastic leukaemia: phase 2 results of the single-arm, open-label, multicentre ZUMA-3 study. Lancet. 2021;398(10299):491–502. Available from: https://doi.org/10.1016/S0140-6736(21)01222-8.
[10] Brudno JN, Kochenderfer JN. Toxicities of chimeric antigen receptor T cells: recognition and management. Blood 2016;127(26):3321–30. Available from: https://doi.org/10.1182/blood-2016-04-703751.
[11] Wang X, Riviere I. Clinical manufacturing of CAR T cells: foundation of a promising therapy. Mol Ther Oncolytics 2016;3:16015. Available from: https://doi.org/10.1038/mto.2016.15.
[12] Locke FL, Ghobadi A, Jacobson CA, et al. Long-term safety and activity of axicabtagene ciloleucel in refractory large B-cell lymphoma (ZUMA-1): a single-arm, multicentre, phase 1–2 trial. Lancet Oncol 2019;20(1):31–42. Available from: https://doi.org/10.1016/S1470-2045(18)30864-7.
[13] Chong EA, Ruella M, Schuster SJ, Lymphoma Program Investigators at the University of Pennsylvania. Five-year outcomes for refractory B-cell lymphomas with CAR T-cell therapy. N Engl J Med 2021;384(7):673–4. Available from: https://doi.org/10.1056/NEJMc2030164.
[14] Abramson JS, Palomba ML, Gordon LI, et al. Pivotal safety and efficacy results from transcend NHL 001, a multicenter phase 1 study of lisocabtagene maraleucel (liso-cel) in relapsed/refractory (R/R) large B cell lymphomas. Blood 2019;134(Supplement_1):241. Available from: https://doi.org/10.1182/blood-2019-127508.
[15] Locke FL, Miklos DB, Jacobson C, et al. Primary analysis of ZUMA-7: a phase 3 randomized trial of axicabtagene ciloleucel (Axi-Cel) vs standard-of-care therapy in patients with relapsed/refractory large B-cell lymphoma. Blood 2021;138(Supplement 1):2. Available from: https://doi.org/10.1182/blood-2021-148039.
[16] Kamdar M, Solomon SR, Arnason JE, et al. Lisocabtagene maraleucel (liso-cel), a CD19-directed chimeric antigen receptor (CAR) T cell therapy, vs standard of care (SOC) with salvage chemotherapy (CT) followed by autologous stem cell transplantation (ASCT) as second-line (2L) treatment in patients (Pts) with relapsed or refractory (R/R) large B-cell lymphoma (LBCL): results from the

randomized phase 3 transform study. Blood 2021;138(Supplement 1):91. Available from: https://doi.org/10.1182/blood-2021-147913.

[17] Maude SL, Laetsch TW, Buechner J, et al. Tisagenlecleucel in children and young adults with B-cell lymphoblastic leukemia. N Engl J Med 2018;378(5):439–48. Available from: https://doi.org/10.1056/NEJMoa1709866.

[18] Park JH, Riviere I, Gonen M, et al. Long-term follow-up of CD19 CAR therapy in acute lymphoblastic leukemia. N Engl J Med 2018;378 (5):449–59. Available from: https://doi.org/10.1056/NEJMoa1709919.

[19] Lee DW, Kochenderfer JN, Stetler-Stevenson M, et al. T cells expressing CD19 chimeric antigen receptors for acute lymphoblastic leukaemia in children and young adults: a phase 1 dose-escalation trial. Lancet. 2015;385(9967):517–28. Available from: https://doi.org/10.1016/S0140-6736(14)61403-3.

[20] Shah BD, Bishop MR, Oluwole OO, et al. KTE-X19 anti-CD19 CAR T-cell therapy in adult relapsed/refractory acute lymphoblastic leukemia: ZUMA-3 phase 1 results. Blood 2021;138(1):11–22. Available from: https://doi.org/10.1182/blood.2020009098.

[21] Romero D. KTE-X19 active in MCL. Nat Rev Clin Oncol 2020;17 (6):336. Available from: https://doi.org/10.1038/s41571-020-0373-3.

[22] Turtle CJ, Hay KA, Hanafi LA, et al. Durable molecular remissions in chronic lymphocytic leukemia treated with CD19-specific chimeric antigen receptor-modified T cells after failure of ibrutinib. J Clin Oncol 2017;35(26):3010–20. Available from: https://doi.org/10.1200/JCO.2017.72.8519.

[23] Porter DL, Hwang WT, Frey NV, et al. Chimeric antigen receptor T cells persist and induce sustained remissions in relapsed refractory chronic lymphocytic leukemia. Sci Transl Med 2015;7(303). Available from: https://doi.org/10.1126/scitranslmed.aac5415 303ra139.

[24] Geyer MB, Riviere I, Senechal B, et al. Safety and tolerability of conditioning chemotherapy followed by CD19-targeted CAR T cells for relapsed/refractory CLL. JCI Insight 2019;5. Available from: https://doi.org/10.1172/jci.insight.122627.

[25] Locke FL, Rossi JM, Neelapu SS, et al. Tumor burden, inflammation, and product attributes determine outcomes of axicabtagene ciloleucel in large B-cell lymphoma. Blood Adv 2020;4(19):4898–911. Available from: https://doi.org/10.1182/bloodadvances.2020002394.

[26] Nastoupil LJ, Jain MD, Feng L, et al. Standard-of-care axicabtagene ciloleucel for relapsed or refractory large B-cell lymphoma: results

from the US lymphoma CAR T consortium. J Clin Oncol 2020;38 (27):3119–28. Available from: https://doi.org/10.1200/JCO.19.02104.
[27] Pasquini MC, Hu ZH, Curran K, et al. Real-world evidence of tisagenlecleucel for pediatric acute lymphoblastic leukemia and non-Hodgkin lymphoma. Blood Adv 2020;4(21):5414–24. Available from: https://doi.org/10.1182/bloodadvances.2020003092.
[28] Iacoboni G, Iraola-Truchuelo J, Martinez-Cibrian N, et al. Real-world evidence of tisagenlecleucel for the treatment of relapsed or refractory large B-cell lymphoma. Blood 2020;136(Supplement 1):19–21. Available from: https://doi.org/10.1182/blood-2020-136201.
[29] Teachey DT, Lacey SF, Shaw PA, et al. Identification of predictive biomarkers for cytokine release syndrome after chimeric antigen receptor T-cell therapy for acute lymphoblastic leukemia. Cancer Discov 2016;6(6):664–79. Available from: https://doi.org/10.1158/2159-8290.CD-16-0040.
[30] Giavridis T, van der Stegen SJC, Eyquem J, Hamieh M, Piersigilli A, Sadelain MCAR. T cell-induced cytokine release syndrome is mediated by macrophages and abated by IL-1 blockade. Nat Med 2018;24 (6):731–8. Available from: https://doi.org/10.1038/s41591-018-0041-7.
[31] Lee DW, Santomasso BD, Locke FL, et al. ASTCT consensus grading for cytokine release syndrome and neurologic toxicity associated with immune effector cells. Biol Blood Marrow Transpl 2019;25(4):625–38. Available from: https://doi.org/10.1016/j.bbmt.2018.12.758.
[32] Porter DL, Levine BL, Kalos M, Bagg A, June CH. Chimeric antigen receptor-modified T cells in chronic lymphoid leukemia. N Engl J Med 2011;365(8):725–33. Available from: https://doi.org/10.1056/NEJMoa1103849.
[33] Pennisi M, Jain T, Santomasso BD, et al. Comparing CAR T-cell toxicity grading systems: application of the ASTCT grading system and implications for management. Blood Adv 2020;4(4):676–86. Available from: https://doi.org/10.1182/bloodadvances.2019000952.
[34] Yakoub-Agha I, Chabannon C, Bader P, et al. Management of adults and children undergoing chimeric antigen receptor T-cell therapy: best practice recommendations of the European Society for Blood and Marrow Transplantation (EBMT) and the Joint Accreditation Committee of ISCT and EBMT (JACIE). Haematologica. 2020;105(2):297–316. Available from: https://doi.org/10.3324/haematol.2019.229781.
[35] Yáñez LA-O, Alarcón A, Sánchez-Escamilla M, Perales MA. How I treat adverse effects of CAR-T cell therapy. ESMO Open. 4, e000746. doi:10.1136/esmoopen-2020-000746

[36] Neelapu SS, Tummala S, Kebriaei P, et al. Chimeric antigen receptor T-cell therapy – assessment and management of toxicities. Nat Rev Clin Oncol 2018;15(1):47–62. Available from: https://doi.org/10.1038/nrclinonc.2017.148.

[37] Brudno JN, Kochenderfer JN. Recent advances in CAR T-cell toxicity: mechanisms, manifestations and management. Blood Rev 2019;34:45–55. Available from: https://doi.org/10.1016/j.blre.2018.11.002.

[38] Shalabi H, Gust J, Taraseviciute A, et al. Beyond the storm – subacute toxicities and late effects in children receiving CAR T cells. Nat Rev Clin Oncol 2021;18(6):363–78. Available from: https://doi.org/10.1038/s41571-020-00456-y.

[39] FDA-NIH Biomarker Working Group. BEST (Biomarkers, EndpointS, and other Tools) resource. Silver Spring, MD: US Food and Drug Administration; 2016.

[40] Paczesny S, Pasquini MC, Pavletic SZ, et al. Blueprint for the discovery of biomarkers of toxicity and efficacy for CAR T cells and T-cell engagers. Blood Adv 2021;5(11):2519–22. Available from: https://doi.org/10.1182/bloodadvances.2021004604.

[41] Gardner RA, Finney O, Annesley C, et al. Intent-to-treat leukemia remission by CD19 CAR T cells of defined formulation and dose in children and young adults. Blood 2017;129(25):3322–31. Available from: https://doi.org/10.1182/blood-2017-02-769208.

[42] Hay KA, Hanafi LA, Li D, et al. Kinetics and biomarkers of severe cytokine release syndrome after CD19 chimeric antigen receptor-modified T-cell therapy. Blood 2017;130(21):2295–306. Available from: https://doi.org/10.1182/blood-2017-06-793141.

[43] Santomasso BD, Park JH, Salloum D, et al. Clinical and biological correlates of neurotoxicity associated with CAR T-cell therapy in patients with B-cell acute lymphoblastic leukemia. Cancer Discov 2018;8(8):958–71. Available from: https://doi.org/10.1158/2159-8290.CD-17-1319.

[44] Turtle CJ, Hanafi LA, Berger C, et al. Immunotherapy of non-Hodgkin's lymphoma with a defined ratio of CD8+ and CD4+ CD19-specific chimeric antigen receptor-modified T cells. Sci Transl Med 2016;8(355):355ra116. Available from: https://doi.org/10.1126/scitranslmed.aaf8621.

[45] Turtle CJ, Hanafi LA, Berger C, et al. CD19 CAR-T cells of defined CD4+:CD8+ composition in adult B cell ALL patients. J Clin Invest 2016;126(6):2123–38. Available from: https://doi.org/10.1172/JCI85309.

[46] Gust J, Hay KA, Hanafi LA, et al. Endothelial activation and blood-brain barrier disruption in neurotoxicity after adoptive immunotherapy with CD19 CAR-T cells. Cancer Discov 2017;7(12):1404–19. Available from: https://doi.org/10.1158/2159-8290.CD-17-0698.

[47] Davila ML, Riviere I, Wang X, et al. Efficacy and toxicity management of 19–28z CAR T cell therapy in B cell acute lymphoblastic leukemia. Sci Transl Med 2014;6(224):224ra25. Available from: https://doi.org/10.1126/scitranslmed.3008226.

[48] Shah NN, Lee DW, Yates B, et al. Long-term follow-up of CD19-CAR T-cell therapy in children and young adults with B-ALL. J Clin Oncol 2021;39(15):1650–9. Available from: https://doi.org/10.1200/JCO.20.02262.

[49] Jae Hong P, Isabelle R, Xiuyan W, Terence P, Michel S, Renier JB. Impact of disease burden on long-term outcome of 19–28z CAR modified T cells in adult patients with relapsed B-ALL. J Clin Oncol 2016;34(15_suppl):7003. Available from: https://doi.org/10.1200/JCO.2016.34.15_suppl.7003.

[50] Karschnia P, Jordan JT, Forst DA, et al. Clinical presentation, management, and biomarkers of neurotoxicity after adoptive immunotherapy with CAR T cells. Blood 2019;133(20):2212–21. Available from: https://doi.org/10.1182/blood-2018-12-893396.

[51] Faramand R, Jain M, Staedtke V, et al. Tumor microenvironment composition and severe cytokine release syndrome (CRS) influence toxicity in patients with large B-cell lymphoma treated with axicabtagene ciloleucel. Clin Cancer Res 2020;26(18):4823–31. Available from: https://doi.org/10.1158/1078-0432.CCR-20-1434.

[52] Deng Q, Han G, Puebla-Osorio N, et al. Characteristics of anti-CD19 CAR T cell infusion products associated with efficacy and toxicity in patients with large B cell lymphomas. Nat Med 2020;26(12):1878–87. Available from: https://doi.org/10.1038/s41591-020-1061-7.

[53] Young HA. Cytokine multiplex analysis. Methods Mol Biol 2009;511:85–105. Available from: https://doi.org/10.1007/978-1-59745-447-6_4.

[54] Morris EC, Neelapu SS, Giavridis T, Sadelain M. Cytokine release syndrome and associated neurotoxicity in cancer immunotherapy. Nat Rev Immunol 2021;. Available from: https://doi.org/10.1038/s41577-021-00547-6.

[55] Norelli M, Camisa B, Barbiera G, et al. Monocyte-derived IL-1 and IL-6 are differentially required for cytokine-release syndrome and

neurotoxicity due to CAR T cells. Nat Med 2018;24(6):739–48. Available from: https://doi.org/10.1038/s41591-018-0036-4.

[56] Strati P, Ahmed S, Kebriaei P, et al. Clinical efficacy of anakinra to mitigate CAR T-cell therapy-associated toxicity in large B-cell lymphoma. Blood Adv 2020;4(13):3123–7. Available from: https://doi.org/10.1182/bloodadvances.2020002328.

[57] Hu Y, Wu Z, Luo Y, et al. Potent anti-leukemia activities of chimeric antigen receptor-modified T cells against CD19 in Chinese patients with relapsed/refractory acute lymphocytic leukemia. Clin Cancer Res 2017;23(13):3297–306. Available from: https://doi.org/10.1158/1078-0432.CCR-16-1799.

[58] Fried S, Avigdor A, Bielorai B, et al. Early and late hematologic toxicity following CD19 CAR-T cells. Bone Marrow Transpl 2019;54(10):1643–50. Available from: https://doi.org/10.1038/s41409-019-0487-3.

[59] Jain T, Knezevic A, Pennisi M, et al. Hematopoietic recovery in patients receiving chimeric antigen receptor T-cell therapy for hematologic malignancies. Blood Adv 2020;4(15):3776–87. Available from: https://doi.org/10.1182/bloodadvances.2020002509.

[60] Rejeski K, Perez Perez A, Sesques P, et al. CAR-HEMATOTOX: a model for CAR T-cell related hematological toxicity in relapsed/refractory large B-cell lymphoma. Blood 2021;. Available from: https://doi.org/10.1182/blood.2020010543.

[61] Major A, Collins J, Craney C, et al. Management of hemophagocytic lymphohistiocytosis (HLH) associated with chimeric antigen receptor T-cell (CAR-T) therapy using anti-cytokine therapy: an illustrative case and review of the literature. Leuk Lymphoma 2021;62(7):1765–9. Available from: https://doi.org/10.1080/10428194.2021.1881507.

[62] Sairah A, Fateeha F, Paolo S, et al. Haemophagocytic lymphohistiocytosis (HLH) in patients with large B-cell lymphoma treated with standard of care (SOC) axicabtagene ciloleucel (Axi-cel). J Clin Oncol 2020;38(15_suppl). Available from: https://doi.org/10.1200/JCO.2020.38.15_suppl.8057.

[63] Sandler RD, Tattersall RS, Schoemans H, et al. Diagnosis and management of secondary HLH/MAS following HSCT and CAR-T cell therapy in adults: a review of the literature and a survey of practice within EBMT Centres on Behalf of the Autoimmune Diseases Working Party (ADWP) and Transplant Complications Working Party (TCWP). Front Immunol 2020;11:524. Available from: https://doi.org/10.3389/fimmu.2020.00524.

[64] Lichtenstein DA, Schischlik F, Shao L, et al. Characterization of HLH-like manifestations as a CRS variant in patients receiving CD22 CAR T-cells. Blood 2021;. Available from: https://doi.org/10.1182/blood.2021011898.

[65] Luft T, Benner A, Jodele S, et al. EASIX in patients with acute graft-vs-host disease: a retrospective cohort analysis. Lancet Haematol 2017;4(9):e414–23. Available from: https://doi.org/10.1016/S2352-3026(17)30108-4.

[66] Pennisi M, Sanchez-Escamilla M, Flynn JR, et al. Modified EASIX predicts severe cytokine release syndrome and neurotoxicity after chimeric antigen receptor T cells. Blood Adv 2021;5(17):3397–406. Available from: https://doi.org/10.1182/bloodadvances.2020003885.

[67] Greenbaum U, Strati P, Saliba RM, et al. CRP and ferritin in addition to the EASIX score predict CAR-T-related toxicity. Blood Adv 2021;5(14):2799–806. Available from: https://doi.org/10.1182/bloodadvances.2021004575.

[68] Rubin DB, Al Jarrah A, Li K, et al. Clinical predictors of neurotoxicity after chimeric antigen receptor T-cell therapy. JAMA Neurol 2020;77(12):1536–42. Available from: https://doi.org/10.1001/jamaneurol.2020.2703.

[69] Kochenderfer JN, Somerville RPT, Lu T, et al. Lymphoma remissions caused by anti-CD19 chimeric antigen receptor T cells are associated with high serum interleukin-15 levels. J Clin Oncol 2017;35(16):1803–13. Available from: https://doi.org/10.1200/JCO.2016.71.3024.

[70] Gauthier J, Chou C, Hirayama AV, et al. High IL-15 serum concentrations are associated with response to CD19 CAR T-cell therapy and robust in vivo CAR T-cell kinetics. Blood 2020;136(Supplement 1):37–8. Available from: https://doi.org/10.1182/blood-2020-140120.

[71] Hirayama AV, Gauthier J, Hay KA, et al. The response to lymphodepletion impacts PFS in patients with aggressive non-Hodgkin lymphoma treated with CD19 CAR T cells. Blood 2019;133(17):1876–87. Available from: https://doi.org/10.1182/blood-2018-11-887067.

[72] Fraietta JA, Lacey SF, Orlando EJ, et al. Determinants of response and resistance to CD19 chimeric antigen receptor (CAR) T cell therapy of chronic lymphocytic leukemia. Nat Med 2018;24(5):563–71. Available from: https://doi.org/10.1038/s41591-018-0010-1.

[73] Hay KA, Gauthier J, Hirayama AV, et al. Factors associated with durable EFS in adult B-cell ALL patients achieving MRD-negative

CR after CD19 CAR T-cell therapy. Blood 2019;133(15):1652–63. Available from: https://doi.org/10.1182/blood-2018-11-883710.

[74] Vercellino L, Di Blasi R, Kanoun S, et al. Predictive factors of early progression after CAR T-cell therapy in relapsed/refractory diffuse large B-cell lymphoma. Blood Adv 2020;4(22):5607–15. Available from: https://doi.org/10.1182/bloodadvances.2020003001.

[75] Yang Y, Kohler ME, Chien CD, et al. TCR engagement negatively affects CD8 but not CD4 CAR T cell expansion and leukemic clearance. Sci Transl Med 2017;9(417). Available from: https://doi.org/10.1126/scitranslmed.aag1209.

[76] Sommermeyer D, Hudecek M, Kosasih PL, et al. Chimeric antigen receptor-modified T cells derived from defined CD8 + and CD4 + subsets confer superior antitumor reactivity in vivo. Leukemia 2016;30(2):492–500. Available from: https://doi.org/10.1038/leu.2015.247.

[77] Garfall AL, Dancy EK, Cohen AD, et al. T-cell phenotypes associated with effective CAR T-cell therapy in postinduction vs relapsed multiple myeloma. Blood Adv 2019;3(19):2812–15. Available from: https://doi.org/10.1182/bloodadvances.2019000600.

[78] Beider K, Besser MJ, Schachter J, et al. Upregulation of senescent/exhausted phenotype of CAR T cells and induction of both treg and myeloid suppressive cells correlate with reduced response to CAR T cell therapy in relapsed/refractory B cell malignancies. Blood 2019;134(Supplement_1):3234. Available from: https://doi.org/10.1182/blood-2019-128068.

[79] Mailankody S, Jakubowiak AJ, Htut M, et al. Orvacabtagene autoleucel (orva-cel), a B-cell maturation antigen (BCMA)-directed CAR T cell therapy for patients (pts) with relapsed/refractory multiple myeloma (RRMM): update of the phase 1/2 EVOLVE study (NCT03430011). J Clin Oncol 2020;38(15):8504. Available from: https://doi.org/10.1200/JCO.2020.38.15_suppl.8504.

[80] Shah NN, Highfill SL, Shalabi H, et al. CD4/CD8 T-cell selection affects chimeric antigen receptor (CAR) T-cell potency and toxicity: updated results from a phase I anti-CD22 CAR T-cell trial. J Clin Oncol 2020;38(17):1938–50. Available from: https://doi.org/10.1200/jco.19.03279.

[81] Bachanova V, Tam CS, Borchmann P, et al. Impact of tisagenlecleucel chimeric antigen receptor (CAR)-T cell therapy product attributes on clinical outcomes in adults with relapsed or refractory diffuse large B-cell lymphoma (r/r DLBCL). Blood. 2019;134

(Supplement_1):242. Available from: https://doi.org/10.1182/blood-2019-128302.

[82] Rossi J, Paczkowski P, Shen YW, et al. Preinfusion polyfunctional anti-CD19 chimeric antigen receptor T cells are associated with clinical outcomes in NHL. Blood 2018;132(8):804–14. Available from: https://doi.org/10.1182/blood-2018-01-828343.

[83] Xue Q, Bettini E, Paczkowski P, et al. Single-cell multiplexed cytokine profiling of CD19 CAR-T cells reveals a diverse landscape of polyfunctional antigen-specific response. J Immunother Cancer 2017;5(1):85. Available from: https://doi.org/10.1186/s40425-017-0293-7.

[84] Zhou CB, Zhou YL, Fang JY. Gut microbiota in cancer immune response and immunotherapy. Trends Cancer 2021;7(7):647–60. Available from: https://doi.org/10.1016/j.trecan.2021.01.010.

[85] Smith M, Littmann E, Slingerland J, et al. Intestinal microbiome analyses identify biomarkers for patient response to CAR T cell therapy. Biol Blood Marrow Transplant 2019;25(3):S177. Available from: https://doi.org/10.1016/j.bbmt.2018.12.319.

[86] Lindo L, Wilkinson LH, Hay KA. Befriending the hostile tumor microenvironment in CAR T-cell therapy. Front Immunol 2020;11:618387. Available from: https://doi.org/10.3389/fimmu.2020.618387.

[87] Scarfo I, Maus MV. Current approaches to increase CAR T cell potency in solid tumors: targeting the tumor microenvironment. J Immunother Cancer 2017;5:28. Available from: https://doi.org/10.1186/s40425-017-0230-9.

[88] Reiss DJ, Do T, Kuo D, et al. Multiplexed immunofluorescence (IF) analysis and gene expression profiling of biopsies from patients with relapsed/refractory (R/R) diffuse large B cell lymphoma (DLBCL) treated with lisocabtagene maraleucel (liso-cel) in transcend NHL 001 reveal patterns of immune infiltration associated with durable response. Blood 2019;134(Supplement_1):202. Available from: https://doi.org/10.1182/blood-2019-127683.

[89] Mancikova V, Peschelova H, Kozlova V, et al. Performance of anti-CD19 chimeric antigen receptor T cells in genetically defined classes of chronic lymphocytic leukemia. J Immunother Cancer 2020;8(1). Available from: https://doi.org/10.1136/jitc-2019-000471.

[90] Shouval R, Alarcon Tomas A, Fein JA, et al. Impact of TP53 genomic alterations in large B-cell lymphoma treated with CD19-chimeric antigen receptor T-cell therapy. J Clin Oncol 2021;. Available from: https://doi.org/10.1200/JCO.21.02143.

[91] Li L, Liu J, Xu M, et al. Treatment response, survival, safety, and predictive factors to chimeric antigen receptor T cell therapy in Chinese relapsed or refractory B cell acute lymphoblast leukemia patients. Cell Death Dis 2020;11(3):207. Available from: https://doi.org/10.1038/s41419-020-2388-1.

[92] Shah NN, Nagle SJ, Torigian DA, et al. Early positron emission tomography/computed tomography as a predictor of response after CTL019 chimeric antigen receptor-T-cell therapy in B-cell non-Hodgkin lymphomas. Cytotherapy 2018;20(12):1415–18. Available from: https://doi.org/10.1016/j.jcyt.2018.10.003.

[93] Dean EA, Mhaskar RS, Lu H, et al. High metabolic tumor volume is associated with decreased efficacy of axicabtagene ciloleucel in large B-cell lymphoma. Blood Adv 2020;4(14):3268–76. Available from: https://doi.org/10.1182/bloodadvances.2020001900.

[94] Sesques P, Tordo J, Ferrant E, et al. Prognostic impact of 18F-FDG PET/CT in patients with aggressive B-cell lymphoma treated with anti-CD19 chimeric antigen receptor T cells. Clin Nucl Med 2021;46(8):627–34. Available from: https://doi.org/10.1097/RLU.0000000000003756.

[95] Jacobson CA, Hunter B, Armand P, et al. Axicabtagene ciloleucel in the real world: outcomes and predictors of response, resistance and toxicity. Blood 2018;132(Supplement 1):92. Available from: https://doi.org/10.1182/blood-2018-99-117199.

[96] Neelapu SS, Locke FL, Go WY. CAR T-cell therapy in large B-cell lymphoma. N Engl J Med 2018;378(11):1065. Available from: https://doi.org/10.1056/NEJMc1800913.

[97] Garcia-Recio M, Wudhikarn K, Pennisi M, et al. The international prognostic index is associated with outcomes in diffuse large B cell lymphoma after chimeric antigen receptor T cell therapy. Transpl Cell Ther 2021;27(3):233–40. Available from: https://doi.org/10.1016/j.jtct.2020.10.022.

CHAPTER

9

Cerebrospinal fluid circulating tumor cells for diagnosis, response evaluation, and molecular profiling of leptomeningeal metastases from solid tumors

Maria Diaz[1], Martin Fleisher[2] and Elena I. Pentsova[1]

[1]*Department of Neurology, Memorial Sloan Kettering Cancer Center, New York, NY, United States*

[2]*Department of Pathology and Laboratory Medicine, Memorial Sloan Kettering Cancer Center, New York, NY, United States*

Objective

- To review the laboratory process of CTC quantification in CSF and describe its clinical context

9.1 Introduction

Leptomeningeal metastases (LM), previously called carcinomatous meningitis or leptomeningeal carcinomatosis, are a rare complication of solid tumors (ST) in which tumoral cells invade the cerebrospinal fluid (CSF)-filled leptomeninges and subarachnoid space of the central nervous system (CNS). LM is relatively uncommon and

Cancer Biomarkers: Clinical Aspects and Laboratory Determination. DOI: https://doi.org/10.1016/B978-0-12-824302-2.00007-2

© 2022 Elsevier Inc. All rights reserved.

diagnosed in 5%–10% of patients with metastatic cancers, although autopsy studies suggest its true prevalence could be up to 30% [1]. Symptoms of LM are devastating and may result in rapid neurological decline, dramatical impact on patient's quality of life and even death. The overall median overall survival remains less than 4 months from LM diagnosis, despite treatment with intrathecal chemotherapy, available targeted therapy, immunotherapy, and radiation therapy [2–4]. Thus, reliable and early detection of LM in ST is urgently needed.

The standard methods to diagnose LM include cytopathological analysis of CSF, magnetic resonance imaging (MRI) of the brain and/or spine and clinical symptoms (Fig. 9.1). CSF cytopathology is the gold standard of LM diagnosis, but its sensitivity is only approximately 50% on the first lumbar puncture (LP). Thus, diagnosis requires several LPs to achieve a consistent threshold up to 90% [5]. The detection of LM by MRIs is challenging; studies have shown high interrater variability when using neuroimaging as a diagnostic tool for LM [6]. Although certain clinical presentations—

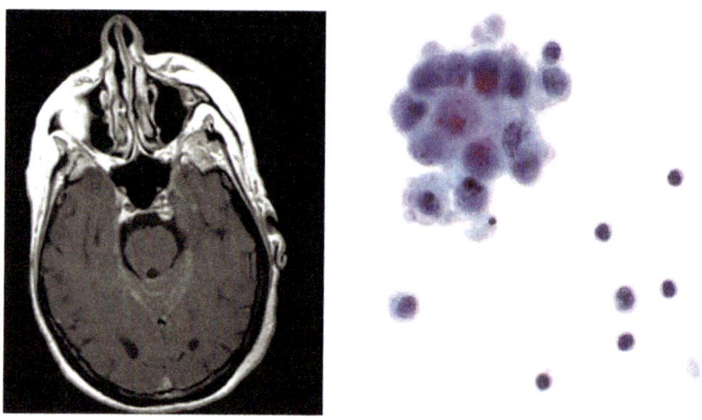

FIGURE 9.1

Standard methods of LM diagnosis. On the left, an axial T1 postcontrast MRI showing linear enhancement within the cerebellar folia, consistent with a diagnosis of leptomeningeal disease in a patient with lung adenocarcinoma. On the right, clump of malignant cells seen on cytopathological exam of a CSF sample, consistent with a diagnosis of leptomeningeal disease. *CSF*, Cerebrospinal fluid; *LM*, leptomeningeal metastases; *MRI*, magnetic resonance imaging.

such as multifocal cranial neuropathies in a patient with metastatic ST—can be very suggestive of LM, the variable neurological manifestations of LM make impossible to diagnose it based on clinical symptoms alone. Overall, establishing the diagnosis of LM can be challenging, particularly at early stages before the patient is disabled. Improved diagnostic tools are required to facilitate early diagnosis when treatment is likely to be the most helpful.

9.2 Rare cell capture technology as a diagnostic tool for leptomeningeal metastases

Rare cell capture technology (RCCT) utilizing immunomagnetic platforms and antibody-covered ferroparticles has emerged as a new tool for capturing circulating tumor cells (CTC), and analysis of peripheral blood CTC has been explored as a prognostic marker of disease and response to anticancer treatments in patients with ST [7]. Although several cell surface antigens can be used to detect and isolate CTC, the most commonly used marker is epithelial cell adhesion molecule (EpCAM), because it has the ability to detect the widest range of carcinoma [8]. Anti-EpCAM-based RCCT (Menarini Silicon BioSystems) is a Food and Drug Administration (FDA)-approved methodology for capturing and enumerating blood CTC in patients with STs, and has been associated with decreased survival in patients with metastatic breast, colorectal and prostate cancer [7,9–12]. This methodology has been applied to other body fluids, in particular to CSF, and has been evaluated in a limited number of patients with LM from breast carcinomas and other types of ST (Table 9.1) [13–15]. Studies have demonstrated RCCT's utility in CSF as a diagnostic tool in this disease, with a sensitivity in LM diagnosis that ranges from 77.8% to 100% and surpasses CSF cytology in all cases (Table 9.1).

This idea led to an Institutional Review Board (IRB)-approved prospective study at Memorial Sloan Kettering Cancer Center, which enrolled 95 patients with epithelial tumors and either with neurological symptoms or MRI findings suspicious for LM who agreed to undergo LP. For this study, a composite definition of LM was based on positive CSF cytology and/or neuroimaging positive

Table 9.1 Published studies investigating CSF-CTC analysis through the CellSearch platform as a diagnostic tool for leptomeningeal metastases.

Study	N	Patient population	CSF-CTC cutoff (cells/mL)	CSF-CTC sensitivity	CSF-CTC specificity	CSF cytology sensitivity	CSF cytology specificity
Nayak [13]	51	Clinical suspicion of LM from solid tumors	>0	100%	97.2%	66.7%	100% (GS)
Lee [30]	38	Confirmed or clinically suspected LM from breast cancer	1.9	80.95%	84.62%	66.67%	100% (GS)
Tu [31]	18	Confirmed LM (in MRI) from lung cancer	0.2	77.8%	100%	44.4%	NR
Jiang [32]	21	Clinical suspicion of LM from NSCLC	>0	95.2%	100%	61.5%	100%
Lin [33]	95	Clinical suspicion of LM from solid tumors	1	93%	95%	29%	NR
Torre [34]	20	Clinical suspicion of LM from solid tumors (18 patients with epithelial tumors, 2 patients with melanoma)	1	88.9%	100%	77.8%	100%

CSF, *Cerebrospinal fluid;* CTC, *circulating tumor cells;* GS, *gold standard;* LM, *leptomeningeal metastases;* N, *number of patients;* NR, *not reported.*

for LM, if MRIs performed within a month of the CSF analysis. The 1-month window was defined arbitrarily to ensure that the diagnosis of LM was new and relevant to the initial CSF-CTC result. At the time of LP, and in addition to CSF cytology, 3 mL of CSF from the same spinal tap were collected in CellSave tubes and analyzed. LM was diagnosed in 30 (32%) patients at initial evaluation: 12 had positive CSF cytology, two had unequivocal MRI findings for LM, and 16 had positive results from both CSF cytology and MRI. The remainder 65 patients were deemed to be negative for LM at the time of initial evaluation. At the initial LP, CSF-CTC were detected in 43 (45%) patients with a median of 19.3 CSF-CTC/mL (range 0.33–66.7). The optimal CSF-CTC cutoff was determined by the area under the receiver operating characteristic (ROC) curve. With the optimal cutoff of ≥1 CSF-CTC/mL, the specificity was 95%, sensitivity was 93%, positive predictive value was 90%, and negative predictive value was 97% (Fig. 9.2). All of these suggest

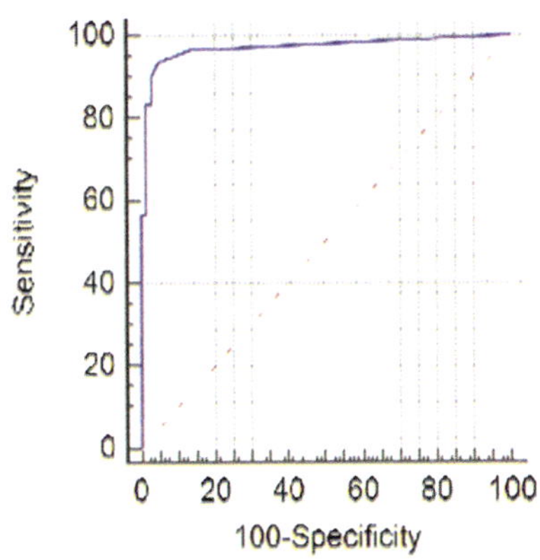

FIGURE 9.2

Diagnostic performance of CSF-CTC testing. Using the pictured ROC curve, optimal cutoff was calculated at ≥3 CSF-CTC/3 mL (or ≥1 CSF-CTC/mL), which provides a specificity of 95%, sensitivity of 93%, positive predictive value of 90% and negative predictive value of 97%. *CSF*, Cerebrospinal fluid; *CTC*, circulating tumor cells; *ROC*, receiver operating characteristic. Lin X, Fleisher M, Rosenblum M, Lin O, Boire A, Briggs S, et al. Cerebrospinal fluid circulating tumor cells: a novel tool to diagnose leptomeningeal metastases from epithelial tumors. Neuro Oncol. 2017;19(9):1248–54 [33].

strong diagnostic performance of CSF-CTC to discriminate between LM and no disease.

9.3 Clinical Laboratory Improvement Act certification

Any change in the approved methods or specimen type—that is, body fluid other than whole blood—used to analyze and enumerate CTC must be validated and submitted for approval by the Clinical Laboratory Improvement Act (CLIA) before the new Laboratory Developed Test (LDT) can be used for clinical management. It was determined that for the validation of CTC in CSF more than 100 patient samples were required in order to definitively differentiate "normal" CTC counts from "abnormal." Therefore based on the aforementioned validation study and prior pilot studies, all results and conclusion were submitted for CLIA certification and were supported by an outline of the validation protocol. The summary addressed how the analytical and clinical performance characteristics were established and raw data was provided using an appropriate number of samples across all representative specimen matrices and expected outcomes. Data was summarized and clearly labeled in tables, figures, and photographs.

Described below is a list of required items for the validation for an LDT to be used for clinical management.

A. Description of the assay, assay principle and clinical validity must be provided.
B. A detailed description of the methods used in performing the assay, including algorithms and flowcharts as necessary in specimen collection and processing. Safety considerations must be provided.
C. A list of equipment/instrumentation essential to the assay.
D. A complete list of all reagents. Preparation and storage stability and handling must be provided including quality control materials used in the assay.

E. Calculation of results (enumeration of CTC). Identify the critical steps in the assay procedure and the quality control measures taken to monitor assay performance.

F. A description of how reference intervals or assignment of cutoff values were determined.

G. Assay interferences and limitations.

9.3.1 Definition of a circulating tumor cell using the CellSearch circulating tumor cell test

A CTC is defined as a cell of epithelial origin that is CD45 negative, EpCAM positive, and cytokeratins 8, 18, and/or 19 positives.

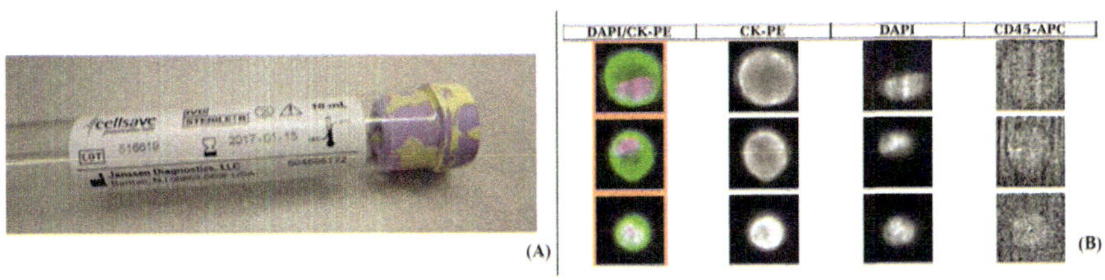

FIGURE 9.3

CellSave tube and CTC gallery. (A) CellSave preservative tube. For CSF samples, 3.0 mL must be collected and transported to the lab at room temperature (15°C–30°C). Sample is stable for 96 h at room temperature; refrigerating samples prior to processing should be avoided as it could adversely affect sample integrity. (B) Fragment of a CTC gallery. Cells are captured through ferromagnetic antibodies directed against EpCAM, and further identified as CTCs (first column) when they are cytokeratin positive (epithelial marker, second column), 4′6-diaminodino-2-phenylindole or DAPI positive (nuclear marker, third column) and CD45 negative (lymphocyte marker, fourth column). All cell galleries are individually inspected by a trained technologist and reviewed and approved by a CLIA certified Laboratory Director. *CLIA*, Clinical Laboratory Improvement Act; *CSF*, cerebrospinal fluid; *CTC*, circulating tumor cells.

9.3.2 Specimen needed for circulating tumor cells enumeration in cerebrospinal fluid

Three microliters of CSF in CellSave Preservative tubes must be collected and transported to the lab at room temperature (15°C–30°C). Refrigerating samples prior to processing could adversely affect sample integrity. Sample is stable for 96 hours at room temperature (Fig. 9.3A).

9.3.3 Interpretation of results

CellSearch automatically identifies CTC based on predetermined cancer cell characteristics, namely CD45 negative, EpCAM positive, and cytokeratins 8, 18, and/or 19 positives. However, all cell galleries are individually inspected by a trained technologist and reviewed and approved by a CLIA certified Laboratory Director (Fig. 9.3B).

9.3.4 Quality control of circulating tumor cells in cerebrospinal fluid

A comprehensive quality control program must be established for an LDT and data over the validation time course must be provided. CellSearch Circulating Tumor Cell Control Kit (Epithelial), Menarini Silicon Biosystems, Inc., (cat#7900003) is indicative of the validity of the overall system performance, including instrument and reagents. CellSearch CTC Control should be run each day of patient testing or when using a new lot of CellSearch Circulating Tumor Cell Reagent Kit. Each single use of 3 mL bottle contains two populations of SK-BR-3 cells at different concentrations (low and high). The two cell populations are distinguished from each other by use of fluorescent dyes that are specific for each population. The Quality Control materials are stored at 2°C–8°C and protected from bright light until the expiration date printed on the bottle.

9.3.5 Reporting results

Results are reported as number of cells per 3.0 mL of CSF (units on the report are "cells/3.0 mL"). This method of reporting is preferred, so that fractions of cells are not reported per 1 mL of CSF.

9.3.6 Reference range

Less than or equal to two cells/3.0 mL of CSF.

9.4 Additional applications of cerebrospinal fluid circulating tumor cells and other techniques

After studies demonstrated the excellent diagnostic utility of the CSF-CTC analysis in patients with LM, other applications of this analysis were explored. Early studies demonstrated dynamic changes in CSF-CTC count during treatment with chemotherapy [14,16], and CSF-CTC analysis has started to be included as an exploratory endpoint in clinical trials [17]. In a phase I/II trial of intrathecal trastuzumab for LM in human epidermal growth factor receptor 2 positive (HER2 +) breast carcinoma, a decrease in CSF-CTC count with treatment, and a later increase in CTCs predating radiographic LM progression, were seen in several patients and indicated that CSF-CTC can be used as a surrogate marker of response to treatment in LM (Fig. 9.4) [18]. However, it is too early to conclude whether CSF-CTC results should influence the decision to discontinue treatment, and the relationship between number of CTC and radiographic disease burden needs to be investigated further.

Retrospective studies have shown that the survival of LM patients with a higher CSF-CTC count at the time of diagnosis is worse than for patients with lower counts, suggesting a potential prognostic role for CSF-CTC testing [19,20].

Beyond the detection of CSF-CTC, the captured cells can also be subjected to molecular and genetic analysis, broadening the possibilities for characterization of CNS metastases. The feasibility of performing genomic sequencing of CSF-CTC has been shown in patients with LM from breast cancer, demonstrating similarities

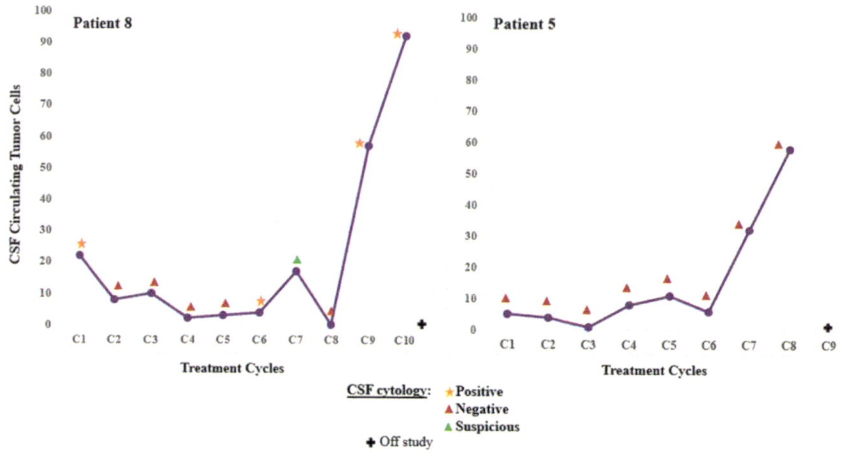

FIGURE 9.4

CSF-CTC changes over time in two patients enrolled in a trial of intrathecal trastuzumab for LM in HER2 + breast carcinoma. Both patients had an initial decrease in CSF-CTC count after treatment initiation, with a later increase after several cycles heralding progression; in patient 8 increase was accompanied by positive CSF cytology, but in patient 5 CSF cytology remained negative despite CSF-CTC count increase and subsequent disease progression. *CSF*, Cerebrospinal fluid; *CTC*, circulating tumor cells; *LM*, leptomeningeal metastases.

Reproduced with author's permission from Malani, R., et al. Cerebrospinal fluid circulating tumor cells as a quantifiable measurement of leptomeningeal metastases in patients with HER2 positive cancer. J Neurooncol 2020;148(3):599–606.

between CSF-CTC and tumor cells from primary site in addition to several genetic alterations private to CSF-CTC [16,18,21]. More detailed characterization of CSF-CTC through single-cell transcriptome analysis has been performed in a small number of patients; [22] these results can not only help improve our understanding of the biology of LM but also have direct clinical implications such as identifying the cell of origin when the primary malignancy is unknown, or monitoring the mutational profile of LM in a given patient over time, potentially guiding changes in targeted therapies.

In addition to immunomagnetic techniques like the CellSearch platform, EpCAM-expressing CTC can also be detected through flow cytometry, using fluorescently labeled EpCAM antibodies to stain and count the cells [23]. Although it has not yet been validated for clinical use, several studies have applied this technology to CSF and reported a sensitivity in the diagnosis of LM in a similar range to studies using the CellSearch platform [24–28]. One of these employed ROC curves to identify the optimal cutoff for the diagnosis of LM, with a result of 0.86 CTC/mL, similar to the result validated for the CellSearch system [28]. CSF-CTC captured by flow cytometry can also be further characterized through genomic profiling [16], which may provide interesting insights regarding the molecular nature of LM cells, but is yet to be clinically validated.

9.5 Limitations of the CellSearch platform

One of the limitations of the CellSearch platform is that it only detects cells from epithelial origin, and therefore cannot be applied to hematological malignancies, primary brain tumors or melanoma. Given melanoma is one of the most common cancer histologies to undergo CNS spread including LM, modified analysis with melanoma markers have been used to capture CTC in CSF [29], but these are not as widely available as the standard CellSearch test and are not clinically validated at this time.

9.6 Summary Points

- Quantification of CTCs in CSF, in particular through the CellSearch platform, has been validated as a sensitive and specific diagnostic method for LM in patients with solid tumors of epithelial origin. In addition to aiding in the diagnosis of LM, CSF-CTC detection allows for serial longitudinal enumeration, which is currently being explored as a biomarker of response to therapies that may have significant clinical applications in the treatment of LM patients in the future.
- Beyond standard EpCAM-based CTC analysis, the ability to assess CSF-CTC for molecular markers or single cell analysis makes this method even more attractive for the management of patients with solid cancers harboring targeted mutations.

9.7 Future directions

CSF-CTC quantification should be prospectively validated as a biomarker of response to therapies and of disease progression. Further testing of CSF-CTCs such as single-cell analyses for now, which are available only in the research setting, should continue to be explored and eventually incorporated into clinical practice. In addition, CSF-CTC detection using alternative markers to EpCAM that could capture other histologies (such as melanoma or primary brain tumors) should also be validated as a diagnostic tool for LM in these malignancies.

References

[1] Chamberlain MC. Leptomeningeal metastasis. Curr Opin Oncol 2010;22(6):627–35.

[2] Le Rhun E, Taillibert S, Chamberlain MC. Carcinomatous meningitis: leptomeningeal metastases in solid tumors. Surg Neurol Int 2013;4 (Suppl 4):S265–88.

[3] Glitza IC, et al. Leptomeningeal disease in melanoma patients: an update to treatment, challenges, and future directions. Pigment Cell Melanoma Res 2020;33(4):527–41.

[4] Devecka M, et al. Craniospinal irradiation(CSI) in patients with leptomeningeal metastases: risk-benefit-profile and development of a prognostic score for decision making in the palliative setting. BMC Cancer 2020;20(1):501.

[5] Boire A, et al. Liquid biopsy in central nervous system metastases: a RANO review and proposals for clinical applications. Neuro Oncol 2019;21(5):571–84.

[6] Le Rhun E, et al. The RANO Leptomeningeal Metastasis Group proposal to assess response to treatment: lack of feasibility and clinical utility and a revised proposal. Neuro Oncol 2019;21(5):648–58.

[7] Cristofanilli M, et al. Circulating tumor cells, disease progression, and survival in metastatic breast cancer. N Engl J Med 2004;351 (8):781–91.

[8] Ferreira MM, Ramani VC, Jeffrey SS. Circulating tumor cell technologies. Mol Oncol 2016;10(3):374–94.

[9] Allard WJ, et al. Tumor cells circulate in the peripheral blood of all major carcinomas but not in healthy subjects or patients with nonmalignant diseases. Clin Cancer Res 2004;10(20):6897–904.

[10] Scher HI, et al. Circulating tumour cells as prognostic markers in progressive, castration-resistant prostate cancer: a reanalysis of IMMC38 trial data. Lancet Oncol 2009;10(3):233–9.
[11] Cohen SJ, et al. Prognostic significance of circulating tumor cells in patients with metastatic colorectal cancer. Ann Oncol 2009;20 (7):1223–9.
[12] Liu MC, et al. Circulating tumor cells: a useful predictor of treatment efficacy in metastatic breast cancer. J Clin Oncol 2009;27 (31):5153–9.
[13] Nayak L, et al. Rare cell capture technology for the diagnosis of leptomeningeal metastasis in solid tumors. Neurology 2013;80(17):1598–605.
[14] Patel AS, et al. Identification and enumeration of circulating tumor cells in the cerebrospinal fluid of breast cancer patients with central nervous system metastases. Oncotarget 2011;2(10):752–60.
[15] Le Rhun E, et al. Development of a new method for identification and quantification in cerebrospinal fluid of malignant cells from breast carcinoma leptomeningeal metastasis. BMC Clin Pathol 2012;12:21.
[16] Magbanua MJ, et al. Molecular profiling of tumor cells in cerebrospinal fluid and matched primary tumors from metastatic breast cancer patients with leptomeningeal carcinomatosis. Cancer Res 2013;73(23):7134–43.
[17] Morikawa A, et al. Phase I study of intermittent high-dose lapatinib alternating with capecitabine for HER2-positive breast cancer patients with central nervous system metastases. Clin Cancer Res 2019;25(13):3784–92.
[18] Malani R, et al. Cerebrospinal fluid circulating tumor cells as a quantifiable measurement of leptomeningeal metastases in patients with HER2 positive cancer. J Neuro Oncol 2020;148(3):599–606.
[19] Nevel KS, et al. A retrospective, quantitative assessment of disease burden in patients with leptomeningeal metastases from non-small-cell lung cancer. Neuro Oncol 2020;22(5):675–83.
[20] Diaz M, et al. Biom-05. circulating tumor cells (CTC) in cerebrospinal fluid (CSF) as a predictor of survival in cns metastases. Neuro-Oncology 2020;22(Suppl 2):ii2.
[21] Li X, et al. Clinical significance of detecting CSF-derived tumor cells in breast cancer patients with leptomeningeal metastasis. Oncotarget 2018;9(2):2705–14.
[22] Ruan H, et al. Circulating tumor cell characterization of lung cancer brain metastases in the cerebrospinal fluid through single-cell transcriptome analysis. Clin Transl Med 2020;10(8):e246.
[23] van Bussel MTJ, et al. EpCAM-based assays for epithelial tumor cell detection in cerebrospinal fluid. J Neurooncol 2018;137(1):1–10.

[24] Subira D, et al. Role of flow cytometry immunophenotyping in the diagnosis of leptomeningeal carcinomatosis. Neuro Oncol 2012;14(1):43–52.
[25] Subira D, et al. Diagnostic and prognostic significance of flow cytometry immunophenotyping in patients with leptomeningeal carcinomatosis. Clin Exp Metastasis 2015;32(4):383–91.
[26] Milojkovic Kerklaan B, et al. EpCAM-based flow cytometry in cerebrospinal fluid greatly improves diagnostic accuracy of leptomeningeal metastases from epithelial tumors. Neuro Oncol 2016;18(6):855–62.
[27] Acosta M, Pereira J, Arroz M. Screening of carcinoma metastasis by flow cytometry: a study of 238 cases. Cytometry B Clin Cytom 2016;90(3):289–94.
[28] van Bussel MTJ, et al. Circulating epithelial tumor cell analysis in CSF in patients with leptomeningeal metastases. Neurology 2020;94(5):e521–8.
[29] Le Rhun E, et al. Detection and quantification of CSF malignant cells by the CellSearch technology in patients with melanoma leptomeningeal metastasis. Med Oncol 2013;30(2):538.
[30] Lee JS, et al. Detection of cerebrospinal fluid tumor cells and its clinical relevance in leptomeningeal metastasis of breast cancer. Breast Cancer Res Treat 2015;154(2):339–49.
[31] Tu Q, et al. CellSearch technology applied to the detection and quantification of tumor cells in CSF of patients with lung cancer leptomeningeal metastasis. Lung Cancer 2015;90(2):352–7.
[32] Jiang BY, et al. Detection of driver and resistance mutations in leptomeningeal metastases of NSCLC by next-generation sequencing of cerebrospinal fluid circulating tumor cells. Clin Cancer Res 2017;23(18):5480–8.
[33] Lin X, Fleisher M, Rosenblum M, Lin O, Boire A, Briggs S, et al. Cerebrospinal fluid circulating tumor cells: a novel tool to diagnose leptomeningeal metastases from epithelial tumors. Neuro Oncol 2017;19(9):1248–54.
[34] Torre M, et al. Integration of rare cell capture technology into cytologic evaluation of cerebrospinal fluid specimens from patients with solid tumors and suspected leptomeningeal metastasis. J Am Soc Cytopathol 2020;9(1):45–54.

CHAPTER

Saliva as a matrix for measurement of cancer biomarkers 10

Lucas Trevisan França de Lima[1,2,3,4], **Juliana Müller Bark**[1,2,3], **Mohammad Rasheduzzaman**[1,2,3], **Chameera Ekanayake Weeramange**[1,2,3,5] **and Chamindie Punyadeera**[1,2,3,6]

[1]Saliva and Liquid Biopsy Translational Laboratory, Centre for Biomedical Technologies, School of Biomedical Sciences, Faculty of Health, Queensland University of Technology, Brisbane, QLD, Australia

[2]Translational Research Institute, Brisbane, QLD, Australia

[3]Griffith Institute for Drug Discovery (GRIDD), Griffith University, Brisbane, Australia

[4]Gallipoli Medical Research Institute, Greenslopes Private Hospital, Brisbane, QLD, Australia

[5]Department of Medical Laboratory Sciences, Faculty of Health Sciences, The Open University of Sri Lanka, Nugegoda, Sri Lanka

[6]Menzies Health Institute Queensland (MIHQ), Griffith University, Queensland, Australia

Objectives

1. Describe the composition of saliva and the transport of molecules from the blood to the oral cavity
2. Understand the techniques and methods used to collect saliva
3. Review salivary biomarkers in different cancer types
4. Understand the limitations of using saliva for the detection and measurement of cancer biomarkers

10.1 Introduction

Cancer is the second most noticeable cause of death worldwide [1]. The number of cases of cancer is expected to increase

Cancer Biomarkers: Clinical Aspects and Laboratory Determination. DOI: https://doi.org/10.1016/B978-0-12-824302-2.00008-4
© 2022 Elsevier Inc. All rights reserved.

approximately by 70% over the next two decades. As such, there is a great need to develop noninvasive, easily accessible methods to early diagnose, prognose, monitor disease progression and/or monitor response to treatment [2]. Furthermore, these noninvasive diagnostic tools integrated with point of care technologies could revolutionize the current clinical practice for managing cancer patients.

Biopsy of tumor tissue, the gold standard for cancer diagnosis, has several limitations. The analysis of tumor tissue is not always feasible or repeatable and it does not capture the complex tumor heterogeneity [3]. Liquid biopsy, the use of biomolecules present in body fluids, is a simple and noninvasive alternative to solid tumor biopsy testing and can predict the outcome of cancer patients that is not available to clinicians using conventional methods. The application of liquid biopsy has emerged as a useful strategy for the detection of cancer-derived biomolecules in a minimally invasive manner [4]. During tumor progression, tumors shed content into the bloodstream, namely circulating tumor cells (CTCs), proteins, circulating tumor DNA (ctDNA), circulating RNAs, and tumor-derived exosomes [4] (Fig. 10.1). These circulating biomarkers are then transported through the lymphatic system and the bloodstream to other biofluids, such as saliva, where they can be noninvasively assessed [5].

Salivary diagnostics is attracting increasing attention due to its noninvasive nature and the ease of sample collection, leading to greater patient compliance compared to blood-based methods [6]. Saliva is also a more stable and less complex matrix compared to blood and as such, is ideal for field testing [7]. The launch of the world's first biosensor to measure glucose levels in type 2 diabetic patients has opened a new era of saliva-based point of care (PoC) diagnostics [8]. The utility of saliva testing is further evidenced by its applications in oral cancer detection. More recently, our team showed the world's first saliva-based test can easily detect 2 mm occult human papillomavirus (HPV) driven head and neck cancer (HNC) based on saliva testing for HPV [9,10]. In addition, in collaboration with Viome, United States, our team has developed a test that can predict the likelihood of developing oral cancer based on

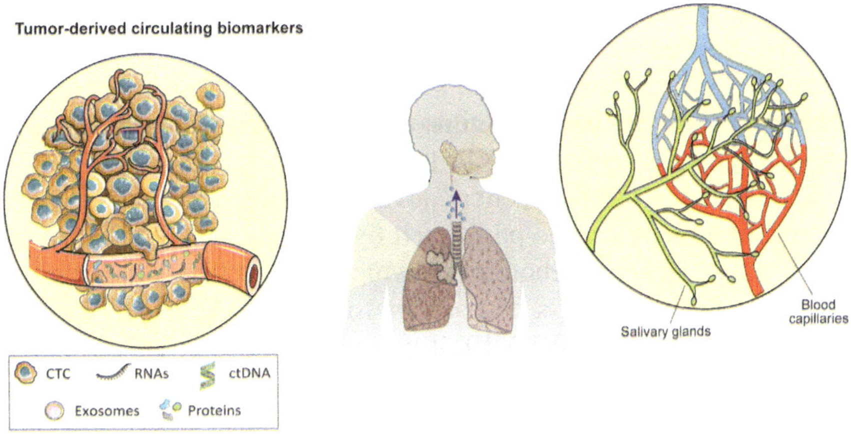

FIGURE 10.1

Cancer-derived circulating biomarkers. During cancer progression, tumors release content into circulation, such as circulating tumor cells (CTCs), circulating tumor DNA (ctDNA), RNAs (microRNA—miRNA, long noncoding RNA—lncRNA, and messenger RNA–mRNA), exosomes and proteins. These molecules can be detected using liquid biopsy, which is often performed using blood samples. However, recently there is a growing interest in exploring less invasive alternatives such as saliva sampling.

the changes in oral microbiome [11]. This saliva test recently received Food and Drug Administration (FDA) approval under the breakthrough device designation. This chapter focuses on biomolecules present in saliva and their potential clinical applications in cancer research.

Compared to traditional tissue biopsy, liquid biopsies present clearer advantages. Firstly, the collection of serum, saliva or other biofluids is less invasive, cheaper and presents fewer risks for patients [12,13]. In addition, while tissue biopsies only present a fraction of the analyzed tumor, liquid biopsy samples can better reflect its genetic heterogeneity. In a clinical setting, a better insight into the tumor genetic landscape is the basis for the development of precision medicine strategies [12]. More importantly, the possibility of performing repeated sampling over the course of the disease can improve the quality of life for an individual and potentially improve outcomes. As an example, ctDNA mutation analysis (T790M) has already entered clinical practice for lung cancer management. Although liquid biopsy is often carried out using blood, recent studies have shown the clinical utility of other biofluids such as saliva, urine, and cerebrospinal fluid [4,13].

Saliva testing offers practical advantages over blood. The collection of the sample can be performed by the patient without the need for a nurse or a phlebotomist, which makes the process cost-effective and causes less discomfort for patients [13,14]. With the current outbreak of COVID-19, it was very clear that we need alternative methods of collection to minimize the spread of the virus in the community, as well as testing options that can be performed in a field setting [15]. Thus, the development of saliva-based diagnosis can positively impact public health.

10.2 Saliva production and secretion

Saliva is a fluid excreted by three major salivary glands (submandibular, sublingual, and parotid) and several minor glands throughout the oral cavity [16]. It is a slightly acidic liquid (pH 6–7) with multiple key functions, including buffering, mineralization, digestion of polysaccharides to disaccharides by salivary alpha-amylase, lubrification, increasing the perception of taste, and protecting against viral, fungal, and bacterial infections [11,17]. The characteristics of the saliva produced by each of the glands are different in many aspects, including the viscosity and protein content. Advances in the gland-specific collection of saliva samples have shown differences in the proteomic and transcriptomic profiles of saliva obtained from a different anatomical location within the oral mucosa [18,19].

The production of saliva is variable between individuals due to factors such as flow rate, type and duration of the stimulus, oral hygiene, circadian rhythm, and physiological/pathological status. The production of saliva in a healthy adult varies between 0.5 to 1.5 L per day [16]. The main components in this biofluid are water (99%), proteins (0.3%), electrolytes, nucleic acids, lipids (0.2%) and cellular components [19] (Fig. 10.2). The proteins present in saliva are mainly antibodies, enzymes, hormones, and inflammatory

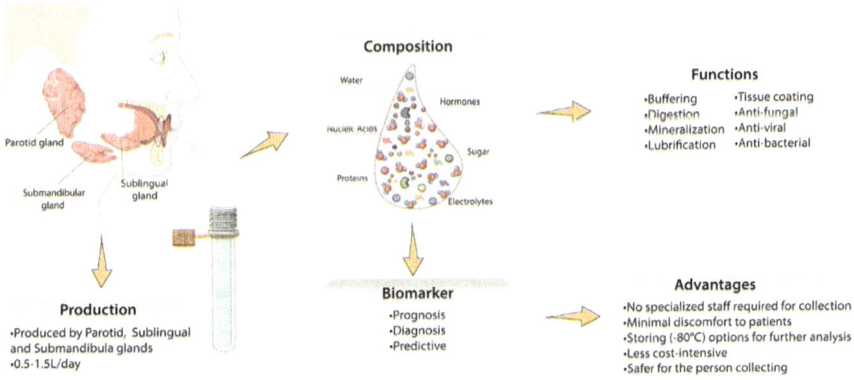

FIGURE 10.2

Production, functions, and clinical utility of saliva for the management of cancer.

markers [20]. The main secretory units in the salivary glands are called acini cells, a cluster of epithelial cells responsible for the production of both a watery fluid, secreted by serous cells and the mucous part of saliva, a mucin-rich fraction responsible for the viscoelastic properties of saliva [7,16]. In addition, part of the saliva is derived from the gingival crevicular fluid. This fluid is a serum transudate or an inflammatory exudate that is originated from the periodontal tissues. Inflammatory markers, locally generated products of tissue breakdown and serum are the main components of the gingival crevicular fluid, which can be transferred to saliva [21].

10.3 Transport of biomolecules from the blood into saliva

Increasing evidence shows that oral health mirrors systemic health, rendering saliva an ideal biofluid to discern the general health of an individual [22]. This is possible because saliva contains molecules that are synthesized by the salivary glands as well as molecules that are transported across the bloodstream into saliva (Fig. 10.3). The transport usually depends on the size and charge of the molecules [19,23]. Small, nonionic molecules are transported from the

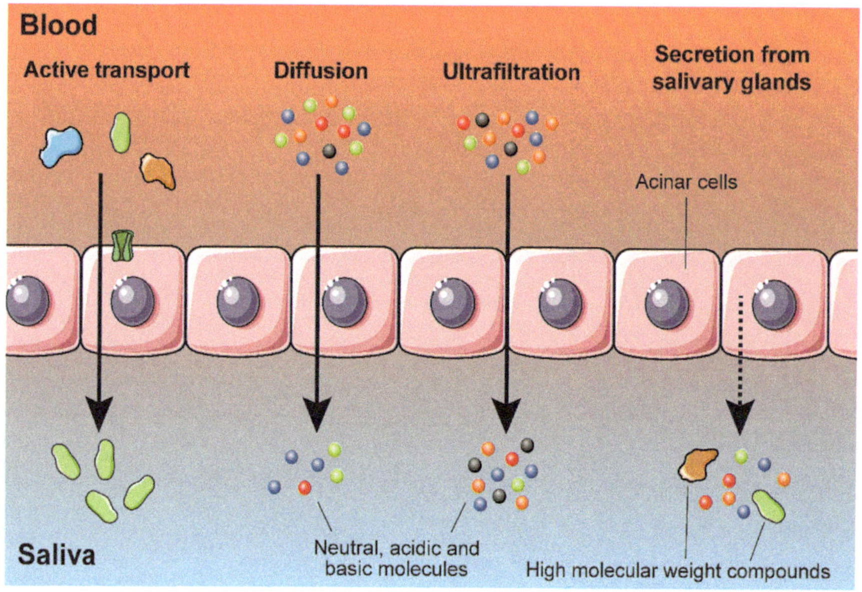

FIGURE 10.3

Schematic representation of the pathways that link blood capillaries and salivary acinar cells. Three main transport pathways are presented: [1] active transport for larger molecules (> 1900 kDa); [2] passive diffusion for the passage of small and neutral molecules through the acinar cells and [3] ultrafiltration of small molecules and ions through the gap junctions between adjacent acinar cells [16,19].

surrounding blood capillaries to the acinar cells in the salivary glands by passive diffusion. This is the case for steroid hormones, which are usually nonpolar due to the high fatty acid content [16]. The active transport is initiated by the interaction of molecules with specific ligands. This is the most common way to transport molecules with a molecular weight higher than 1900 kDa, such as immunoglobulins secreted by B-lymphocyte cells [24]. In addition, molecules smaller than 1900 kDa, such as ions, water, and several neutral steroid hormones pass through the gap junctions between the acinar cells in a process called ultrafiltration [16]. Also, molecules get transported through the gingival crevicular fluid. When these molecules reach saliva their concentration can be reduced by 100- to 1000-fold in comparison to the concentrations in blood [25].

Therefore the wide use of salivary diagnosis remains highly dependent on the development of highly sensitive detection platforms [25].

10.4 Saliva collection methods and saliva types

Saliva sampling can be influenced by different factors such as salivary flow rate, saliva composition, type of saliva collected and collection method/device [26]. Briefly, saliva can be collected under unstimulated (i.e., resting) or stimulated conditions. Common methods of increasing saliva production include: chewing on paraffin wax or rubber bands (mechanical induction) and applying an 0.1–0.2 mol/L solution of citric acid to the tongue [6]. Furthermore, gland-specific collection can be performed by cannulation of the salivary gland ducts or using collection devices specifically designed for those purposes [26]. The saliva collection methods are summarized in Table 10.1.

There are commercially available saliva collection devices, which include: Super Sal and Versi Sal (Oasis Diagnostic Corporation, https://4saliva.com/); Quantisal (Immunalysis Corporation, https://immunalysis.com/); Intercept (Orasure Technologies Inc., https://www.orasure.com/); Salivette (Sarstedt, https://www.sarstedt.com/); SalivaBio Oral Swab (Salimetrics, https://salimetrics.com/); OraQuick Advance HIV-1/2 (Orasure Technologies Inc., http://www.oraquick.com/); and BBL CultureSwab (BD, https://www.bd.com/). The main function for these saliva collection devices to date are for the detection of illicit and prescription drugs (swab based Quantisal and Orasure). The Salivatte is the most used swab-based device, which has been previously used for the detection of salivary cortisol [23]. The Oasis Diagnostic Versi Sal and Super Sal are universal samplers for biofluids (i.e., saliva, urine, amniotic, and vaginal fluids). They have been used for the detection of antimicrobial proteins (secretory IgA, α-amylase, and lysozyme) and cortisol in humans, as well as validated to be used in several other species such as cows, horses, pigs, cats, dogs, and nonhuman primates [36–38]. Saliva collection devices need to be standardized in the future prior to clinical implementation.

Table 10.1 Description of saliva types and collection methods

Saliva types	Collection method	References
Unstimulated whole mouth saliva	Passive drool: The individual should lean forward, accumulate saliva in their mouth and expectorate it into a vial without the use of oral motility	[27]
	Spitting method: Instruct the individual to spit into a vial.	[28]
Stimulated whole mouth saliva	This method uses physical (chewing different substances including natural gum or a piece of paraffin wax) and chemical (citric acids, and powdered drink crystals) stimulants.	[29]
Parotid gland	The parotid glands are found beneath and in front of both ears, with the gland duct opening near the first and second molars. Parotid saliva can be collected using intraductal cannulation, which is invasive and time-consuming. A noninvasive alternative is the Lashley cup. Using suction, saliva is collected through a metal cannulae attached to a cup. Although not invasive, this method requires skilled personnel and is also time-consuming.	[30,31]
Submandibular/ sublingual gland	These glands are found beneath the floor of the mouth. The procedure for saliva collection is similar to the parotid gland, which can be performed using intraductal cannulation, Lashey cups, or custom-made Wolf saliva collectors. However, a difference between this method and the parotid gland is that the first 2 mL of saliva collected from submandibular glands are discarded to avoid contamination.	[18,32,33]
Minor glands	Kutscher et al. (1967) employed capillary tubes to collect saliva from small glands on the lower lips everted surface. Homogeneous samples from minor glands are challenging to obtain and the development of custom-made devices is often necessary.	[34,35]

10.5 Salivary biomarkers for cancer

10.5.1 Saliva-based nucleic acid biomarkers for cancer detection

Detection of tumor-derived nucleic acids is one of the most common approaches to liquid biopsy-based cancer identification. A range of genomic, epigenomic, and transcriptomic biomarkers can be described under this category. These tumor-derived nucleic acids are released into the tumor microenvironment mainly as a consequence of frequent necrosis and apoptosis of tumor cells [39]. However, several other mechanisms of nucleic acid release such as secretion in extracellular vesicles (EV) and active release also play a significant role [39,40]. These molecules released into the tumor microenvironment find their way to surrounding body fluids and into circulation. Similar to other body fluids, saliva is also a rich medium for tumor-derived nucleic acids.

Even though the characterization of tumor-derived DNA in blood has become a major field of cancer research, limited attention has been paid to the detection of these molecules in other body fluids such as saliva. The detection of tumor mutations using saliva samples has been reported for head and neck cancers (i.e., oral cavity, oropharynx, larynx, and hypopharynx) [41]. In addition, saliva has also been shown to contain mutated DNA molecules that originated from cancers in distant organs, for example, lung cancer [42]. These DNA molecules can also carry signature epigenetic modifications, such as aberrant methylation patterns [43]. The methylation of the DNA is one of the most common mechanisms to regulate gene expression. Aberrant DNA methylation patterns have been described in cancer, with the activation of tumor-promoting genes and the silencing of tumor suppressors. The clinical utility of DNA methylation changes detected in saliva, however, was only investigated for HNCs with a focus on HPV-driven tumors [44]. Studies investigating tumor-derived DNA as biomarkers for cancer are summarized in Table 10.2.

Nucleic acids from tumors can also be detected in the form of RNA molecules, which comprise messenger RNAs (mRNAs) and noncoding RNAs (e.g., micro RNAs). Messenger RNAs are single-stranded RNA molecules that are complementary to one of the DNA strands of a gene and used for protein synthesis [60]. The

Table 10.2 Salivary nucleic acid-based biomarkers potentially useful for cancer detection.

Study	Cancer type	Samples		Fraction of saliva	Biomarker investigated	Methods	Findings
		Cancer	Controls				
Somatic mutations							
Liao et al. [45]	OSCC	10	27	Whole saliva	DNA–p53 mutation	PCR and sequencing	p53 codon 63 mutation was detected in 5/8 OSCC patients in saliva
Wang et al. [41]	HNSCC	93	-	Whole saliva, Oral rinse	DNA–somatic mutations in common cancer-associated genes	Multiplex PCR, Massively parallel sequencing	Tumor DNA was detected in 100% of oral cancer patients and 70% of other HNSCC patients' saliva
Izumchenko et al. [46]	OSCC	91	-	Not mentioned	DNA–somatic mutations in seven most frequently mutated genes	NGS	90.9% agreement between saliva and tumor for somatic mutations
Shanmugam et al. [47]	OSCC	121	-	Oral rinse	DNA–somatic mutations in the coding regions of seven most frequently mutated genes	NGS	93% agreement between saliva and tumor for investigated mutations
Wei et al. [42]	NSCLC	62	-	Cell-free supernatant of saliva	DNA–EGFR mutations (exon 19 deletion and p. L858 mutations)	EFIRM	EGFR L858R mutation and exon 19 deletions in NSCLC patients can be detected in saliva (AUC - 0.925)
Pu et al. [48]	NSCLC	17	-	Cell-free supernatant of saliva	DNA–EGFR mutations (exon 19 deletion and p. L858 mutations)	EFIRM	EGFR L858R mutation and exon 19 deletions in NSCLC patients can be detected in saliva (AUC-1.00)

DNA methylation changes							
Guerrero-Preston et al. [49]	Head and neck cancer (HNC)	OSCC-16, OPSCC-16	19	Oral rinse	DNA methylation - 27,578 CpG sites from 14,495 protein-coding gene promoters and 110 mRNA gene promoters	Microarray, qMSP	DNA promoter methylation changes in HOXA9 (AUC- 0.75) and NID2 (AUC - 0.73) in saliva can be used for HNC detection
Ovchinnikov et al. [44]	HNSCC	143	31	Whole saliva and buccal cell scrape	DNA methylation - RASSF1A, DPK1, p16INK4a	Methylation-specific PCR	RASSF1A, DPK1 and p16INK4a gene promoters are hypermethylated in HNSCC patients' saliva. A salivary methylation panel consisting of these markers could detect HNSCC with an overall accuracy of 81%.
Ovchinnikov et al. [50]	HNSCC	Tissue-6, Saliva-5′-CpG site – 49, 3′-CpG site - 45	5′-CpG–49, 3′-CpG - 45	Cellular deposit of saliva.	DNA methylation - MED15/ PCQAP (5′-CpG and 3′-CpG)	Sequencing and methylation-specific PCR	MED15/ PCQAP gene promoter is hypermethylated in HNSCC patients' saliva and can be used as a methylation marker to detect HNSCC (5' CpG; AUC-0.7, 3' CpG; AUC-0.63)

(Continued)

Table 10.2 Salivary nucleic acid-based biomarkers potentially useful for cancer detection. *Continued*

Study	Cancer type	Samples		Fraction of saliva	Biomarker investigated	Methods	Findings
		Cancer	Controls				
Lim et al. [51]	HNC	133	122	Whole saliva	DNA methylation - RASSF1α, p16INK4a, TIMP3, PCQAP 5′ and PCQAP 3′	Methylation-specific PCR	RASSF1α, PCQAP-5′, and PCQAP-3′ are hypermethylated in the saliva of HPV negative HNC and p16INK4a, PCQAP 5′and PCQAP 3′ are hypomethylated in the saliva of HPV positive HNC. These salivary methylation panels could detect HPV negative HNC (AUC-0.86) and HPV positive HNC (AUC-0.80)
Liyanage et al. [52]	OC, OPC	OC-54OPC-34	60	Whole saliva	DNA methylation - RASSF1α, p16INK4a, TIMP3, PCQAP 5′ and PCQAP 3′	Methylation-specific PCR	RASSF1A, TIMP3, and PCQAP/MED15 are hypermethylated in OC and OPC. All four targets tested had a combined AUC of 0.97 for OC and 0.96 for OPC
Giuliano et al. [53]	OPC	101	101	Oral rinse	DNA methylation - EPB41L3, HPV16 L1, L2, and E2 genes	Pyrosequencing	Salivary DNA methylation levels of EPB41L3 gene and HPV 16 L1, L2, and E2 genes can be used for OPC detection (AUC-0.885)

Messenger RNA (mRNA)							
Li et al. [54]	OSCC	32	32	Cell-free supernatant of saliva	mRNA	Microarray, RT-qPCR	DUSP1, H3F3A, IL1B, IL8, OAZ1, S100P, and SAT expression changes in saliva can be used for OSCC detection (AUC-0.95)
Brinkmann et al. [55]	OSCC	35	51	Cell-free supernatant of saliva	mRNA –DUSP1, IL8, IL1B, OAZ1, SAT1, and S100P	RT-qPCR	Salivary levels of IL8, IL1B, SAT1, and S100P were significantly elevated in OSCC patients
Horváth et al. [56]	OSCC	31	59	Cell-free supernatant of saliva	mRNA –DUSP1, IL8, IL1B, OAZ1, SAT1, S100P, and IL6	RT-qPCR	Salivary IL6 is downregulated in OSCC patients
Zhang et al. [57]	Lung cancer	42	74	Cell-free supernatant of saliva	mRNA	Microarray, RT-qPCR	Salivary BRAF, CCNI, EGFR, FGF19, FRS2, GREB1, and LZTS1 expression changes can be used for lung cancer detection (AUC-0.925)
Zhang et al. [58]	Pancreatic cancer	42	Chronic pancreatitis-30, healthy-42	Cell-free supernatant of saliva	mRNA	Microarray, RT-qPCR	Salivary KRAS, MBD3L2, ACRV1, and DPM1 expression changes can be used for pancreatic cancer detection (AUC-0.971)
Zhang et al. [59]	Breast cancer	40	73	Cell-free supernatant of saliva	mRNA	Microarray, RT-qPCR	CSTA, TPT1, IGF2BP1, GRM1, GRIK1, H6PD, MDM4, S100A8, CA6 mRNA are differentially expressed in the saliva of breast cancer patients

(Continued)

Table 10.2 Salivary nucleic acid-based biomarkers potentially useful for cancer detection. *Continued*

Study	Cancer type	Samples		Fraction of saliva	Biomarker investigated	Methods	Findings
		Cancer	Controls				
Lee et al. [60]	Ovarian cancer	32	46	Cell-free supernatant of saliva	mRNA	Microarray, RT-qPCR	Salivary H3F3A, SRGN, B2M, BASP1, AGPAT1, IL1B, and IER3 mRNA levels are downregulated in ovarian cancer patients. Salivary RNA panel consisting of B2M, BASP1, AGPAT1, IL1B, and IER3 can detect ovarian cancer (AUC-0.909)
Micro RNA (miRNA)							
Park et al. [61]	OSCC	50	50	Whole saliva and cell-free supernatant of saliva	miRNA (314 types)	RT-qPCR	Salivary miR-125a and miR-200a expression is downregulated in OSCC patients
Liu et al. [62]	OSCC	45	Verrucous leukoplakia-10, healthy-24	Cell free supernatant of saliva	miRNA–miRNA-31	RT-qPCR	Increased levels of miRNA-31 can be detected in the saliva of OSCC patients and can be used to diagnose OSCC (AUC-0.82)
Salazar et al. [63]	HNSCC	61	61	Whole saliva	miRNA (84 types)	PCR array, RT-qPCR	miR-9 and miR-134 are differently expressed in the saliva of HNSCC patients (AUC-0.85)

Romani et al. [64]	OSCC	89	58	Cell free supernatant of saliva	miRNA (1361 types)	Microarray, RT-qPCR	Salivary miR-106b-5p, miR-423–5p and miR-193b-3p can be used for OSCC detection (AUC - 0.98). Overexpression of miR-423–5p is associated with poor prognosis
Wan et al. [65]	HNC	101	113	Whole saliva	miRNA (9 types)	RT-qPCR	miR-9, miR-127, miR-134, miR-191, miR-222 and miR-455 differentially expressed in HPV negative HNC (AUC-0.82). miR-9, miR-134, miR-196b, miR-210, and miR-455 differentially expressed in HPV positive HNC (AUC-0.8)
Xie et al. [66]	Esophageal cancer	46	22	Whole saliva and cell-free supernatant of saliva collected under citric acid stimulation	miRNA (923 types)	Microarray, RT-qPCR	miR-10b, miR-144, and miR-451 are upregulated in the whole saliva and miR-10b, miR-144, miR-21, and miR-451 are upregulated in saliva supernatant of esophageal cancer patients
Fendereski et al. [67]	Esophageal cancer	5	5	Not mentioned	miRNA - miR-196a	RT-qPCR	Salivary miR-196a is upregulated in esophageal cancer patients

(Continued)

Table 10.2 Salivary nucleic acid-based biomarkers potentially useful for cancer detection. *Continued*

Study	Cancer type	Samples		Fraction of saliva	Biomarker investigated	Methods	Findings
		Cancer	Controls				
Xie et al. [68]	Pancreatic cancer	48	Benign-20, healthy-48	Cell-free supernatant of saliva collected with citric acid stimulation	miRNA (2006 types)	Microarray, RT-qPCR	Salivary miR-3679–5p is downregulated (AUC-0.673) and miR-940 is upregulated (AUC-0.680) in pancreatic cancer patients
Sazanov et al. [69]	Colorectal cancer	34	34	Cell-free supernatant of saliva	miRNA–miR-21	RT-qPCR	Salivary miR-21 is upregulated in colorectal cancer patients (Sensitivity-97%, Specificity-91%)
Long noncoding RNA (lncRNA)							
Tang et al. [70]	OSCC	20		Cell-free supernatant of saliva	lncRNA	RT-qPCR	HOTAIR was detected in a higher frequency in patients with metastasis
Xie et al. [71]	Pancreatic cancer	55	Benign-20, healthy-55		lncRNA (lncRNA) H19, HOTAIR, HOTTIP, MALAT1, PVT1	RT-qPCR	Salivary HOTAIR and PVT1 is upregulated in pancreatic cancer patients and can be used for pancreatic cancer detection (AUC-0.909)

EFIRM, *Electric field-induced release and measurement;* HNSCC, *head and neck squamous cell carcinoma;* NGS, *next generation sequencing; NSCLC, non-small cell lung carcinoma;* OC, *oral cancer;* OPC, *oropharyngeal cancer;* OPSCC, *oropharyngeal squamous cell carcinoma;* OSCC, *oral squamous cell carcinoma;* qMSP, *quantitative methylation-specific PCR;* RT-qPCR, *quantitative reverse transcription.*

majority of RNA-based diagnostic biomarkers are mRNAs, which are usually combined in gene expression panels. Gene expression changes with diagnostic potential have been reported for oral [72], lung [57], pancreas [58], breast [59], and ovarian cancer [60], among other malignancies. Noncoding RNAs are RNA molecules that are not translated into proteins, but play a role in the regulation of gene expression [73,74]. Due to their dynamic engagement in cellular pathways, active or passive changes in noncoding RNA expression are often observed in cancer [75]. Over the last two decades, studies have gathered much evidence suggesting the utility of noncoding RNA as diagnostic and therapeutic targets for a wide range of cancers. Among noncoding RNAs, micro RNAs (miRNA) are a group of approximately 22-nucleotide long, small RNA molecules known for their involvement in post transcription gene expression regulation [76]. They are believed to have control over 60% of the protein-coding genes in the human genome and hence their expression alters alongside almost every physiological or pathological change [75,76]. Cancer is one of such diseases where miRNA expression changes are well documented [76,77]. More importantly, these are not always just passive changes, as certain miRNA play a direct role in tumorigenesis and tumor suppression [77,78]. Parallel to expression changes in tumor tissue, miRNA expression in body fluids has also been shown to alter, allowing them to be used in liquid biopsy-based cancer detection [63,65]. Several studies have identified potentially diagnostic salivary miRNAs for head and neck cancers (i.e., miR-31, miR-9, miR-134) [62,63]. These authors also identified specific salivary miRNAs associated with patient overall survival and tumor stage [61]. More recently, salivary miRNA expression changes have also been identified for several other cancer types, such as esophageal [66,67], pancreatic [68], and colorectal cancer [69]. Changes in salivary miRNA expression in oral and systemic cancers are summarized in Table 10.2.

10.5.2 Protein-based cancer biomarkers in saliva

The salivary proteome, defined as the complete content of proteins present in the oral cavity, can provide information on oral and systemic health [72]. Detailed profiling of salivary peptides is the first step towards understanding disease pathogenesis and developing robust platforms for the detection of clinically relevant proteins [16,79]. From a biochemical perspective, proteins represent the most relevant molecules in saliva and are responsible for most of its functions: histatins and lysozymes present antimicrobial properties; salivary amylases are part of the digestive process; carbonic anhydrases promote buffering to maintain homeostasis, and immunoglobulins offer mucosal immunity [19]

In the past, the proteome of serum and plasma samples have been more widely investigated than saliva. There are 22 abundant proteins in serum and plasma samples, including albumin and immunoglobulins, and together they make up for approximately 99% of the total blood proteome [80]. The variation of protein concentrations in this biofluid can reach 10 orders of magnitude, and disease-specific biomarkers are more likely to be found in the lower concentration ranges. The over-representation of these few proteins limits the ability to detect lower abundant proteins that may have clinical potential. As a result, several methods have been developed to deplete highly abundant proteins in serum/plasma samples [81,82]. In contrast, the 20 most abundant proteins in human saliva correspond to only 40% of the total salivary proteome. Therefore the dynamic range of protein concentration does not pose as much difficulty as for blood, and expensive depletion methods are not required [80,83]. Table 10.3 summarizes studies investigating protein biomarkers for different cancer types.

Due to the close proximity to the tumors in the oral cavity, there is a plethora of proteomics data regarding oral cancers such as oral squamous cell carcinoma (OSCC). Many studies have focused on developing protein biomarkers panels that could diagnose cancer or be used to determine patients prognosis. For example, Chi and coauthors investigated the differences between the salivary proteome of OSCC patients and healthy controls. Out of 30 differentially abundant proteins, five protein candidates – MMP-1, PADI1, TNC, CSTA, and MMP3 – were used in a panel to diagnose OSCC with an AUC of 0.77–0.91 [84]. Interestingly, in this study, the authors showed that the same biomarkers were not detected in paired serum samples, highlighting their clinical relevance to the disease of the oral cavity. Recent evidence suggests an association between immune response and the tumor

Table 10.3 Protein biomarkers potentially useful for cancer detection

Study	Cancer type	Samples		Fraction of saliva	Methods	Findings
		Cancer	Controls			
Chi et al. [84]	OSCC	30	30	Unstimulated	MRM-MS	Higher levels of salivary MMF1, PADI1, TNC, CSTA and MMP3 in saliva samples from OSCC patients compared to saliva from controls
Feng et al. [85]	OSCC	12	4	Stimulated	Human protease arrayELISA	Cathepsin V, kallikrein 5 and ADAM9 were increased in the saliva of OSCC patients (AUC 0.77) compared to control saliva samples
Rajkumar et al. [86]	OSCC	100	100	Unstimulated	ELISA	Interleukin-8 increase is correlated with tumor stage, and salivary IL-8 has better diagnostic power than serum (AUC 0.951 vs AUC 0.821)
Chu et al. [87]	OSCC	117	115	Unstimulated	iTRAQMRM-MsELISA	Concentrations of CFH, FGA, and SERPINA1 are significantly correlated with advanced stages of OSCC
Van Der Merwe et al. [88]	HNSSC	58	38	Unstimulated	Bio-Plex Pro human cancer biomarker Panel 1	First evidence that angiogenic factors in saliva can be used as biomarkers forehead and neck cancer
Brooks et al. [89]	BC	49	49	Unstimulated	ELISA	VEGF, EGF and CEA elevated in the saliva of breast cancer patients, and a panel combining those markers had an AUC of 0.84
Wood and Streckfus [90]	BC	16	16	Stimulated	HPLC-MsWestern blot	Higher concentrations of LRP present in the saliva of breast cancer patients
Laidi et al. [91]	BC	29	31	Unstimulated	ELISA	Positive correlation of CA15−3 concentrations in serum and saliva, useful for monitoring BC progression

(*Continued*)

Table 10.3 Protein biomarkers potentially useful for cancer detection *Continued*

Study	Cancer type	Samples		Fraction of saliva	Methods	Findings
		Cancer	Controls			
Wu et al. [92]	Gastric	18	18	Unstimulated	MALDI-TOF	Identification of a mass fingerprint of the salivary proteins for gastric patients in comparison to controls, with four significantly altered peaks
Xiao et al. [93]	Gastric	20	20	Unstimulated	LC-Ms/ MSELISA	Salivary biomarker panel—CSTB, TPI1, DMBT1—to distinguish gastric cancer patients from controls with an AUC of 0.93
He et al. [94]	Gastric	149	124	Unstimulated	ECL immunoassay	CEA, AFP and CA19–9 increased in the saliva of gastric cancer patients, with a combined AUROC of 0.785
Xiao et al. [93]	NSCLC	36	36	Unstimulated	2D-DIGEMALDI-TOF	Three out of 16 potential salivary biomarkers were validated. Panel (calprotein, AZGP1, and HP) showed AUC of 0.90 for the detection of lung cancer
Ding et al. [95]	HCC	15	15	Unstimulated	iTRAQELISA	Validation of SOD2 as a potential diagnostic biomarker for HCC (AUC 0.90)
Chen et al. [96]	Ovarian	41	92	Unstimulated	ELISA	Measurement of CA125 in saliva had a better performance than serum for the diagnosis of ovarian cancer

AFP, *Alpha-fetoprotein;* AZGP1, *alpha-2-glycoprotein 1;* BC, *breast cancer;* CA15–3, *cancer antigen 15–3;* CEA, *carcinoembryonic antigen;* CTSB, *cystatin-B;* CSTA, *cystatin-A;* CFH, *complement factor H;* CA125, *cancer antigen 125;* 2D-DIGE, *two-dimensional difference gel electrophoresis;* DMBT1, *deleted in malignant brain tumors 1;* EGF, *epithelial growth factor;* ELISA, *enzyme-linked immunosorbent assay;* FGA, *fibrinogen alpha chain;* HPLC, *high-performance liquid chromatography;* iTRAQ, *isobaric tag for relative and absolute quantification;* LRP, *lung resistant protein;* LC-Ms/MS, *liquid chromatography coupled with mass spectrometry;* MMP-1, *matrix metalloproteinase-1;* MRM-Ms, *multiple-reaction monitoring mass spectrometry;* OSCC, *oral squamous cell carcinoma;* PADI1, *peptidyl arginine deiminase 1;* SERPINA1, *serpin family a member 1;* SOD2, *superoxide dismutase 2;* TNC, *tenascin;* TPI1, *triosephosphate isomerase 1;* VEGF, *vascular endothelial growth factor.*

microenvironment in the progression of cancer [97]. This led to several studies proposing salivary cytokines and chemokines as potential biomarkers for oral cancers [86,88,98,99]. Interleukin-6 (IL-6) and interleukin-8 (IL-8) were elevated in the saliva of OSCC patients compared to individuals with chronic periodontitis and healthy controls [100]. Furthermore, an increase in salivary IL-6 has been linked to the progression to OSCC in individuals with oral leukoplakia, a premalignant condition, whereas salivary IL-8 can differentiate OSCC from potentially malignant lesions, with better performance over serum IL-8 (AUC 0.951 vs AUC 0.821) [86,99]. However, the main challenge in the use of inflammation markers for the diagnosis of oral cancer is the presence of these proteins in other oral conditions such as chronic periodontitis and lichen planus [98].

Regarding cancer in distant organs, a lot of efforts have been focused on the detection of previously described serum-based biomarkers in saliva. For example, cancer antigen 15−3 (CA15−3), an FDA-approved biomarker for monitoring metastatic breast cancer, was detected in higher concentrations in the saliva of breast cancer patients, as well as vascular endothelial growth factor (VEGF), epidermal growth factor (EGF), and carcinoembryonic antigen (CEA) [89,91,101]. Furthermore, the combination of salivary VEGF and EGF was able to discriminate the breast cancer patient group from a healthy group with a sensitivity of 83% and specificity of 74% (AUC, 0.84) [89]. In addition, cancer biomarkers have been identified in the saliva of patients with gastric [102], ovarian [96], prostate [103], and hepatobiliary cancers [95]. For these cancers, a potentially important finding in the field of saliva diagnosis was achieved with the detection of salivary α-fetoprotein (AFP) and prostate-specific antigen (PSA) in the saliva of hepatocellular carcinoma (HCC) and prostate cancers, respectively, with a positive correlation between serum and salivary concentrations [103,104]. The application of these findings, however, has not yet been translated to the clinical management of these malignancies.

10.5.3 Glycans and glycoproteomic biomarkers for cancer

Glycosylation is the attachment of sugar groups to proteins, which plays a critical role in protein structure, function, and stability.

Alterations or modifications in the glycosylation machinery are involved in health and pathological conditions, including cancers [105]. Recent reports showing changes in glycoprotein levels in saliva associated with cancer are summarized in Table 10.4. Most studies regarding glycoconjugates are focused on oral cancers. For these malignancies, several glycoproteins with diagnostic potential have been identified, such as salivary N-acetylglucosamine (GlcNAc) or Gal1–4GlcNAc-containing glycans, which were significantly increased in the saliva of OSCC patients [106]. Several reports also indicate a higher proportion of salivary N-glycan, a class of glycoproteins, in these patients (79%) compared to healthy individuals (68%) [107]. However, recent reports show that the glycoconjugate composition in saliva varies according to cancer types, with fucosylated proteins being more abundant in gastric cancers, but not relevant in HCC.

Regarding cancers in distant organs, a number of proteins that are currently used for diagnosis are glycoproteins, such as cancer antigen 19–9 (CA 19–9) and CEA. Both CA19–9 and CEA are significantly increased in breast cancer patients' saliva (29.4 μ/mL and 279 ng/mL, respectively) compared to saliva from healthy controls (12.4 μ/mL and 155.3 ng/mL, respectively) [112]. In addition, salivary zinc-alpha-2-glycoprotein (AZGP1) is a promising biomarker for the detection of lung, prostate, pancreatic, and breast cancer. Liu et al. performed in-depth characterization of AZGP1 glycan structures, with over 300 structures identified. Interestingly, five glycan structures were uniquely found in the saliva collected from lung cancer patients. The authors suggest that a cancer-specific glycan signature can enhance the diagnostic potential for lung cancer [115]. These studies show the potential of saliva glycobiology to improve the early diagnosis and monitoring of cancer patients.

10.5.4 Exosome based salivary biomarkers for cancer

Exosomes are small EVs (30–200 nm) surrounded by a lipid bilayer membrane and released via exocytosis from cells [121,122]. Nearly all eukaryotic cells release exosomes under physiological and pathological conditions [123]. These vesicles transport several biomolecules as cargo (DNA, RNA, proteins, and lipids) which can be transferred between cells [124]. Since exosomes can cross epithelial

Table 10.4 Summary of salivary glycan biomarkers found in cancer patients

Study	Cancer type	Samples		Fraction of saliva	Biomarker investigated	Methods	Findings
		Cancer	Controls				
[106]	OSCC	18	5	Unstimulated whole mouth saliva	N glycans	Hydrophilic interaction ultra-performance liquid chromatography (HILIC–UPLC) for glycan identification; Mann–Whitney testing and principle component analysis (PCA) for differential N glycomic profile	11 distinctive N glycan peaks were identified between the patient saliva and healthy control that could carry biomarker potential
[107]	ESCC	16	25	Whole mouth saliva	N-acetylglucosamine (GlcNAc) or Gal1–4GlcNAc-containing N-glycans	Lectin Datura stramonium (DSA) and MALDI-TOF/TOF-MS	GlcNAc or Galβ1–4GlcNAc-containing N-glycans increased in ESCC patients (79.04%) compared with healthy control (63.20%)
[108]	Breast cancer	66 stage I, 62 stage II, 65 benign breast cyst	66	Whole mouth saliva	Galactosylated N- and O-glycans	*Bandeiraea simplicifolia* Lectin I and MALDI-TOF/TOF-MS	Distinctive variation of specific galactosylated N- and O-glycans between stages of cancer and healthy volunteers were observed

(*Continued*)

Table 10.4 Summary of salivary glycan biomarkers found in cancer patients *Continued*

Study	Cancer type	Samples		Fraction of saliva	Biomarker investigated	Methods	Findings
		Cancer	Controls				
[109], [110]	Gastric cancer	64 and 51	30 and 51	Whole mouth saliva	Alterations of N- and O-linked glycans	*Aleuria aurantia* Lectin (AAL)-magnetic particle, MALDI-TOF/TOF-MS	Mannose, T-antigen and blood group B antigen significantly increased and both fucosylated N- and O-glycans decreased
[111]	Hepatocellular carcinoma	60	50	Whole mouth saliva	Alterations of N glycans	MALDI-TOF/TOF-MS	Proportion of fucosylated N-glycans increased (84.8%) compared to healthy volunteers (73.1%), while sialylated N-glycans decreased (12.1% and 17.23% respectively)
[112]	Ductal adenocarcinoma	5	5	Whole mouth saliva	CA19–9 and CEA	Biochemical assay	Both CA19–9 and CEA increased significantly in patients saliva compared to saliva from healthy control

[113]	OSCC	60	60	Whole mouth saliva	Glycoprotein CA 125	ELISA	Mean CA concentration significantly increased in patients saliva (320.25) compared to healthy control (33.14)
[114]	Breast cancer	26	35	Unstimulated whole mouth saliva	Glycoprotein CA-15–3	ELISA	Salivary CA15–3 concentration was significantly high compared to healthy control ($P < 0.001$)
[115]	Lung cancer	10	20	Whole mouth saliva	N glycans	Lectin affinity chromatography	Unique glycopattern observed from salivary AZGP1 that could carry biomarker potential
[116]	OSCC	63	60	Whole mouth saliva	EGFR	ELISA	Salivary EGFR concentrations were higher in patient's saliva compared to healthy control ($P = 0014$)
[117]	OSCC	20	20	Whole mouth saliva	TSA	Biochemical assay	The level of TSA significantly increased in patient's saliva sample compared to healthy control

(Continued)

Table 10.4 Summary of salivary glycan biomarkers found in cancer patients *Continued*

Study	Cancer type	Samples		Fraction of saliva	Biomarker investigated	Methods	Findings
		Cancer	Controls				
[118]	OC	20	20	Unstimulated whole mouth saliva	Ratio between sialic acid and fucose	Biochemical assay	OC patients had significant elevation in the levels of salivary sialic acid, fucose, and their ratios compared to healthy control
[119]	Breast cancer	100	106	Whole mouth saliva	Sialic acid	Surface-enhanced Raman spectroscopy	Sialic acid concentration in saliva significantly increased in cancer patients compared to healthy control with sensitivity/ specificity 94%/ 98%
[120]	Ovarian cancer	15	37	Whole mouth saliva	Sialic acid	Surface-enhanced Raman spectroscopy	Sialic acid concentration increased with sensitivity/ specificity of 80%/ 100% in benign/ cancer patients

barriers (e.g., blood-brain barrier) and travel through the circulation, they play a significant role in communication not only between neighboring cells but also across cells in distant sites [125].

Although cells are continuously releasing exosomes, emerging evidence has shown that this release pattern can be altered under stress conditions [126,127]. In this context, the potential clinical relevance of exosomes has been explored in several diseases, including cancer. The concentration of tumor-derived exosomes from ovarian [128] and breast cancer cells [129] has been reported to be increased in comparison to normal controls. In addition, exosomes favor cancer progression [130] and have been associated with several cellular processes, such as proliferation, invasion, angiogenesis, and also with other "hallmarks of cancer" [131,132]. *In vitro* studies have shown that cells exposed to tumor cell-derived exosomes presented higher proliferation rates than control cells [133]. Exosomes play a role in stimulating angiogenesis [134], invasion [135], and metastasis, [136] facilitating the establishment of a premetastatic niche [137]. Also, exosomes have been described as promising biomarkers in the liquid biopsy field [138]. These discoveries reinforce the clinical potential of the use of exosomes in cancer research and highlight the need for further studies in the field.

Several techniques can be used to isolate and characterize exosomes [139–146]. A summary of these techniques and possible applications in the cancer field is shown in Fig. 10.4. The selected method influences the recovery and specificity of the isolated fraction. However, there is no individual optimal method for isolation. A number of considerations, including downstream applications, must be taken into account when determining the selected method [146]. In order to help researchers perform the most appropriate technique and to standardize currently used approaches in the field of EVs, the International Society for Extracellular Vesicles (ISEV) has published a guideline for 'minimal information for studies of extracellular vesicles ("MISEV")' in 2014 [147], which was updated in 2018 [146].

Exosomes can be found in several body fluids, such as blood, saliva, urine, cerebrospinal fluid, breast milk, semen, [121] and cell-conditioned media [145,148]. Currently, the most investigated source of exosomes in the literature is blood. However, saliva has been

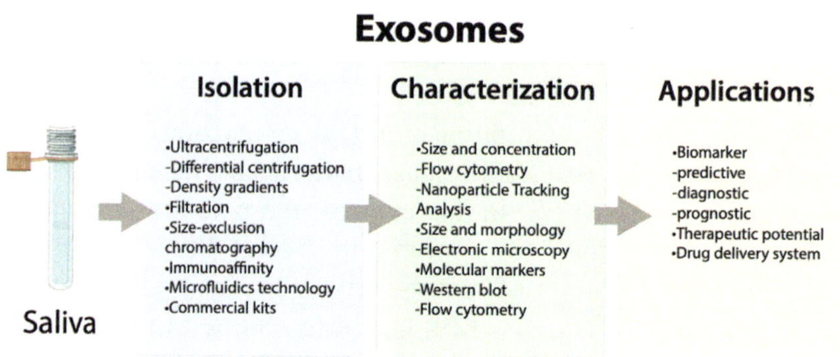

FIGURE 10.4

Schematic workflow summarizing the isolation, characterization and applications of salivary exosomes.

gaining attention as a matrix for cancer biomarkers, mainly because of the overlap of proteins and other molecules [16,149–151]. In addition, whole mouth saliva may contain highly abundant proteins, such as amylase, which may mask other less abundant proteins. [152]. However, this issue is addressed when using the exosome fraction which is purified and better characterized [153]. Salivary exosomes have been studied as biomarkers in different cancer types, including OSSC [154,155], pancreatic cancer [156,157], and lung cancer [158,159]. These studies using salivary exosomes for cancer detection are summarized in Table 10.5.

Even though salivary exosomes are promising biomarker candidates, some limitations in the field need to be addressed before translating these findings from basic research to a clinical setting. One of the main roadblocks to the translation of exosomes to the clinic is the lack of standardization of the techniques that are used to isolate and characterize these vesicles. Since they can be isolated from different biofluids, an appropriate standardization needs to be elucidated. In addition, most studies conducted using saliva have used small cohorts of patients and larger multicenter research must be conducted to validate those findings. Further studies in the salivary exosome field are essential to conclusively elucidate their role as biomarkers in cancer.

Table 10.5 Summary of studies on salivary exosomes as potential cancer biomarkers

Study	Cancer Type	Samples	Fraction of Saliva	Biomarker Investigated	Methods	Findings
Tang et al. [155]	HPV-driven OPC	10 p16INK4a (p16)-positive patients20 healthy controls	Unstimulated whole saliva	Protein	Isolation Ultracentrifugation Characterization NTA TEM WB Mass spectrometry	More abundant in OPC patients: ALDOA, GAPDH, A/B LDHA and LDHB, PGK1, PKM
Winck et al. [160]	OSCC	24 OSCC patients10 healthy individuals	Unstimulated whole saliva	Protein	Isolation Ultracentrifugation Characterization NTA TEM WB Mass spectrometry	More abundant n OSCC patients: A2M, HPA, MUC5B, LGALS3BP, IGHA1, PIP, PKM, and GAPDH
Sun et al. [159]	Lung cancer	Three lung cancer patientsSix healthy individuals	Unstimulated whole saliva	Protein	Isolation Ultracentrifugation Characterization NTA Mass spectrometry	Exclusively found in lung cancer patients: ANXA1, ANXA2, ANXA3, ANXA5, ANXA6, ANXA11, NPRL2, CEACAM-1, MUC1, PROM1, H4HIST1H4A, and TNFAIP3.
Sun et al. [158]	Lung cancer	Six lung cancer patientsSix healthy individuals	Unstimulated whole saliva	Protein	Isolation Ultracentrifugation Characterization NTA TEM Mass spectrometry	BPIFA1, CRNN, MUC5B, and IQGAP
Lau et al. [157]	Pancreatic cancer	C57BL/6 mice tumor and control group (22 animals per group).	Unstimulated pooled whole saliva	mRNA	Isolation Magnetic bead-based exosome extraction Characterization TEM	Upregulated in pancreatic cancer: Apbb1ip, Aspn, BCO31781, Daf2, Foxp1, Gng2, Incenp

(Continued)

Table 10.5 Summary of studies on salivary exosomes as potential cancer biomarkers *Continued*

Study	Cancer Type	Samples	Fraction of Saliva	Biomarker Investigated	Methods	Findings
Machida et al. [156]	Pancreaticobiliary cancer	12 patients with pancreatobiliary tract cancers13 healthy individuals	Unstimulated whole saliva	miRNA	Isolation total exosome isolation reagent (invitrogen)	Upregulated in pancreaticobiliary cancer: miR-1246 and miR-4644
Langevin et al. [154]	HNSCC	Five patients with incident primary HNSCCFive cancer-free controls	Unstimulated whole saliva	miRNA	Isolation Ultracentrifugation Characterization NTA TEM	Upregulated in HNSCC: miR-486–5p, miR-486–3p, and miR-10b-5p
Farag et al. [161]	HNSCC	14 patients with OSCC6 healthy individuals17 smokers	Unstimulated whole saliva	miRNA	Isolation Ultracentrifugation Characterization TEM	Upregulated in HNSCC: microRNA200a and microRNA-134
He et al. [162]	OSCC	Four OSCC patientsFour healthy individuals	Unstimulated whole saliva	miRNA	Isolation ExoQuick-TC™ (SBI) Characterization NTA TEM WB	Upregulated in OSCC: MiR-24–3p

ALDOA, *aldolase;* ANXA1, ANXA2, ANXA3, ANXA5, ANXA6, ANXA11, *Annexin family members;* A2M, *alpha-2-macroglobulin;* CEACAM-1, *carcinoembryonic antigen-related cell adhesion molecule 1'* MUC1, *mucin 1;* GAPDH, *glyceraldehyde-3-phosphate dehydrogenase;* H4HIST1H4A, *histone;* HNSCC, *head and neck squamous cell carcinomas;* HPA, *haptoglobin alpha chain;* IGHA1, *immunoglobulin alpha-1 chain c region;* LDHA and LDHB, *lactate dehydrogenase A/B;* LGALS3BP, *galectin3-binding protein;* MUC5B, *mucin-5B;* NTA, *nanoparticle tracking analysis;* NPRL2, *nitrogen permease regulator 2-like protein;* OSCC, *oral squamous cell carcinoma;* OPC, *oropharyngeal cancer;* PIP, *prolactin-inducible protein;* PGK1, *phosphoglycerate kinase 1;* PKM, *pyruvate kinase isozymes M1/M2;* PROM1, *prominin-1;* TEM, *transmission electron microscopy;* TNFAIP3, *tumor necrosis factor-alpha-induced protein 3;* WB, *western blot.*

10.6 Oral microbiome as a biomarker for cancer

The oral microbiome represents all the microorganisms that reside in the oral cavity [163]. The understanding of the human microbiome and its role in human health has been increasing after the completion of the Human Microbiome Project in 2009 [164]. Since then, changes in the oral microbiome have been linked to several systemic conditions, for example, diabetes mellitus, cardiovascular diseases, inflammatory diseases, and cancer [165]. The first evidence of the involvement of microorganisms in cancer was reported by Marshall et al. in the 80s. They observed that patients with persistent *Helicobacter pylori* infection had a higher incidence of gastric ulcers and cancer [166]. Until recently, the study of the microbiome was limited to bacterial culture and identification. However, this method ignores a large proportion of "non-culturable" bacteria present in the oral cavity, which can reach upto 80% of the microbiome [163]. Therefore next-generation sequencing is currently the most reliable tool to assess the human microbiome.

Due to the proximity, the oral microbiome has been linked to cancers in the oral cavity. For instance, persistent HPV infection is a major cause of OSCC. Approximately 25% of OSCC patients are positive for HPV, and the infection with this virus can increase the risk of developing oral cancer by upto threefold [51,167,168]. Additionally, coinfection with other viruses have also been reported as potential biomarkers for head and neck cancers, such as Epstein–Barr virus (EBV) [169], Polyoma BK Virus (BKPyV) [170] and herpes simplex virus (HSV) [171]. The majority of the studies on the oral microbiome and cancers are limited to the description of the microbial communities in the oral cavity. During the progression of OSCC, Yang et al. observed an increase in the genus *Fusobacterium* [172]. They developed a panel—*Fusobacterium periodonticum*, *Streptococcus mitis*, and *Porphyromonas pasteri*—that had an AUC of 0.956 in discriminating OSCC stage 4 from healthy individuals. Another group of bacteria with carcinogenic properties is *Porphyromonas gingivalis*, which is a common cause of periodontal disease. Using a chemically-induced carcinogenesis model in mice, Wen et al. reported that infection with *P. gingivalis* increased oral lesions multiplicity and size in OSCC [173]. They observed that infected lesions presented a cancer-promoting tumor microenvironment due to immune modulation. Furthermore, a *meta*-analysis conducted by Corbella et al. showed a correlation between periodontitis and gastric, pancreatic, prostate, breast cancers, Non-Hodgkin lymphoma, and other hematological neoplasms, indicating the importance of the understanding of the role of the oral microbiome in cancer [174]. Other signature changes in the oral microbiota of oral cancer have been reported. For instance, an increase in the Lactobacillus genus is observed in the saliva of patients with higher TNM stages for HNSCC [175]. Furthermore, an oral microbiome panel developed by Lim et al. showed an excellent performance to discriminate cancers in the oral cavity from oropharyngeal tumors with an AUC of 0.974 [167].

Studies on the use of oral microbiome for the diagnosis of tumors outside of the oral cavity are still in their initial stages. Three common strategies are applied when investigating changes in microbial composition: [1] overall changes in microbial diversity; [2] combination of changes in the abundance of specific organisms; [3] combination of changes in abundance of different organisms to generate a panel. As for overall changes in diversity, Chen et al. observed a widespread decrease in the microbial diversity in the saliva of esophageal squamous cell carcinoma patients compared to controls [176]. However, more recent reports indicate an increase in diversity for other cancer types, for example, gastric cancer, suggesting that more studies are necessary to clarify the role of cancer in the diversity of organisms in the oral cavity [177]. An increase in the salivary abundance of specific organisms has been reported for colorectal cancer (*Fusobacterium nucleatum*), pancreatic cancer (*P. gingival*is and *Fusobacterium spp.*), and lung cancer (*Capnocytophaga sp. and Veillonela sp.*) [178–180]. Finally, due to the number of colonies of microorganisms in the oral cavity, combinatorial analysis of taxa has also been used to diagnose cancer. A two-taxon model—*Streptococcus mitis* and *Neisseria elongate*, developed by Farrel et al. showed an AUC of 0.90 to distinguish pancreatic cancer patients from healthy controls [181]. Similarly, Wu et al. identified a bacterial cluster composed of mainly *Prevotella sp* that had a 0.76 AUC for the diagnosis of gastric cancer in a high-risk population [182]. Together, these studies suggest that the oral microbiome represents an informative tool for cancer diagnosis. The 16S rRNA gene profiling of microorganisms using NGS has allowed for a breakthrough in this field of study, but there are still some challenges to be overcome. For example, the standardization of DNA isolation methods and bioinformatics analysis is crucial for clinical purposes. Furthermore, recent advances in omics technologies—metabolomics, proteomics, genomics, will allow for a more in-depth and detailed study regarding the organisms that colonize the mouth and their relationship with cancers.

10.7 Salivary metabolites for cancer diagnosis

Metabolites are small products generated by the body when breaking down drugs, food, or its own tissue [183]. The study of these molecules, that is, metabolomics consists of a sensitive approach that enables the monitoring of alterations in biological systems. The complete set of metabolites (metabolome) interacts with and dynamically influences other omics levels (genomics, transcriptomics and proteomics) [184]. The detection of metabolites is a valuable tool to investigate key evidence on drivers of physiological and pathological biological processes, including cancer [184]. Metabolic alteration is known as a cancer hallmark [131]. Cancer cells undergo metabolic reprogramming to promote and maintain growth and survival [185]. Therefore metabolomics studies have been explored in the cancer field [186], particularly to discover new biomarkers for future translation into clinical practice [187].

Metabolites can be assessed in several body fluids, including saliva [183,188]. The most common fraction of saliva collected for metabolomics studies is unstimulated whole mouth saliva. Currently, the majority of studies using saliva as a source of metabolites for cancer research are related to oral [189] and breast cancers [190]. Nevertheless, studies have also been reported on gastric [191], head and neck [192], and pancreatic [193] cancers, in addition to glioblastoma [194]. A summary of studies on salivary metabolomics in cancer patients is shown in Table 10.6. All studies reviewed in Table 10.6 have used unstimulated whole saliva collected using the spit technique, except for Xavier Assad et al. [198] in which the authors have used a cotton swab (Salivette).

Table 10.6 Summary of studies on salivary metabolites found in cancer patients.

Study	Cancer type	Samples	Fraction of saliva	Methods	Findings
Lohavanichbutr et al. [189]	Oral	101 OCC58 OPC patients35 healthy individuals	Unstimulated whole saliva	Nuclear magnetic resonance spectroscopy and Mass spectrometry	Decreased levels of glycine and proline in OCC compared to healthy individualsNo differences in concentration of metabolites between OCC patients presenting or not nodal metastasis and between OPC patients with healthy individuals.
Ishikawa et al. [195]	Oral cancer	26 oral cancer patients44 healthy individuals	Unstimulated whole saliva	Mass spectrometry	11 metabolites levels correlated with PET/CT scans of patients with oral cancer and were able to differentiate cancer patients from healthy individuals.
Mikkonen et al. [192]	Head and neck	45 head and neck cancer patients30 healthy individuals	Unstimulated whole saliva	Nuclear magnetic resonance spectroscopy	The combination of fucose, glycine, methanol, and proline showed promising results as a diagnostic panel for head and neck cancer.
Ohshima et al. [196]	Oral	22 oral cancer patients21 healthy individuals	Unstimulated whole saliva	Mass spectrometry	25 metabolites were able to distinguish oral cancer patients from healthy individuals, namely choline, p-hydroxyphenylacetic acid, and 2-hydroxy-4-methylvaleric acid; valine, 3-phenyllactic acid, leucine, hexanoic acid, octanoic acid, terephthalic acid, γ-butyrobetaine, and 3-(4-hydroxyphenyl)propionic acid; and isoleucine, tryptophan, 3-phenylpropionic acid, 2-hydroxyvaleric acid, butyric acid, cadaverine, 2-oxoisovaleric acid, N6,N6,N6-trimethyllysine, taurine, glycolic acid, 3-hydroxybutyric acid, heptanoic acid, alanine, and urea

Takayama et al. [197]	Breast	111 breast cancer patients61 healthy individuals	Unstimulated whole saliva	Mass spectrometry	The combination of six polyamines (N8-Ac-SPD, N1-Ac-SPD, CAD, DAc-SPD, PUT, and Ac-PUT) showed promising potential as a panel to diagnose breast cancer compared to healthy controls
Xavier Assad et al. [198]	Breast	23 breast cancer patients35 healthy individuals	Unstimulated whole saliva	Mass spectrometry	31 upregulated metabolites in patients with breast cancer compared to healthy individuals.
Sugimoto et al. [199]	Oral Pancreatic Breast	69 Oral cancer patients18 pancreatic cancer patients30 breast cancer patients87 healthy induviduals	Unstimulated whole saliva	Mass spectrometry	28 metabolites were increased in samples from patients with breast cancer, 48 for pancreatic cancer and 28 for oral cancer.
Asai et al. [193]	Pancreatic	39 pancreatic cancer patients26 healthy individuals	Unstimulated whole saliva	Mass spectrometry	Increased levels of polyamines (spermine; N1-acetylspermidine; N1-acetylspermine; 2-aminobutanoate) in cancer patients
Garcia-Villaescusa et al. [194]	Glioblastoma	10 glioblastoma patients120 healthy individuals	Unstimulated whole saliva	Nuclear magnetic resonance spectroscopy	Increased levels of leucine, valine, isoleucine, propionate, alanine, acetate, ethanolamine, and sucrose in glioblastoma patients
Chen et al. [191]	Gastric	104 gastric cancer patients220 healthy individuals	Unstimulated whole saliva	Mass spectrometry	The combination of taurine, glycine, glutamine, ethanolamine, histidine, alanine, glutamic acid, hydroxylysine, proline, lysine shows a promising panel for the diagnosis of gastric cancer

Ac-PUT; Ac-SPM; CAD, *cadaverine; N1-acetyl-putrescine; N1-acetyl-spermidine; N1-Ac-SPD; N8-acetyl-spermidine; N8-Ac-SPD; N1-acetyl-spermine, N1N12-diacetyl-spermine, DAc-SPD; 1,6-Diaminohexane,* DAH; *PET/CT, 18F-fluorodeoxyglucose (18F-FDG) positron emission tomography (PET)/ computed tomography (CT);* OCC, *oral cavity squamous cell carcinoma;* OPC, *oropharyngeal squamous cell carcinoma;* SPD, *spermidine.*

The most common techniques for the generation of metabolomics data are spectroscopic (nuclear magnetic resonance spectroscopy) and spectrometric methodologies (mass spectrometry). The latter is commonly coupled to chromatographic systems, such as gas or liquid chromatography [200]. Two different methods can be performed for the analysis of metabolites, an untargeted and a targeted methodology [201]. The technique should be selected according to the aim of the study. The untargeted approach consists of a global profiling of all metabolites in the system, while the targeted approach analyses specific metabolic pathways or validates results obtained from the untargeted approach [202]. Different reports have suggested the use of salivary metabolites for the diagnosis and prognosis of different cancer types and are summarized in Table 10.6.

Although the analysis of salivary metabolome in cancer patients shows promising data, there is still the need for further studies and validations to be conducted using larger cohorts. In general, the combination of metabolites in a panel generates superior specificity and sensitivity values, when compared to single molecules, thus resulting in a higher potential for future applications. Also, despite the knowledge that saliva's composition is a reflection of overall health, its use for distant diseases from the oral cavity has not been explored as much as for oral diseases. There is increasing attention in the metabolomics field using saliva samples for different cancer types. Several reports have suggested the use of these biomarkers as a potential diagnostic tool with high specificity and sensitivity. Therefore further research must be conducted for future implementation in a clinical setting.

10.8 Limitations of salivary biomarkers

The current pipeline for the development of novel saliva-based diagnostic markers often includes a discovery and a validation phase [203]. During the discovery phase, unbiased and semiquantitative techniques such as liquid chromatography–mass spectrometry (LC-Ms) are applied to identify potential biomarkers in complex samples [204]. Following the discovery stage, a selected subgroup of molecules enters the verification phase, where orthogonal targeted measurement techniques are used to confirm the abundance of candidate biomarkers in normal and diseased specimens [205]. Finally, the verified candidates are used in high throughput and multiplexing techniques (e.g., automated immunoassays) and the clinical performance of these biomarkers is assessed in a larger number of human samples [204]. The main limitation for the use of saliva diagnosis involves the lack of robust, standard operating procedures (SOPs) for the collection, storage, preparation, and analysis of saliva. This causes most saliva-based tests to be designed, manufactured and validated within a single laboratory, therefore classified as laboratory developed tests (LDTs), which are rarely submitted for further approval for general clinical use [206].

10.9 Summary Points

- The clinical relevance of salivary diagnostics gained attention in the current SARS-CoV-2 global pandemic. The collection of samples in a timely, cost-effective, accurate, non-invasive manner became paramount to combat the rapidly rising spread of the virus.
- Human saliva is a biofluid that can be sampled non-invasively and contains biomarkers of clinical relevance for oral and systemic diseases such as cancer. Diagnosis, prognosis, and monitoring of cancer patients frequently require painful, invasive procedures (e.g., tumor tissue biopsies) and repeated blood collections.
- The discovery of saliva-based biomarkers or salivary 'omics' offers unique opportunities to bypass invasive measurements by utilizing oral fluids. In this context, significant advances have been achieved in the field of saliva omics.
- The accurate and reliable detection of early-stage cancers combined with non-invasive modes of sample collection is the holy grail of molecular diagnostics.

10.10 Future perspective

Detecting cancer at an early stage of tumorigenesis can significantly affect patient discomfort, prognosis, therapeutic intervention, survival rates, and recurrence. The translation of salivary biomarkers for clinical applications is a challenging process with very few successful implementations, at this stage. This is because several limitations need to be overcome prior to salivary diagnostics entering clinical practice: the lack of standardized saliva sample collection devices; biomarker levels are affected by diurnal variations; biomarkers/analytes are present at low concentration and requires advanced, highly sensitive, and high-throughput techniques (e.g., EFIRM, NGS); and environmental factors affecting saliva production. Its successful application in clinical practice will depend on multidisciplinary, collaborative partnerships across continents with a relevant well defined clinical question and with well-defined parameters of recruitment. Thus, the use of saliva as a diagnostic fluid will be increasingly accepted as a biofluid of choice because it provides a safe and effective means by which to evaluate patients and personalize their treatment.

Acknowledgments

CP is currently receiving funding from Cancer Australia (APP1145657, 2012560), NIH R21, NHMRC Ideas Grant (APP 2002576 and APP) and RBWH Foundation Project Grants. She has one patent on heart failure licensed to ESN Cleer where she receives upfront payments and has assigned her IP to Viome as an upfront payment. Figures were created using Servier Medical Art templates, licensed under a Creative Commons Attribution 3.0 Unported License; https://smart.servier.com.

References

[1] Siegel RL, Miller KD, Fuchs HE, Jemal A. Cancer statistics, 2021. CA: A Cancer J Clin 2021;71(1):7–33.

[2] Schiffman JD, Fisher PG, Gibbs P. Early detection of cancer: past, present, and future. Am Soc Clin Oncol Educ Book 2015;35:57–65.

[3] Fisher BM, Tang K, Warkiani M, Punyadeera C, Batstone M. A pilot study for presence of circulating tumour cells in adenoid cystic carcinoma. Int J Oral Maxillofac Surg 2021;50(8):994–8.

[4] Kuhn P, Bethel K. A fluid biopsy as investigating technology for the fluid phase of solid tumors. Phys Biol 2012;9(1):010301.

[5] Punyadeera C, Dimeski G, Kostner K, Beyerlein P, Cooper-White J. One-step homogeneous C-reactive protein assay for saliva. J Immunol Methods 2011;19–25.

[6] Pfaffe T, Cooper-White J, Beyerlein P, Kostner K, Punyadeera C. Diagnostic potential of saliva: current state and future applications. Clin Chem. 2011;675–87.

[7] Ovchinnikov DA, Cooper MA, Pandit P, Coman WB, Cooper-White JJ, Keith P, et al. Tumor-suppressor gene promoter hypermethylation in saliva of head and neck cancer patients. Transl Oncol. 2012. p. 321–6.

[8] Zhang W, Du Y, Wang ML. Noninvasive glucose monitoring using saliva nano-biosensor. Sens Bio-Sensing Res 2015;4:23–9.

[9] Tang KD, Vasani S, Menezes L, Taheri T, Walsh LJ, Hughes BG, et al. Oral HPV16 DNA as a screening tool to detect early oropharyngeal squamous cell carcinoma. Cancer Sci 2020;111(10):3854.

[10] Tang KD, Vasani S, Taheri T, Walsh LJ, Hughes BG, Kenny L, et al. An occult HPV-driven oropharyngeal squamous cell carcinoma discovered through a saliva test. Front Oncol 2020;10:408.

[11] Banavar G, Ogundijo O, Toma R, Rajagopal S, Lim Y, Tang KD, et al. The salivary metatranscriptome as an accurate diagnostic indicator of oral cancer. npj Genomic Medicine 2021;6:105.

[12] Keller L, Pantel K. Unravelling tumour heterogeneity by single-cell profiling of circulating tumour cells. Nat Rev Cancer 2019;19(10):553–67.

[13] Siravegna G, Marsoni S, Siena S, Bardelli A. Integrating liquid biopsies into the management of cancer. Nat Rev Clin Oncol 2017;14(9):531–48.

[14] Carlton M, Voisey J, Jones L, Parker TJ, Punyadeera C, Cuttle L. An exploratory study demonstrating that salivary cytokine profiles are altered in children with small area thermal injury. J Burn Care Res 2021;.

[15] Kevadiya BD, Machhi J, Herskovitz J, Oleynikov MD, Blomberg WR, Bajwa N, et al. Diagnostics for SARS-CoV-2 infections. Nat Mater 2021;20(5):593–605.

[16] Pfaffe T, Cooper-White J, Beyerlein P, Kostner K, Punyadeera C. Diagnostic potential of saliva: current state and future applications. Clin Chem 2011;57(5):675–87.

[17] Dawes C, Pedersen AML, Villa A, Ekström J, Proctor GB, Vissink A, et al. The functions of human saliva: a review sponsored by the World Workshop on Oral Medicine VI. Arch Oral Biol 2015;60(6):863–74.

[18] Denny P, Hagen FK, Hardt M, Liao L, Yan W, Arellanno M, et al. The proteomes of human parotid and submandibular/sublingual gland salivas collected as the ductal secretions. J Proteome Res 2008;7(5):1994–2006.
[19] Kaczor-Urbanowicz KE, Wei F, Rao SL, Kim J, Shin H, Cheng J, et al. Clinical validity of saliva and novel technology for cancer detection. Biochim Biophys Acta Rev Cancer 2019;1872(1):49–59.
[20] Chiappin S, Antonelli G, Gatti R, De Palo EF. Saliva specimen: a new laboratory tool for diagnostic and basic investigation. Clinica Chim Acta 2007;383(1):30–40.
[21] Subbarao KC, Nattuthurai GS, Sundararajan SK, Sujith I, Joseph J, Syedshah YP. Gingival crevicular fluid: an overview. J Pharm Bioallied Sci 2019;11(Suppl 2):S135–9.
[22] Sun CX, Bennett N, Tran P, Tang KD, Lim Y, Frazer I, et al. A pilot study into the association between oral health status and human papillomavirus—16 infection. Diagnostics. 2017;7(1):11.
[23] Bellagambi FG, Lomonaco T, Salvo P, Vivaldi F, Hangouët M, Ghimenti S, et al. Saliva sampling: methods and devices. An overview. TrAC Trends Anal Chem 2020;124:115781.
[24] Carpenter GH, Proctor GB, Anderson LC, Zhang XS, Garrett JR. Immunoglobulin A secretion into saliva during dual sympathetic and parasympathetic nerve stimulation of rat submandibular glands. Exp Physiol 2000;85(3):281–6.
[25] Aro K, Wei F, Wong DT, Tu M. Saliva liquid biopsy for point-of-care applications. Front Public Health 2017;5 77-.
[26] Navazesh M. Methods for collecting saliva. Ann N Y Acad Sci 1993;694:72–7.
[27] Michishige F, Kanno K, Yoshinaga S, Hinode D, Takehisa Y, Yasuoka S. Effect of saliva collection method on the concentration of protein components in saliva. J Med Invest 2006;53(1-2):140–6.
[28] Khurshid Z, Zohaib S, Najeeb S, Zafar MS, Slowey PD, Almas K. Human saliva collection devices for proteomics: an update. Int J Mol Sci 2016;17(6).
[29] Gröschl M, Rauh M. Influence of commercial collection devices for saliva on the reliability of salivary steroids analysis. Steroids. 2006;71(13-14):1097–100.
[30] Carlson AJ, Crittenden AL. The relation of ptyalin concentration to the diet and to the rate of salivary secretion. Proc Soc Exp Biol Med 1909;7(2):52–4.
[31] Ryu O, Atkinson J, Hoehn G, Illei G, Hart T. Identification of parotid salivary biomarkers in Sjögren's syndrome by surface-enhanced laser

desorption/ionization time-of-flight mass spectrometry and two-dimensional difference gel electrophoresis. Rheumatology. 2006;45 (9):1077–86.

[32] Truelove EL, Bixler D, Merritt AD. Simplified method for collection of pure submandibular saliva in large volumes. J Dent Res 1967;46 (6):1400–3.

[33] Navazesh M, Kumar SK. Measuring salivary flow: challenges and opportunities. J Am Dental Assoc 2008;139:35S–40S.

[34] Kutscher AH, Mandel ID, Zegarelli EV, Denning C, Eriv A, Ruiz L, et al. A technique for collecting the secretion of minor salivary glands: I. Use of capillary tubes. J Oral Ther Pharmacol 1967;3(5):391–2.

[35] Wang Z, Shen M-M, Liu X-J, Si Y, Yu G-Y. Characteristics of the saliva flow rates of minor salivary glands in healthy people. Arch oral Biol 2015;60(3):385–92.

[36] Fortes MB, Diment BC, Di Felice U, Walsh NP. Dehydration decreases saliva antimicrobial proteins important for mucosal immunity. Appl Physiol Nutr Metab 2012;37(5):850–9.

[37] Rapp-Santos KJ, Altamura LA, Norris SL, Lugo-Roman LA, Rico PJ, Hofer CC. Comparison of saliva collection methods for the determination of salivary cortisol levels in Rhesus macaques (*Macaca mulatta*), Cynomolgus Macaques (*Macaca fascicularis*), and African Green monkeys (*Chlorocebus aethiops*). J Am Assoc Lab Anim Sci 2017;56(2):181–9.

[38] Rai B, Kaur J, Foing BH. Stress, workload and physiology demand during extravehicular activity: a pilot study. N Am J Med Sci 2012;4 (6):266–9.

[39] Schwarzenbach H, Hoon DSB, Pantel K. Cell-free nucleic acids as biomarkers in cancer patients. Nat Rev Cancer 2011;11(6):426–37.

[40] Thierry AR, El Messaoudi S, Gahan PB, Anker P, Stroun M. Origins, structures, and functions of circulating DNA in oncology. Cancer Metastasis Rev 2016;35(3):347–76.

[41] Wang Y, Springer S, Mulvey CL, Silliman N, Schaefer J, Sausen M, et al. Detection of somatic mutations and HPV in the saliva and plasma of patients with head and neck squamous cell carcinomas. Sci Transl Med 2015;7(293):293ra104–293ra1ra104.

[42] Wei F, Lin C-C, Joon A, Feng Z, Troche G, Lira ME, et al. Noninvasive saliva-based EGFR gene mutation detection in patients with lung cancer. Am J Respir Crit Care Med 2014;190(10):1117–26.

[43] Pan Y, Liu G, Zhou F, Su B, Li Y. DNA methylation profiles in cancer diagnosis and therapeutics. Clin Exp Med 2018;18(1):1–14.

[44] Ovchinnikov DA, Cooper MA, Pandit P, Coman WB, Cooper-White JJ, Keith P, et al. Tumor-suppressor gene promoter hypermethylation in saliva of head and neck cancer patients. Transl Oncol 2012;5 (5):321–6.
[45] Liao PH, Chang YC, Huang MF, Tai KW, Chou MY. Mutation of p53 gene codon 63 in saliva as a molecular marker for oral squamous cell carcinomas. Oral Oncol 2000;36(3):272–6.
[46] Izumchenko E, Hasina R, Hariharan AK, Shanmugam A, Irusappan S, Maji S, et al. Detection of somatic mutations in saliva of patients with oral cavity squamous cell carcinoma. J Clin Oncol 2020;38 (15_suppl):6562.
[47] Shanmugam A, Hariharan AK, Hasina R, Nair JR, Katragadda S, Irusappan S, et al. Ultrasensitive detection of tumor-specific mutations in saliva of patients with oral cavity squamous cell carcinoma. Cancer. 2021;127(10):1576–89.
[48] Pu D, Liang H, Wei F, Akin D, Feng Z, Yan Q, et al. Evaluation of a novel saliva-based epidermal growth factor receptor mutation detection for lung cancer: a pilot study. Thorac Cancer 2016;7 (4):428–36.
[49] Guerrero-Preston R, Soudry E, Acero J, Orera M, Moreno-López L, Macía-Colón G, et al. NID2 and HOXA9 promoter hypermethylation as biomarkers for prevention and early detection in oral cavity squamous cell carcinoma tissues and saliva. Cancer Prev Res 2011;4 (7):1061–72.
[50] Ovchinnikov DA, Wan Y, Coman WB, Pandit P, Cooper-White JJ, Herman JG, et al. DNA methylation at the novel CpG sites in the promoter of MED15/PCQAP gene as a biomarker for head and neck cancers. Biomarker Insights 2014;9 BMI.S16199.
[51] Lim Y, Wan Y, Vagenas D, Ovchinnikov DA, Perry CF, Davis MJ, et al. Salivary DNA methylation panel to diagnose HPV-positive and HPV-negative head and neck cancers. BMC Cancer 2016;16(1):749.
[52] Liyanage C, Wathupola A, Muraleetharan S, Perera K, Punyadeera C, Udagama P. Promoter hypermethylation of tumor-suppressor genes p16INK4a, RASSF1A, TIMP3, and PCQAP/MED15 in salivary DNA as a quadruple biomarker panel for early detection of oral and oropharyngeal cancers. Biomolecules. 2019;9(4):148.
[53] Giuliano AR, Nedjai B, Lorincz AT, Schell MJ, Rahman S, Banwait R, et al. Methylation of HPV 16 and EPB41L3 in oral gargles: associations with oropharyngeal cancer detection and tumor characteristics. Int J Cancer 2019;.

[54] Li St. Y, John MAR, Zhou X, Kim Y, Sinha U, Jordan RCK, et al. Salivary transcriptome diagnostics for oral cancer detection. Clin Cancer Res 2004;10(24):8442–50.

[55] Brinkmann O, Kastratovic DA, Dimitrijevic MV, Konstantinovic VS, Jelovac DB, Antic J, et al. Oral squamous cell carcinoma detection by salivary biomarkers in a Serbian population. Oral Oncol 2011;47(1):51–5.

[56] Horváth J, Szabó A, Tar I, Dezső B, Kiss C, Márton I, et al. Oral health may affect the performance of mRNA-based saliva biomarkers for oral squamous cell cancer. Pathol Oncol Res 2018;24:833–42 (4).

[57] Zhang L, Xiao H, Zhou H, Santiago S, Lee JM, Garon EB, et al. Development of transcriptomic biomarker signature in human saliva to detect lung cancer. Cell Mol Life Sci 2012;69(19):3341–50.

[58] Zhang L, Farrell JJ, Zhou H, Elashoff D, Akin D, Park NH, et al. Salivary transcriptomic biomarkers for detection of resectable pancreatic cancer. Gastroenterology. 2010;138(3):949–57 e7.

[59] Zhang L, Xiao H, Karlan S, Zhou H, Gross J, Elashoff D, et al. Discovery and preclinical validation of salivary transcriptomic and proteomic biomarkers for the non-invasive detection of breast cancer. PLoS One 2010;5(12):e15573.

[60] Lee Y-H, Kim JH, Zhou H, Kim BW, Wong DT. Salivary transcriptomic biomarkers for detection of ovarian cancer: for serous papillary adenocarcinoma. J Mol Med 2012;90(4):427–34.

[61] Park NJ, Zhou H, Elashoff D, Henson BS, Kastratovic DA, Abemayor E, et al. Salivary microRNA: discovery, characterization, and clinical utility for oral cancer detection. Clin Cancer Res 2009;15(17):5473–7.

[62] Liu C-J, Lin S-C, Yang C-C, Cheng H-W, Chang K-W. Exploiting salivary miR-31 as a clinical biomarker of oral squamous cell carcinoma. Head Neck 2012;34(2):219–24.

[63] Salazar C, Nagadia R, Pandit P, Cooper-White J, Banerjee N, Dimitrova N, et al. A novel saliva-based microRNA biomarker panel to detect head and neck cancers. Cell Oncol (Dordr) 2014;37(5):331–8.

[64] Romani C, Salviato E, Paderno A, Zanotti L, Ravaggi A, Deganello A, et al. Genome-wide study of salivary miRNAs identifies miR-423-5p as promising diagnostic and prognostic biomarker in oral squamous cell carcinoma. Theranostics 2021;11(6):2987–99.

[65] Wan Y, Vagenas D, Salazar C, Kenny L, Perry C, Calvopina D, et al. Salivary miRNA panel to detect HPV-positive and HPV-negative head and neck cancer patients. Oncotarget. 2017;8(59):99990–100001.

[66] Xie Z, Chen G, Zhang X, Li D, Huang J, Yang C, et al. Salivary microRNAs as promising biomarkers for detection of esophageal cancer. PLoS One 2013;8(4):e57502.

[67] Fendereski M, Zia MF, Shafiee M, Safari F, Saneie MH, Tavassoli M. MicroRNA-196a as a potential diagnostic biomarker for esophageal squamous cell carcinoma. Cancer Investig 2017;35(2):78–84.

[68] Xie Z, Yin X, Gong B, Nie W, Wu B, Zhang X, et al. Salivary microRNAs show potential as a noninvasive biomarker for detecting resectable pancreatic cancer. Cancer Prev Res 2015;8(2):165–73.

[69] Sazanov AA, Kiselyova EV, Zakharenko AA, Romanov MN, Zaraysky MI. Plasma and saliva miR-21 expression in colorectal cancer patients. J Appl Genet 2017;58(2):231–7.

[70] Tang H, Wu Z, Zhang J, Su B. Salivary lncRNA as a potential marker for oral squamous cell carcinoma diagnosis. Mol Med Rep 2013;7(3):761–6.

[71] Xie Z, Chen X, Li J, Guo Y, Li H, Pan X, et al. Salivary HOTAIR and PVT1 as novel biomarkers for early pancreatic cancer. Oncotarget. 2016;7(18):25408–19.

[72] Zhang X, Kulasinghe A, Karim RS, Punyadeera C. Saliva diagnostics for oral diseases. Advances in Salivary Diagnostics. Berlin: Springer; 2015. p. 131–56.

[73] Wei JW, Huang K, Yang C, Kang CS. Non-coding RNAs as regulators in epigenetics (Review). Oncol Rep 2017;37(1):3–9.

[74] Dhanoa JK, Sethi RS, Verma R, Arora JS, Mukhopadhyay CS. Long non-coding RNA: its evolutionary relics and biological implications in mammals: a review. J Anim Sci Technol 2018;60:25.

[75] Esteller M. Non-coding RNAs in human disease. Nat Rev Genet 2011;12(12):861–74.

[76] Sannigrahi MK, Sharma R, Panda NK, Khullar M. Role of non-coding RNAs in head and neck squamous cell carcinoma: a narrative review. Oral Dis 2018;24(8):1417–27.

[77] Rupaimoole R, Calin GA, Lopez-Berestein G, Sood AK. miRNA deregulation in cancer cells and the tumor microenvironment. Cancer Discov 2016;6(3):235–46.

[78] Khan A, Ahmed E, Elareer N, Junejo K, Steinhoff M, Uddin S. Role of miRNA-regulated cancer stem cells in the pathogenesis of human malignancies. Cells. 2019;8(8):840.

[79] Zhang X, Sadowski P, Punyadeera C. Evaluation of sample preparation methods for label-free quantitative profiling of salivary proteome. J Proteom 2020;210:103532.

[80] Loo JA, Yan W, Ramachandran P, Wong DT. Comparative human salivary and plasma proteomes. J Dent Res 2010;89 (10):1016–23.

[81] Issaq HJ, Xiao Z, Veenstra TD. Serum and plasma proteomics. Chem Rev 2007;107(8):3601–20.

[82] Whiteaker JR, Zhang H, Eng JK, Fang R, Piening BD, Feng LC, et al. Head-to-head comparison of serum fractionation techniques. J Proteome Res 2007;6(2):828–36.

[83] Wren ME, Shirtcliff EA, Drury SS. Not all biofluids are created equal: chewing over salivary diagnostics and the epigenome. Clin Ther 2015;37(3):529–39.

[84] Chi L-M, Hsiao Y-C, Chien K-Y, Chen S-F, Chuang Y-N, Lin S-Y, et al. Assessment of candidate biomarkers in paired saliva and plasma samples from oral cancer patients by targeted mass spectrometry. J Proteom 2020;211:103571.

[85] Feng Y, Li Q, Chen J, Yi P, Xu X, Fan Y, et al. Salivary protease spectrum biomarkers of oral cancer. Int J Oral Sci 2019;11(1):7.

[86] Rajkumar K, Nandhini G, Ramya R, Rajashree P, Kumar AR, Anandan SN. Validation of the diagnostic utility of salivary interleukin 8 in the differentiation of potentially malignant oral lesions and oral squamous cell carcinoma in a region with high endemicity. Oral Surg Oral Med Oral Pathol Oral Radiol 2014;118 (3):309–19.

[87] Chu H-W, Chang K-P, Hsu C-W, Chang IY-F, Liu H-P, Chen Y-T, et al. Identification of salivary biomarkers for oral cancer detection with untargeted and targeted quantitative proteomics approaches*[S]. Mol Cell Proteom 2019;18(9):1796–806.

[88] Van Der Merwe L, Wan Y, Cheong H, Perry C, Punyadeera C. A pilot study to profile salivary angiogenic factors to detect head and neck cancers. BMC cancer 2018;18(1):1–8.

[89] Brooks MN, Wang J, Li Y, Zhang R, Elashoff D, Wong DT. Salivary protein factors are elevated in breast cancer patients. Mol Med Rep 2008;1(3):375–8.

[90] Wood N, Streckfus CF. The expression of lung resistance protein in saliva: a novel prognostic indicator protein for carcinoma of the breast. Cancer Invest 2015;33(10):510–15.

[91] Laidi F, Bouzianc A, Lakhdar A, Khabouze S, Amrani M, Rhrab B, et al. Significant correlation between salivary and serum Ca 15-3 in healthy women and breast cancer patients. Asian Pac J Cancer Prev 2014;15(11):4659–62.

[92] Wu Z-Z, Wang J-G, Zhang X-L. Diagnostic model of saliva protein finger print analysis of patients with gastric cancer. World J Gastroenterol: WJG 2009;15(7):865.

[93] Xiao H, Zhang Y, Kim Y, Kim S, Kim JJ, Kim KM, et al. Differential proteomic analysis of human saliva using tandem mass tags quantification for gastric cancer detection. Sci Rep 2016;6(1):1–13.

[94] He C-Z, Zhang K-H, Li Q, Liu X-H, Hong Y, Lv N-H. Combined use of AFP, CEA, CA125 and CAl9-9 improves the sensitivity for the diagnosis of gastric cancer. BMC Gastroenterol 2013;13 (1):1–5.

[95] Ding F, Sun K, Sun N, Jiang Q, Cao M, Wu Z. iTRAQ-based proteomics reveals SOD2 as a potential salivary biomarker in liver cancer. Int J Biol Markers 2019;34(3):221–31.

[96] Chen DX, Schwartz PE, Li FQ. Saliva and serum CA 125 assays for detecting malignant ovarian tumors. Obstet Gynecol 1990;75(4):701–4.

[97] Lambert AW, Pattabiraman DR, Weinberg RA. Emerging biological principles of metastasis. Cell. 2017;168(4):670–91.

[98] Babiuch K, Kuśnierz-Cabala B, Kęsek B, Okoń K, Darczuk D, Chomyszyn-Gajewska M. Evaluation of proinflammatory, NF-kappaB dependent cytokines: IL-1α, IL-6, IL-8, and TNF-α in tissue specimens and saliva of patients with oral squamous cell carcinoma and oral potentially malignant disorders. J Clin Med 2020;9(3).

[99] Panneer Selvam N, Sadaksharam J. Salivary interleukin-6 in the detection of oral cancer and precancer. Asia Pac J Clin Oncol 2015;11(3):236–41.

[100] Lisa Cheng YS, Jordan L, Gorugantula LM, Schneiderman E, Chen HS, Rees T. Salivary interleukin-6 and -8 in patients with oral cancer and patients with chronic oral inflammatory diseases. J Periodontol 2014;85(7):956–65.

[101] Assad DX, Mascarenhas ECP, Normando AGC, Chardin H, Barra GB, Pratesi R, et al. Correlation between salivary and serum CA15-3 concentrations in patients with breast cancer. Mol Clin Oncol 2020;13(2):155–61.

[102] Wu Z-Z, Wang J-G, Zhang X-L. Diagnostic model of saliva protein finger print analysis of patients with gastric cancer. World J Gastroenterol 2009;15(7):865–70.

[103] Farahani H, Alaee M, Amri J, Baghinia MR, Rafiee M. Serum and saliva concentrations of biochemical parameters in men with prostate cancer and benign prostate hyperplasia. Lab Med 2020;51 (3):243–51.

[104] You XY, Jiang J, Yin FZ. Preliminary observation on human saliva alpha-fetoprotein in patients with hepatocellular carcinoma. Chin Med J (Engl) 1993;106(3):179–82.

[105] Thomas D, Rathinavel AK, Radhakrishnan P. Altered glycosylation in cancer: a promising target for biomarkers and therapeutics. Biochimica et Biophysica Acta (BBA) - Rev Cancer 2021;1875(1):188464.

[106] Sinevici N, Mittermayr S, Davey GP, Bones J, O'Sullivan J. Salivary N-glycosylation as a biomarker of oral cancer: a pilot study. Glycobiology. 2019;29(10):726–34.

[107] Shu J, Ma J, Ren X, Wang J, Wang Y, Zhang K, et al. The abnormal glycopatterns of salivary glycoproteins in esophageal squamous cell carcinoma patients. Front Chem 2021;9(62).

[108] Yang J, Liu X, Shu J, Hou Y, Chen M, Yu H, et al. Abnormal galactosylated-glycans recognized by *Bandeiraea simplicifolia* Lectin I in saliva of patients with breast cancer. Glycoconj J 2020;37(3):373–94.

[109] Shu J, Yu H, Li X, Zhang D, Liu X, Du H, et al. Salivary glycopatterns as potential biomarkers for diagnosis of gastric cancer. Oncotarget. 2017;8(22):35718–27.

[110] Shu J, Yu H, Du H, Zhang J, Zhang K, Li X, et al. Identification of N- and O-linked glycans recognized by AAL in saliva of patients with atrophic gastritis and gastric cancer. Cancer Biomark 2018;22 (4):669–81.

[111] Qin Y, Zhong Y, Ma T, Zhang J, Yang G, Guan F, et al. A pilot study of salivary N-glycome in HBV-induced chronic hepatitis, cirrhosis, and hepatocellular carcinoma. Glycoconj J 2017;34(4):523–35.

[112] Ben-Ishay O, Haim SO, Kaplan M, Kluger Y. CA19-9 levels in saliva and urine of patients with ductal adenocarcinoma of the pancreas correlates with serum CA19-9 levels - a prospective preliminary study. J Pancreas 2016;17:66–9.

[113] Balan JJ, Rao RS, Premalatha BR, Patil S. Analysis of tumor marker CA 125 in saliva of normal and oral squamous cell carcinoma patients: a comparative study. J Contemp Dent Pract 2012;13(5):671–5.

[114] Agha-Hosseini F, Mirzaii-Dizgah I, Rahimi A. Correlation of serum and salivary CA15-3 levels in patients with breast cancer. Med Oral Patol Oral Cir Bucal [Internet] 2009;14(10). Available from: http://europepmc.org/abstract/MED/19680209, https://doi.org/10.4317/medoral.14.e521.

[115] Liu S, Jiang X, Shang Z, Ji Y, Wang H, Wang Z, et al. N-glycan structures of target cancer biomarker characterized by two-

dimensional gel electrophoresis and mass spectrometry. Analytica Chim Acta 2020;1123:18–27.

[116] Zanotti L, Paderno A, Piazza C, Pagan E, Bignotti E, Romani C, et al. Epidermal growth factor receptor detection in serum and saliva as a diagnostic and prognostic tool in oral cancer. Laryngoscope 2017;127(11):E408–14.

[117] Jacob T, Ramesh M, Murali S, Ramesh K, Sanjay P, Abraham P. A non-invasive study to estimate and compare salivary sialic acid level as tumor marker in patients with pre-cancer and oral cancer. J Cancer Res Therapeutics 2016;12(2):634–9.

[118] Arthisri A, Sathiyamoorthy A, Meenakshi B, Chandran C. Ratio of salivary sialic acid to fucose as tumor markers in potentially malignant disorders and oral cancer. Contemporary Clin Dent 2020;11 (2):131–5.

[119] Hernández-Arteaga A, de Jesús Zermeño Nava J, Kolosovas-Machuca ES, Velázquez-Salazar JJ, Vinogradova E, José-Yacamán M, et al. Diagnosis of breast cancer by analysis of sialic acid concentrations in human saliva by surface-enhanced Raman spectroscopy of silver nanoparticles. Nano Res 2017;10(11):3662–70.

[120] Zermeño-Nava JdJ, Martínez-Martínez MU, Rámirez-de-Ávila AL, Hernández-Arteaga AC, García-Valdivieso MG, Hernández-Cedillo A, et al. Determination of sialic acid in saliva by means of surface-enhanced Raman spectroscopy as a marker in adnexal mass patients: ovarian cancer vs benign cases. J Ovarian Res 2018;11(1):61.

[121] Colombo M, Raposo G, Thery C. Biogenesis, secretion, and intercellular interactions of exosomes and other extracellular vesicles. Annu Rev Cell Dev Biol 2014;30:255–89.

[122] Raposo G, Stoorvogel W. Extracellular vesicles: exosomes, microvesicles, and friends. J Cell Biol 2013;200(4):373–83.

[123] Nair S, Tang KD, Kenny L, Punyadeera C. Salivary exosomes as potential biomarkers in cancer. Oral Oncol 2018;84:31–40.

[124] Li S-P, Lin Z-X, Jiang X-Y, Yu X-Y. Exosomal cargo-loading and synthetic exosome-mimics as potential therapeutic tools. Acta Pharmacol Sin 2018;39(4):542–51.

[125] Meldolesi J. Exosomes and ectosomes in intercellular communication. Curr Biol 2018;28(8) R435-r44.

[126] Harmati M, Gyukity-Sebestyen E, Dobra G, Janovak L, Dekany I, Saydam O, et al. Small extracellular vesicles convey the stress-induced adaptive responses of melanoma cells. Sci Rep 2019;9(1):15329.

[127] de Jong OG, Verhaar MC, Chen Y, Vader P, Gremmels H, Posthuma G, et al. Cellular stress conditions are reflected in the protein and RNA content of endothelial cell-derived exosomes. J Extracell Vesicles 2012;1 10.3402/jev.v1i0.18396.

[128] Taylor DD, Gercel-Taylor C. MicroRNA signatures of tumor-derived exosomes as diagnostic biomarkers of ovarian cancer. Gynecol Oncol 2008;110(1):13–21.

[129] Riches A, Campbell E, Borger E, Powis S. Regulation of exosome release from mammary epithelial and breast cancer cells—a new regulatory pathway. Eur J Cancer 2014;50(5):1025–34.

[130] Whiteside TL. Tumor-derived exosomes and their role in cancer progression. Adv Clin Chem 2016;74:103–41.

[131] Hanahan D, Weinberg Robert A. Hallmarks of cancer: the next generation. Cell. 2011;144(5):646–74.

[132] Meehan K, Vella LJ. The contribution of tumour-derived exosomes to the hallmarks of cancer. Crit Rev Clin Lab Sci 2016;53 (2):121–31.

[133] Huang J, Ding Z, Luo Q, Xu W. Cancer cell-derived exosomes promote cell proliferation and inhibit cell apoptosis of both normal lung fibroblasts and non-small cell lung cancer cell through delivering alpha-smooth muscle actin. Am J Transl Res 2019;11 (3):1711–23.

[134] Aslan C, Maralbashi S, Salari F, Kahroba H, Sigaroodi F, Kazemi T, et al. Tumor-derived exosomes: implication in angiogenesis and antiangiogenesis cancer therapy. J Cell Physiol 2019;.

[135] Emmanouilidi A, Paladin D, Greening DW, Falasca M. Oncogenic and non-malignant pancreatic exosome cargo reveal distinct expression of oncogenic and prognostic factors involved in tumour invasion and metastasis. Proteomics. 2019;e1800158.

[136] Mao Y, Wang Y, Dong L, Zhang Y, Zhang Y, Wang C, et al. Hypoxic exosomes facilitate angiogenesis and metastasis in esophageal squamous cell carcinoma through altering the phenotype and transcriptome of endothelial cells. J Exp Clin Cancer Res 2019;38 (1):389.

[137] Costa-Silva B, Aiello NM, Ocean AJ, Singh S, Zhang H, Thakur BK, et al. Pancreatic cancer exosomes initiate pre-metastatic niche formation in the liver. Nat Cell Biol 2015;17(6):816–26.

[138] Zhou B, Xu K, Zheng X, Chen T, Wang J, Song Y, et al. Application of exosomes as liquid biopsy in clinical diagnosis. Signal Transduct Target Ther 2020;5(1):144.

[139] Li P, Kaslan M, Lee SH, Yao J, Gao Z. Progress in exosome isolation techniques. Theranostics. 2017;7(3):789–804.

[140] Greening DW, Xu R, Ji H, Tauro BJ, Simpson RJ. A protocol for exosome isolation and characterization: evaluation of ultracentrifugation, density-gradient separation, and immunoaffinity capture methods. Methods Mol Biol 2015;1295:179–209.

[141] Gardiner C, Di Vizio D, Sahoo S, Théry C, Witwer KW, Wauben M, et al. Techniques used for the isolation and characterization of extracellular vesicles: results of a worldwide survey. J Extracell Vesicles 2016;5:32945.

[142] Livshits MA, Khomyakova E, Evtushenko EG, Lazarev VN, Kulemin NA, Semina SE, et al. Isolation of exosomes by differential centrifugation: theoretical analysis of a commonly used protocol. Sci Rep 2015;5:17319.

[143] Gurunathan S, Kang M-H, Jeyaraj M, Qasim M, Kim J-H. Review of the isolation, characterization, biological function, and multifarious therapeutic approaches of exosomes. Cells. 2019;8(4):307.

[144] Zhu L, Sun H-T, Wang S, Huang S-L, Zheng Y, Wang C-Q, et al. Isolation and characterization of exosomes for cancer research. J Hematol Oncol 2020;13(1):152.

[145] Tauro BJ, Greening DW, Mathias RA, Ji H, Mathivanan S, Scott AM, et al. Comparison of ultracentrifugation, density gradient separation, and immunoaffinity capture methods for isolating human colon cancer cell line LIM1863-derived exosomes. Methods. 2012;56(2):293–304.

[146] Théry C, Witwer KW, Aikawa E, Alcaraz MJ, Anderson JD, Andriantsitohaina R, et al. Minimal information for studies of extracellular vesicles 2018 (MISEV2018): a position statement of the International Society for Extracellular Vesicles and update of the MISEV2014 guidelines. J Extracell Vesicles 2018;7 (1):1535750.

[147] Lötvall J, Hill AF, Hochberg F, Buzás EI, Di Vizio D, Gardiner C, et al. Minimal experimental requirements for definition of extracellular vesicles and their functions: a position statement from the International Society for Extracellular Vesicles. J Extracell vesicles 2014;3 26913-.

[148] Skog J, Würdinger T, van Rijn S, Meijer DH, Gainche L, Sena-Esteves M, et al. Glioblastoma microvesicles transport RNA and proteins that promote tumour growth and provide diagnostic biomarkers. Nat Cell Biol 2008;10(12):1470–6.

[149] Malamud D. Saliva as a diagnostic fluid. Dent Clin North Am 2011;55(1):159–78.

[150] Zhang C-Z, Cheng X-Q, Li J-Y, Zhang P, Yi P, Xu X, et al. Saliva in the diagnosis of diseases. Int J Oral Sci 2016;8(3):133–7.

[151] Nonaka T, Wong DTW. Saliva-exosomics in cancer: molecular characterization of cancer-derived exosomes in saliva. Enzymes 2017;42:125–51.

[152] Bailey U-M, Punyadeera C, Cooper-White JJ, Schulz BL. Analysis of the extreme diversity of salivary alpha-amylase isoforms generated by physiological proteolysis using liquid chromatography–tandem mass spectrometry. J Chromatogr B 2012;911:21–6.

[153] Han Y, Jia L, Zheng Y, Li W. Salivary exosomes: emerging roles in systemic disease. Int J Biol Sci 2018;14(6):633–43.

[154] Langevin S, Kuhnell D, Parry T, Biesiada J, Huang S, Wise-Draper T, et al. Comprehensive microRNA-sequencing of exosomes derived from head and neck carcinoma cells in vitro reveals common secretion profiles and potential utility as salivary biomarkers. Oncotarget. 2017;8(47):82459–74.

[155] Tang KD, Wan Y, Zhang X, Bozyk N, Vasani S, Kenny L, et al. Proteomic alterations in salivary exosomes derived from human papillomavirus-driven oropharyngeal cancer. Mol Diagnosis Ther 2021;.

[156] Machida T, Tomofuji T, Maruyama T, Yoneda T, Ekuni D, Azuma T, et al. miR-1246 and miR-4644 in salivary exosome as potential biomarkers for pancreatobiliary tract cancer. Oncol Rep 2016;36 (4):2375–81.

[157] Lau C, Kim Y, Chia D, Spielmann N, Eibl G, Elashoff D, et al. Role of pancreatic cancer-derived exosomes in salivary biomarker development. J Biol Chem 2013;288(37):26888–97.

[158] Sun Y, Huo C, Qiao Z, Shang Z, Uzzaman A, Liu S, et al. Comparative proteomic analysis of exosomes and microvesicles in human saliva for lung cancer. J Proteome Res 2018;17(3):1101–7.

[159] Sun Y, Xia Z, Shang Z, Sun K, Niu X, Qian L, et al. Facile preparation of salivary extracellular vesicles for cancer proteomics. Sci Rep 2016;6 24669-.

[160] Winck FV, Prado Ribeiro AC, Ramos Domingues R, Ling LY, Riaño-Pachón DM, Rivera C, et al. Insights into immune responses in oral cancer through proteomic analysis of saliva and salivary extracellular vesicles. Sci Rep 2015;5 16305.

[161] Farag AF, Sabry D, Hassabou NF, Alaa EL-Din Y. MicroRNA-134/MicroRNA-200a derived salivary exosomes are novel

diagnostic biomarkers of oral squamous cell carcinoma. Egypt Dental J 2021;67(Issue 1):367–77 - January (Oral Medicine, X-Ray, Oral Biology & Oral Pathology)).

[162] He L, Ping F, Fan Z, Zhang C, Deng M, Cheng B, et al. Salivary exosomal miR-24-3p serves as a potential detective biomarker for oral squamous cell carcinoma screening. Biomed Pharmacother 2020;121:109553.

[163] Lim Y, Totsika M, Morrison M, Punyadeera C. Oral microbiome: a new biomarker reservoir for oral and oropharyngeal cancers. Theranostics. 2017;7(17):4313.

[164] Peterson J, Garges S, Giovanni M, McInnes P, Wang L, Schloss JA, et al. The NIH human microbiome project. Genome Res 2009;19(12):2317–23.

[165] Irfan M, Delgado RZR, Frias-Lopez J. The oral microbiome and cancer. Front Immunol 2020;11.

[166] Marshall B, J.B W, Blincow ED. Prospective double-blind trial of duodenal ulcer relapse after eradication of Campyobacfer py/ori. Lancet. 1988;2:1437–42.

[167] Lim Y, Fukuma N, Totsika M, Kenny L, Morrison M, Punyadeera C. The performance of an oral microbiome biomarker panel in predicting oral cavity and oropharyngeal cancers. Front Cell Infect Microbiol 2018;8:267.

[168] Ekanayake Weeramange C, Liu Z, Hartel G, Li Y, Vasani S, Langton-Lockton J, et al. Salivary high-risk human papillomavirus (HPV) DNA as a biomarker for HPV-driven head and neck cancers. J Mol Diagnostics 2021;23(10):1334–42.

[169] Carpén T, Syrjänen S, Jouhi L, Randen-Brady R, Haglund C, Mäkitie A, et al. Epstein–Barr virus (EBV) and polyoma viruses are detectable in oropharyngeal cancer and EBV may have prognostic impact. Cancer Immunol, Immunother 2020;69(8):1615–26.

[170] Drop B, Strycharz-Dudziak M, Kliszczewska E, Polz-Dacewicz M. Coinfection with Epstein–Barr virus (EBV), human papilloma virus (HPV) and polyoma BK virus (BKPyV) in laryngeal, oropharyngeal and oral cavity cancer. Int J Mol Sci 2017;18(12):2752.

[171] Jalouli J, Jalouli MM, Sapkota D, Ibrahim SO, Larsson P-A, Sand L. Human papilloma virus, herpes simplex virus and Epstein–Barr virus in oral squamous cell carcinoma from eight different countries. Anticancer Res 2012;32(2):571–80.

[172] Yang C-Y, Yeh Y-M, Yu H-Y, Chin C-Y, Hsu C-W, Liu H, et al. Oral microbiota community dynamics associated with oral squamous cell carcinoma staging. Front Microbiol 2018;9:862.

[173] Wen L, Mu W, Lu H, Wang X, Fang J, Jia Y, et al. Porphyromonas gingivalis promotes oral squamous cell carcinoma progression in an immune microenvironment. J Dental Res 2020;99(6):666–75.

[174] Corbella S, Veronesi P, Galimberti V, Weinstein R, Del Fabbro M, Francetti L. Is periodontitis a risk indicator for cancer? A *meta*-analysis. PLoS One 2018;13(4):e0195683.

[175] Guerrero-Preston R, White JR, Godoy-Vitorino F, Rodríguez-Hilario A, Navarro K, González H, et al. High-resolution microbiome profiling uncovers Fusobacterium nucleatum, Lactobacillus gasseri/johnsonii, and Lactobacillus vaginalis associated to oral and oropharyngeal cancer in saliva from HPV positive and HPV negative patients treated with surgery and chemo-radiation. Oncotarget. 2017;8(67):110931.

[176] Chen X, Winckler B, Lu M, Cheng H, Yuan Z, Yang Y, et al. Oral microbiota and risk for esophageal squamous cell carcinoma in a high-risk area of China. PLoS One 2015;10(12):e0143603.

[177] Sun JH, Li XL, Yin J, Li YH, Hou BX, Zhang Z. A screening method for gastric cancer by oral microbiome detection. Oncol Rep 2018;39(5):2217–24.

[178] Mitsuhashi K, Nosho K, Sukawa Y, Matsunaga Y, Ito M, Kurihara H, et al. Association of Fusobacterium species in pancreatic cancer tissues with molecular features and prognosis. Oncotarget. 2015;6(9):7209.

[179] Yan X, Yang M, Liu J, Gao R, Hu J, Li J, et al. Discovery and validation of potential bacterial biomarkers for lung cancer. Am J Cancer Res 2015;5(10):3111.

[180] Michaud DS, Izard J. Microbiota, oral microbiome, and pancreatic cancer. Cancer J (Sudbury, Mass) 2014;20(3):203.

[181] Farrell JJ, Zhang L, Zhou H, Chia D, Elashoff D, Akin D, et al. Variations of oral microbiota are associated with pancreatic diseases including pancreatic cancer. Gut. 2012;61(4):582–8.

[182] Wu MZ, Li WN, Cha N, Tian LX, Zhang YI, Wu X, et al. Diagnostic utility of HPV16 E6 mRNA or E7 mRNA quantitative expression for cervical cells of patients with dysplasia and carcinoma. Cell Transpl 2018;27(9):1401–6.

[183] National Cancer Institute N. NCI Dictionary of Cancer Terms. 2020.

[184] Rinschen MM, Ivanisevic J, Giera M, Siuzdak G. Identification of bioactive metabolites using activity metabolomics. Nat Rev Mol Cell Biol 2019;20(6):353–67.

[185] Ahmad F, Cherukuri MK, Choyke PL. Metabolic reprogramming in prostate cancer. Br J Cancer 2021;.

[186] Frezza C. Metabolism and cancer: the future is now. Br J Cancer 2020;122(2):133–5.

[187] Beger RD. A review of applications of metabolomics in cancer. Metabolites. 2013;3(3):552–74.

[188] Griffin JL, Shockcor JP. Metabolic profiles of cancer cells. Nat Rev Cancer 2004;4(7):551–61.

[189] Lohavanichbutr P, Zhang Y, Wang P, Gu H, Nagana Gowda GA, Djukovic D, et al. Salivary metabolite profiling distinguishes patients with oral cavity squamous cell carcinoma from normal controls. PLoS One 2018;13(9):e0204249.

[190] Assad DX, Mascarenhas ECP, de Lima CL, de Toledo IP, Chardin H, Combes A, et al. Salivary metabolites to detect patients with cancer: a systematic review. Int J Clin Oncol 2020;25(6):1016–36.

[191] Chen Y, Cheng S, Zhang A, Song J, Chang J, Wang K, et al. Salivary analysis based on surface enhanced Raman scattering sensors distinguishes early and advanced gastric cancer patients from healthy persons. J Biomed Nanotechnol 2018;14(10):1773–84.

[192] Mikkonen JJW, Singh SP, Akhi R, Salo T, Lappalainen R, González-Arriagada WA, et al. Potential role of nuclear magnetic resonance spectroscopy to identify salivary metabolite alterations in patients with head and neck cancer. Oncol Lett 2018;16(5):6795–800.

[193] Asai Y, Itoi T, Sugimoto M, Sofuni A, Tsuchiya T, Tanaka R, et al. Elevated polyamines in saliva of pancreatic cancer. Cancers (Basel) 2018;10(2):43.

[194] García-Villaescusa A, Morales-Tatay JM, Monleón-Salvadó D, González-Darder JM, Bellot-Arcis C, Montiel-Company JM, et al. Using NMR in saliva to identify possible biomarkers of glioblastoma and chronic periodontitis. PLoS One 2018;13(2):e0188710.

[195] Ishikawa S, Hiraka T, Kirii K, Sugimoto M, Shimamoto H, Sugano A, et al. Relationship between standard uptake values of positron emission tomography/computed tomography and salivary metabolites in oral cancer: a pilot study. J Clin Med 2020;9(12).

[196] Ohshima M, Sugahara K, Kasahara K, Katakura A. Metabolomic analysis of the saliva of Japanese patients with oral squamous cell carcinoma. Oncol Rep 2017;37(5):2727–34.

[197] Takayama T, Tsutsui H, Shimizu I, Toyama T, Yoshimoto N, Endo Y, et al. Diagnostic approach to breast cancer patients based on target metabolomics in saliva by liquid chromatography with tandem mass spectrometry. Clin Chim Acta 2016;452:18–26.

[198] Xavier Assad D, Acevedo AC, Cançado Porto Mascarenhas E, Costa Normando AG, Pichon V, Chardin H, et al. Using an untargeted metabolomics approach to identify salivary metabolites in women with breast cancer. Metabolites. 2020;10(12).

[199] Sugimoto M, Wong DT, Hirayama A, Soga T, Tomita M. Capillary electrophoresis mass spectrometry-based saliva metabolomics identified oral, breast and pancreatic cancer-specific profiles. Metabolomics. 2010;6(1):78–95.

[200] Sas KM, Karnovsky A, Michailidis G, Pennathur S. Metabolomics and diabetes: analytical and computational approaches. Diabetes. 2015;64(3):718–32.

[201] Almontashiri NAM, Zha L, Young K, Law T, Kellogg MD, Bodamer OA, et al. Clinical validation of targeted and untargeted metabolomics testing for genetic disorders: a 3 year comparative study. Sci Rep 2020;10(1):9382.

[202] Zhang X, Zhu X, Wang C, Zhang H, Cai Z. Non-targeted and targeted metabolomics approaches to diagnosing lung cancer and predicting patient prognosis. Oncotarget. 2016;7(39):63437–48.

[203] Pepe MS, Feng Z, Janes H, Bossuyt PM, Potter JD. Pivotal evaluation of the accuracy of a biomarker used for classification or prediction: standards for study design. J Natl Cancer Inst 2008;100(20):1432–8.

[204] Sugimoto M. Salivary metabolomics for cancer detection. Expert Rev Proteom 2020;17(9):639–48.

[205] Rifai N, Gillette MA, Carr SA. Protein biomarker discovery and validation: the long and uncertain path to clinical utility. Nat Biotechnol 2006;24(8):971–83.

[206] Gug IT, Tertis M, Hosu O, Cristea C. Salivary biomarkers detection: analytical and immunological methods overview. TrAC Trends Anal Chem 2019;113:301–16.

CHAPTER

Emerging technologies in cancer detection

11

Zvi Yaari[1,2], Christopher P. Horoszko[1,3], Merav Antman-Passig[1], Mijin Kim[1], Freddy T. Nguyen[4,5] and Daniel A. Heller[1,3,6]

[1]*Memorial Sloan Kettering Cancer Center, New York, NY, United States*
[2]*School of Pharmacy, Faculty of Medicine, The Hebrew University of Jerusalem, Israel*
[3]*Department of Pharmacology, Weill Cornell Medicine, New York, NY, United States*
[4]*Institute for Medical Engineering and Science, Massachusetts Institute of Technology, Cambridge, MA, United States*
[5]*Department of Pathology, Molecular and Cell-Based Medicine, Icahn School of Medicine at Mount Sinai, New York, NY, United States*
[6]*Department of Physiology and Biophysics, Weill Cornell Medicine, New York, NY, United States*

Objectives

- Present innovative technologies that will impact laboratory medicine practice for cancer diagnosis in the next 5–10 years.
- Disseminate information about new types of modalities and mechanisms in biomarker detection.
- Highlight the clinical potential of artificial intelligence and its integration into diagnostic technologies.

11.1 Introduction

Laboratory medicine technologies have undergone rapid improvements in the past few decades, especially related to cancer care. Despite recent technological progress, significant challenges persist along the road toward more accessible, precise, inexpensive, rapid,

Cancer Biomarkers: Clinical Aspects and Laboratory Determination. DOI: https://doi.org/10.1016/B978-0-12-824302-2.00011-4
© 2022 Elsevier Inc. All rights reserved.

automated, and patient-facing methods. Fortunately, many emerging technologies in this field have the potential to revolutionize laboratory medicine methodologies. This chapter highlights new and emerging technologies for cancer diagnosis. Subsections are based on the detection methods—electrochemical, optical, magnetic, mass, mechanical, and others. In each subsection, novel platforms are described, emphasizing the point of care, dimensions, robustness, and simplicity. We predict that the technologies represented in this chapter will model the evolving needs of the laboratory medicine community over the coming few years and may stimulate the field of laboratory medicine to pursue novel technologies. For example, emerging technologies offer new methods to measure dozens of biomarkers for cancer care rapidly. Implementing these technologies can boost cancer research, diagnostics, and treatment, especially since early detection is critical.

Additionally, current clinical technologies generate an enormous volume of data. The noteworthy progress in artificial intelligence (AI) and its integration into the medical field is an area of importance. A subsection of this chapter is dedicated to detailing the progress in AI and other data analytics technologies for cancer detection. We believe that significant potential for AI exists for integrating biological data, intelligent processing, and predictive algorithms. As a result, it can reduce errors, costs, and delays.

11.2 Electrochemical techniques

Among the instrumental assays most used in the clinical laboratory—spectrophotometric, chromatographic, radiometric, and electrochemical—the latter holds significant leverage against common measurement challenges associated with analyte classes, concentrations, and matrices. A substantial branch of analytical chemistry, electrochemical analysis has evolved rapidly with concomitant miniaturization and scaling of the electrode apparatus historically employed in clinical and pharmaceutical studies of gas and ion reactions. Later, biosensors emerged when biochemistry merged with analytical and physical chemistry principles and biological macromolecules were incorporated into electrochemical substrates. This

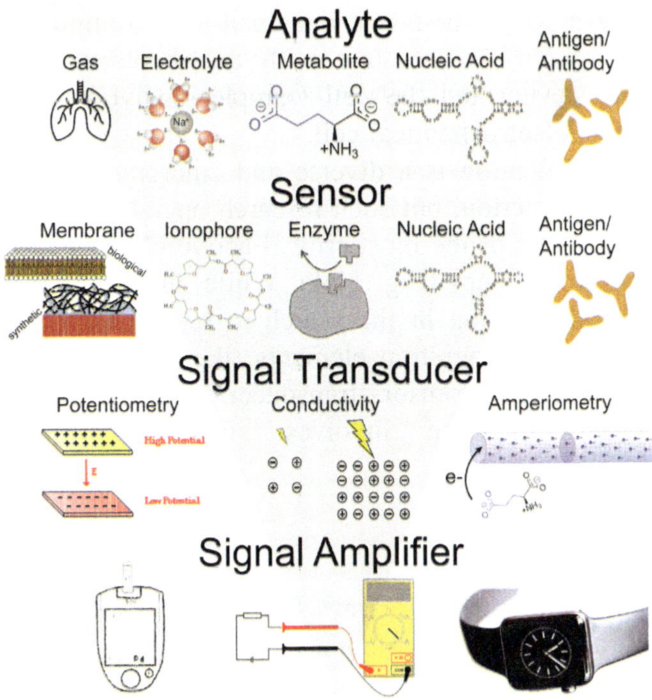

FIGURE 11.1

Design and signal flow of an electrochemical device.

simple addition revolutionized clinical analysis and increased the measurability of human physiology [1]. Subsequently, a wide variety of electrochemical signaling transduction routes became available—potentiometry, amperometry, voltammetry, and impedance—each adds a unique metrological dimension uniquely suited to situations that involve low volume, flowing samples, and point-of-care environments.

11.2.1 Principles

The miniaturization of the circuity that translates electronics into quantitative (discreet, continuous) or qualitative (threshold descriptor) information is a significant advance. Fig. 11.1 summarizes the molecular workflow of an electrochemical assay. In keeping with the importance of phases and interfaces, the depicted workflow cartoon, in proper form, should be imagined as two miniaturized compartments brought into contact with one another: the electrode (sensor–transducer) and the cell (sample–analyte) together make up a complete electrochemical cell.

Signal amplification is a diverse and emerging area outside the purview of this section, but such research operates with its own set of constraints. One of the most critical activities occurs between the sensor–transducer steps (Fig. 11.1). While innovative methods and ideas are rarely lacking in the search to bring analytes of interest into contact with recognition elements (the sensor) that are highly discriminatory, the sensor–transducer step at the electrode–electrolyte interface involves significant physiochemical complexities.

11.2.2 Emerging trends

Translating electrochemical assays from the bench into the clinic at the bedside or in a nearby laboratory is a challenge. However, the biggest hurdle is the instrumentation used to translate electrical signals into a readable output format. This led to research to integrate currently available point-of-care instruments, for example, blood sugar devices [2–5], into biosensor workflow. Virtually all biosensor efforts aim to improve essential analytical sensitivity and specificity for laboratory investigators that use simple or complex samples. It is essential to understand that laboratory experimentation is discreet from clinical validation; clinical specificity and sensitivity found in biomarker studies seeking utility cannot be assessed through analytical research to design and optimize the assay.

Analytical improvements are made on the material composition of the device, the chemistry the device performs, the lock-and-key sensor the device uses, the analyte barrier(s) the device contains, or a combination. Over the past decades of electrochemical research, most improvements have been dedicated to transducer physics or stepwise improvements to well-studied sensing elements. For high-volume usage in automated laboratories, durability and workflow integration are crucial considerations better left to the R&D expertise of industry-level projects. On the other hand, innovative improvements that seek to expand analyte accessibility and analyte sensitivity/specificity by designing new or existing sensing combinations are presented below—with an eye toward the coming decade.

11.2.3 Sensor trends: aptamers

The unique two- and three-dimensional configuration of nucleic acids have attracted significant attention as biomarker binding elements in electronic sensors. These aptamer templates are designed and synthesized relatively inexpensively and are semipredictive methods that improve affinity to an analyte class. By screening against an analyte (or complex target) to enrich the resulting output, aptamers are designed to contain an optimal set of binding variables—bonding residues, conformation, flexibility, size, etc.—within the specific testing matrix. Even in crude, minimally processed blood samples, improved detection limits are possible when coupling the enriched aptamer candidate to an electrochemical transducer. As an example of the progression of aptamer sensing beyond academic interests [6–8], commercial and startup ventures have been expanding into this space.

For cancer diagnostics, a major roadblock is ultra-low biomarker concentrations found in the blood due to limited tissue leakage or low numbers of circulating tumor cells. As a recent proof of concept for the power of aptamer workflow in cancer diagnostics, a DNA sensor was designed using postoperative lung tumor tissue with two goals: to test device sensitivity and specificity and to identify blood biomarkers specific to the lung cancer [9]. Using gold electrodes and a ferro(i)cyanide couple, the authors quantified and characterized a lung cancer target, that is, a histone protein, in crude blood plasma. Notably, the square-wave voltammetric measurement could be enhanced by optimizing their assay with silica-coated iron oxide bead "sandwiches" decorated with hydrophobic linkers. An impressive 100-fold increase in the limit of detection resulted.

A new lung cancer biomarker can enter clinical validation and a high-affinity sensor detecting the low-level antigen in minimally prepped blood samples is beneficial. In comparison, prior improvements were achieved a decade earlier using a redox-dye apta-sensor binding PDGF at a limit of 50 pM in serum [10]. Low detection limits are commonly reported for other analytical laboratory platforms. However, electrochemical analysis currently stands out as a pragmatic route to achieve low-volume sample analysis with readily available materials. The aptamer selection process is amenable to linear workflows, suggesting good in-house time-to-development/calibration within moderate to high complexity laboratories [6–8,11,12].

11.2.4 Transducer trends: low-cost paper devices

A significant push toward economies of scale and space, embodied in part by emerging markets and the developing world, has catalyzed new laboratory priorities. For example, microfluidics can reduce the laboratory real estate necessary for testing, coopt or combine testing steps to reduce operational complexity, and increase the types of environments amenable to operation. One of the more promising avenues in this regard is the microfluidic paper-based platform and printed electrodes for targeted chemistries. Paper-based tech is worth a brief mention because one significant issue with electrochemical enzyme-linked immunosorbent assay (ELISA) assays is electrode fouling and repetitive wash steps. This issue is seemingly eliminated by separating the chemical reaction from the electrical detection step [13]. The emerging utility of paper-based devices is also highlighted by their ability to integrate into off the shelf meters for tumor cell detection with horseradish peroxidase (HRP) electrochemistry (aptamer capture of tumor cell antigen) [14], and rapid point-of-care opportunities for low-cost acute care settings. In this situation, integrated microchemistry and electrical detection take advantage of easily sourced films [15], strips [16], and screen prints [17].

11.2.5 Amplifier trends: wearable and mobile devices

A barrier in laboratory science concerns measurement time—monitoring and measuring health attributes as close to real-time as possible. In addition, noninvasiveness or minimal invasiveness is a top priority. Electronic consumer goods can be engineered to detect and read analytical inputs. Smartphones can integrate external circuitry after a suitable user interface is designed for the device [18]. Wearable devices in contact with the skin are of significant interest in this space. Two critical concepts are access to the sample and chosen analytical technique. Concerning sample access, research tends to focus on semiinvasive devices that access the dermis or epidermis unless noninvasive optical methods are employed.

Most analytes of interest require sampling on, in, or under the skin. Because flexible materials can integrate microelectrochemical cells and circuits, they are perfect for addressing this challenge. For example, microelectrodes in contact with the skin can continuously assess calcium ions (Ca^{+2}) and protons (H^{+}) in sweat in a reproducible and accurate manner. Using amperometry, this sampling concept can avoid artifacts and inconsistencies inherent in colorimetric detection. Detection limits down to 50 μM were achieved [19]. The most impressive aspect of this research is the full integration of skin-appropriate substrates, channels, and pumps into a single unit. While mobile meters and wearables evolve separately, the last 5 years have shown that real-time measurement can be achieved.

Further development will hopefully answer how accessible and practical more invasive transdermal wearables are and how many different chemistries can be performed on one device. It is good to see that both patient comfort [20] and cost [21] are actively considered in this research area.

11.3 Optical techniques

Optical measurements have long been a mainstay of pathology, starting with the basic optical microscope to the more common tools in the clinical laboratories that employ underlying light-based techniques such as light scattering, fluorescence, absorption, and Raman spectroscopic techniques. Newer advances in optical technologies have led to tremendous leaps forward in new core technologies, a combination of multimodal techniques, and the development and commercialization of more portable, affordable, and accessible versions of these techniques. These tools are being developed both for clinical laboratories and point-of-care applications. One of the main advantages of optical spectroscopic methods is its versatility as a noninvasive nondestructive probing method that can be easily integrated into current medical devices using fiber optics to guide, deliver, and collect the light through existing microscopes, catheters, endoscopes, or biopsy needle channels.

11.3.1 Light scattering

Scattering of light is often used to quantify and qualify the size of cells and subcellular features but has also been used to describe the texture or granularity of cells. Light scattering is elastic scattering where the energy of the incident particles is conserved. Although these techniques have long been available, recent advances in optical coherence tomography (OCT) [22–24] or low coherence interferometry continue to bring these tools to the forefront and, more importantly, at the patient's side as a point-of-care tool. The optical window for interrogating biological tissues is best in the visible and near-infrared (NIR) region of the light where tissue is most optically clear or when the tissue absorbs the least amount of light, allowing it for maximal scattering events.

OCT images are created through the interference of the backscattered light with a reference light wave and, like ultrasound, create an A-Scan. With a light source in the NIR region, the penetration depth of OCT ranges between 2 and 3 mm depending on the tissue type limiting the reach of OCT to mucosal surfaces and the skin. OCT can be incorporated into fiber-optic probes, biopsy needles, catheters, and endoscopes. In comparison, OCT images are collected without any physical destruction, sectioning, or tissue staining. This technology allows real-time probing of cellular and subcellular dynamics, such as identifying tumors often characterized by a high cellular density and a high nuclear to cytoplasmic ratio.

Clinical OCT systems have several oncological applications ranging from tumor margin assessment [25–27] to lymph nodes assessment [28,29] to circulating tumor cells or extracellular vesicles. OCT has mainly been explored in the context of solid cancers, in particular breast, brain, lung, gastrointestinal (GI), oral, bladder, laryngeal, and skin cancers [22,23,30–34], where there is the ease of access to the tissue surface intraoperatively, via endoscopes, and noninvasively. The development of contrast agents for OCT and its modalities have extended the capabilities of OCT by improving the sensitivity of the OCT signal or by adding molecular or chemical contrast [35].

Additional advances on the processing and computational side have also enhanced the visualization afforded by OCT by correcting for optical aberrations [36–38] and applying machine learning (ML) or AI algorithms to better identify and label features in OCT images [39–41].

11.3.2 Fluorescence-based techniques

Fluorescence spectroscopy measures the fluorescence emitted by chromophores or the endogenous chemicals (fluorophores) naturally occurring in tissue, including nicotinamide adenine dinucleotide with hydrogen (NADH), flavin adenine dinucleotide (FAD), collagen fibers, elastin, and calcifications. Earlier work in the field had focused on the study of intrinsic fluorescence for cancer diagnostics in the skin, GI, and cervical cancers [42–46]. Over the years, the field has continued to evolve in exploring the role of combining intrinsic fluorescence with other techniques such as diffuse reflectance spectroscopy, light scattering spectroscopy, and Raman spectroscopy [47–51]. Simultaneously, label-free autofluorescence multiharmonic (SLAM) imaging was developed at the University of Illinois at Urbana-Champaign. SLAM is a label-free method to image NADH, FAD, second harmonic generation, and third harmonic generation of tumors in vivo and extracellular vesicles [52,53]. Fluorescence lifetime imaging microscopy can also be used to measure NADH in extracellular vesicles [54].

Single-walled carbon nanotubes (SWCNTs) have been developed as NIR fluorescent nanosensors over the last couple of decades. SWCNT nanosensors have been demonstrated in vitro and in vivo in mice for more than 400 days [55]. The advantages of SWCNT include resistance to photobleaching [56–58] and NIR fluorescence, allowing for deeper penetration in biological tissue [59]. The unique fluorescence signature from each SWCNT chirality and the diversity of wrappings used to encapsulate the SWCNT open the door for the multiplexing capabilities of SWCNT nanosensors. These SWCNT-based nanosensors have been developed to detect diverse analytes such as saccharides [60,61], dopamine [62], reactive oxygen species [55,63,64], nucleic acids [65], protein biomarkers [66,67], and enzymatic activity/deactivation [68]. The interaction between the analyte of interest and SWCNT wrapping induces a wavelength shift or amplitude change in the SWCNT NIR fluorescence due to changes in the local redox and electrostatic/dielectric environment [55,62–64,69–81]. On the basis of this SWCNT platform technology, researchers developed in vivo implantable sensors for the measurement of miRNA and protein biomarkers [65,66,82].

11.3.3 Raman-based techniques

Raman-based spectroscopic techniques probe the molecular vibrations and rotations of chemical bonds [83–85]. Compared to the light scattering techniques discussed earlier on elastic scattering, the Raman spectrum can provide information on the molecular composition of water, protein, lipids, and nucleic acid content, making it an appealing technique for characterizing biological cell content.

In ovarian cancer, Raman bands (785 nm excitation) corresponding to protein and lipid changes can differentiate between normal fallopian tubes, malignant ovaries, and omental metastasis [85]. Ex vivo studies showed more proteins, lipids, and DNA in malignant tissues compared to normal tissue with nearly 100% specificity and sensitivity [86]. Clinical applications of Raman to prostate cancer have included screening, biopsy, margin assessment, and therapy monitoring [85]. Stimulated Raman in the intraoperative diagnosis of pediatric type brain tumors had a high diagnostic concordance to H&E histology with 92%–96% accuracy when using a quantitative histologic method based on image feature extraction. With detection limits as low as 10 cells/mL in SERS, initial studies showed 99.7% confidence in spiked buffer solutions, 92.4% prediction accuracy, and a false-positive rate of $<0.5\%$ in spiked blood serological samples [85]. SERS has also been reported to study extracellular vesicles to detect exosomal miRNAs [87].

Raman spectroscopy is limited by the low signal and potential interference by tissue autofluorescence, leading to increased acquisition times or increased incident light power. There is a need for more standardization in the data collection and preprocessing of Raman spectra. The use of principal components analysis and ML algorithms can potentially improve the sensitivity and specificity of the classification techniques.

11.3.4 Other modalities and multiplexing

Photoacoustic imaging (PAI) allows imaging of optical absorbers by acoustic detectors (light in—sound out) [88,89]. The strengths of PAI include the high spatial/temporal resolution, clinically relevant penetration depth, endogenous and exogenous chromophores, and the absence of ionizing radiation. Under PAI, one can monitor the water, oxyhemoglobin, deoxyhemoglobin, melanin, and lipid content of cells. PAI enhancement can be made with exogenous contrast agents, especially those with a large optical absorption cross-section. The major contrast agents are metallic nanostructures, fluorescent dyes (indocyanine green (ICG), methylene blue), and carbon nanotubes. PAI has been demonstrated in the thyroid, breast, dermatological, pancreatic, gynecological, urological, eye, and gastrointestinal (GI) cancers. Tissue penetration depth in PAI is in the 2–3 cm range. It has been used for imaging blood vessels and estimating blood oxygenation levels. Most of the developments are still at the preclinical stage, and systems are not readily commercially available.

11.4 Magnetic techniques

Harnessing the power of magnetism in scientific research and clinical diagnostics has resulted in powerful technologies such as nuclear magnetic resonance (NMR) and magnetic resonance imaging (MRI). Among the many advantages in using these technologies for diagnostics, nontoxic and radiation-less operation, high-resolution imaging, and broad imaging access due to the involvement of ubiquitous water molecules for imaging throughout the body, there is virtually no need for additional additives or equipment in assay or analysis. Despite significant advances in this field, considerable challenges exist with the instruments' size and cost. Here, we describe the most advanced and emerging technologies in the magnetic field that can potentially revolutionize utility in medical diagnostics.

11.4.1 Miniature nuclear magnetic resonance

One of the most challenging aspects of magnetic-based instruments in clinical diagnostics is size. Typically, these instruments occupy an entire room. In addition, because of the sensitivity of the instruments, the outside environment can influence readings. Therefore instrument location must be carefully chosen and often specifically engineered to isolate the setup from external disturbances (such as vibrations and magnetic radiation). Miniature NMR instruments operate with low-field (LF) (below 100 MHz) and contain permanent magnets, unlike the large high-field (HF) superconducting instruments commonly found in analytical characterization facilities [90]. Even though LF instruments have lower sensitivity than HF, with a series of optimization steps such as solvent suppression [91] and magnet arrays [92], it is possible for these instruments to detect dozens of metabolites in human biofluids successfully. Although detection limits restrict the use of miniature NMRs, several advantages exist, such as point-of-care detection [93].

Haun et al. developed an even smaller NMR called micro-NMR (μNMR) [94]. Their current device is the third generation (DMR-3 [97,98]), which contains multiple microchannels, allowing rapid and quantitative analysis of multiple protein markers.

The μNMR's dimensions are only 10 cm × 10 cm. The device contains a microfluidic specimen delivery system, feedback control for temperature variations, and smartphone software. μNMR was designed to perform the measurements ex vivo for biofluid assays and detect cancer cells and secreted proteins [99]. To assess clinical utility, the researchers collected fine-needle aspirates from 50 patients with various breast lesions, female reproductive organs, lungs, pancreas, GI tract, urinary tract, and others and compared results to samples from an independent cohort of 20 patients. The aspirate samples yielded enough cancer cells for detection. Together with the NMR nanoparticle labels, μNMR was able to detect and quantify four different cancer markers simultaneously: Mucin 1 (MUC-1), human epidermal growth factor receptor 2 (HER2), endothelial growth factor receptor (EGFR), and epithelial cellular adhesion molecule (EpCAM). Compared to traditional clinical assays such as ELISA, FACS, and IHC, μNMR achieved 96% accuracy. Due to the rapid testing capability, results showed that protein expression patterns decayed over time, highlighting the benefit of continuous sampling at the patient bedside. The researchers discovered a high degree of protein expression heterogeneity across and within the same tumor sample based on their observations.

11.4.2 Implantable multiplexed sensor

Daniel et al. successfully developed an implantable device based on magnetic relaxation switches (Mrs) [95]. Mrs relies on magnetic nanoparticles that show transverse relaxation times (T_2) and, as a result, shortened T_2 in the presence of a specific analyte [100]. Most previous platforms that succeeded in detecting various biomarkers such as small molecules, oligonucleotides, proteins, and antibodies have not approached the low femtomolar limit.

The primary mechanism of the T_2 Mrs platform is based on the orientation of the magnetic nanoparticles in the test solution. When the nanoparticles are dispersed, they result in high T_2, whereas the addition of the target analyte causes the nanoparticles to aggregate into clusters, which results in lower T_2. The T_2 shift is due to the aggregate nanoparticle tendency to diphase the spins surrounding water protons. Newer Mrs platforms comprise a multireservoir device made of semipermeable polycarbonate membranes (with 10 nm cutoff) that allow the diffusion of the target analyte into the reservoirs and prevent the nanoparticles (Fig. 11.2A). In addition, this construct was shown to significantly extend nanoparticle stability in a physiological environment suggesting some relevance to in vivo applications. Researchers intend to implant the device in vivo and analyze the aggregation state of the nanoparticles in the presence of the analyte using standard MRI.

A notable advantage of a multireservoir platform is the multiplexing potential. As a proof of concept, researchers modified the device to detect the soluble marker hCG-β, which can be elevated in patients with specific types of testicular and ovarian cancers. Promising results were demonstrated in vitro after incubation in

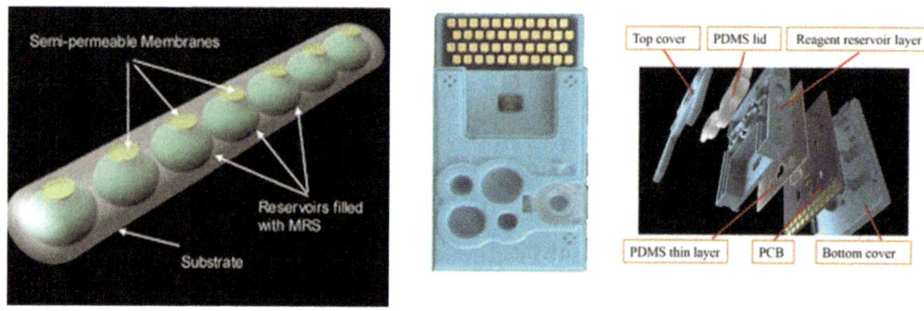

FIGURE 11.2 Emerging magnetic technologies for cancer biomarkers.

(A) Multireservoir device for detecting a soluble cancer biomarker. (B) Multiarray measurement of 12 tumor markers using a GMR immunoassay biosensor.

(A) Reprinted from Daniel KD, Kim GY, Vassiliou CC, Jalali-Yazdi F, Langer R, Cima MJ. Multi-reservoir device for detecting a soluble cancer biomarker. Lab Chip 2007;7(10):1288–93 [95]; and (B) Reprinted from Gao Y, Huo W, Zhang L, Lian J, Tao W, Song C, et al. Multiplex measurement of twelve tumor markers using a GMR multi-biomarker immunoassay biosensor. Biosens Bioelectron 2019;123:204–10 [96].

solutions containing 0.5–5 μg/mL hCG-β. The decrease in T_2 recorded was between 9% and 27%, respectively (compared to control devices), after approximately 28 hours. Although this device has not been tested in vivo, it demonstrates a milestone toward implantable Mrs utility for soluble cancer biomarkers.

11.4.3 Giant magnetoresistance sensor

The Giant Magnetoresistance (GMR) immunoassay technology has been available for the last 6 years. It is characterized by several advantages including low production costs, portability, high sensitivity, high signal-to-noise ratio (SNR), and real-time signal readout. Gao et al. developed a novel magnetic-based device that enables several important capabilities. These include microscale dimensions, fast read times (<15 minutes), multiplexing, and portability [101] using microfluidic device engineering to promote efficient reagent consumption. Device reaction rate and efficiency inside the microfluidic channels are high, suggesting low maintenance and automation are possible.

GMR (Fig. 11.2B) can perform multiarray detection—up to 12 different tumor markers. The team achieved this by integrating a GMR sensor chip and microfluidic magnetic nanobead label into a double antibody sandwich immunoassay. The resulting device was evaluated by screening serum samples from lung, liver, digestive tract, and prostate cancers. Results showed excellent sensitivity, accuracy, precision, and stability. Due to its portability, rapid detection, low cost, high efficacy, and multiarray detection, the GMR device is a notable candidate for point-of-care use.

11.5 Mass spectrometry

In recent years, mass spectrometry (MS) has become an integral analytical technique in cancer research, biomarker discovery, drug formulation development, and the diagnosis and prognosis of disease [102,103]. Emerging MS technologies are becoming increasingly important for cancer diagnosis.

11.5.1 Biomarker discovery and validation with mass spectrometry

Desorption electrospray ionization (DESI), together with mass spectrometry imaging (MSI), which permits identification of (m/z) distributions with spatial resolution, opens the possibility of biomarker research and clinical utility where direct use of primary patient tissue is available (Fig. 11.3A) [104]. Examples of biomarkers discovered with MS are OVA1 [105], an FDA-approved multivariate index assay to detect ovarian cancer risk, and (R)-2-hydroxybutyrate [106], a tumorigenesis marker for gliomas and acute myeloid leukemias. The field of oncological biomarkers is rapidly advancing with MS-based discoveries.

As biomarker assays shift from single analytes (e.g., immunochemistry) to multianalyte analysis and demand ultra-low concentration detection, new MS assays capable of sensitive and multivariate indexing are likely to impact and penetrate the clinical laboratory. As such, MS technologies may become powerful tools in clinical laboratories [107]. There are still challenges in transferring MS

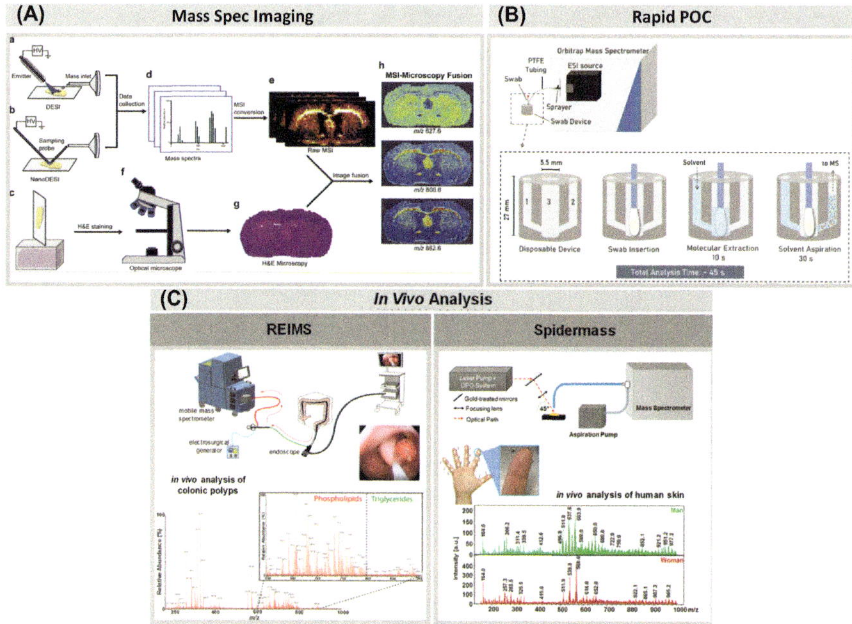

FIGURE 11.3 Illustration of emerging MS technologies.

(A) Illustration of the workflow to ambient ionization MSI for biomarker discovery. Schematic of a representative DESI (desorption electrospray ionization) setup sampling tissue sections (later H&E stained for optical microscopy). MS used ambient ionization. (B) Schematic of the rapid point-of-care (MasSpec Pen-ESI-Ms) system used to diagnose COVID-19 infection from swabs. (C) In vivo analysis using ambient ionization MS techniques. The left panel shows an endoscopic version of REIMS called *iEndoscope*. The right panel shows the use of *SpiderMass*, a resonant IR laser ablation device, in the analysis of human skin.

(A) Adapted from Lin L-E, Chen C-L, Huang Y-C, Chung H-H, Lin C-W, Chen K-C, et al. Precision biomarker discovery powered by microscopy image fusion-assisted high spatial resolution ambient ionization mass spectrometry imaging. Anal Chim Acta 2020;1100:75–87. (B) Adapted from Garza KY, Silva AAR, Rosa JR, Keating MF, Povilaitis SC, Spradlin M, et al. Rapid screening of COVID-19 directly from clinical nasopharyngeal swabs using the MasSpec pen. Anal Chem 2021. (C) Left panel: Reprinted from Alexander J, Gildea L, Balog J, Speller A, McKenzie J, Muirhead L, et al. A novel methodology for in vivo endoscopic phenotyping of colorectal cancer based on real-time analysis of the mucosal lipidome: a prospective observational study of the iKnife. Surg Endosc 2017;31(3):1361–70; Center panel: Adapted with permission from Fatou B, Saudemont P, Leblanc E, Vinatier D, Mesdag V, Wisztorski M, et al. In vivo real-time mass spectrometry for guided surgery application. Sci Rep 2016;6:25919.

technologies geared toward biomarker assay into medical laboratories, such as reproducibility, method transfer, and suitable personnel for the high-complexity training; however, the analytical need for highly quantitative diagnostic tools is expected to advance MS's role within routine clinical chemistry operations.

11.5.2 Precision medicine

One driving motivation for developing clinical laboratory technologies is a need for treatment strategies that take individual variability into account, a concept known as precision medicine or personalized medicine. Precision medicine requires analytical tools capable of mapping the genotype and phenotype of patients so that clinicians may determine an optimal therapeutic strategy [108]. Personalized oncology relies on identifying the tumor's genetic, cellular, and molecular features, as well as the tumor microenvironment. Therapeutic drug monitoring is critical to understanding a drug's metabolism, effectiveness, and treatment protocol.

Gas chromatography–MS (GC–MS) and liquid chromatography–MS (LC–MS) are today's leading analytical tools for detecting drugs and metabolites in patient blood. With other techniques such as whole-genome sequencing, MS enables high throughput data acquisition in various Omics workflows [109,110]. Tandem MS simultaneously analyses a wide array of analytes within a short brief time. It detects deficient drug levels in the blood [111]. Advanced analytical separation techniques, such as capillary electrophoresis, enhance the selectivity and efficiency of sample runs [112]. These techniques can be exploited to provide in-depth MS-based-Omics [113,114].

11.5.3 Mass spectrometry for point-of-care diagnostics

Point-of-care technologies aim to provide rapid information at the care site. New developments in ambient ionization MS, which requires minimal sample preparation and preserves samples in their original state, open new possibilities for MS in point-of-care situations [115]. Ambient ionization MS technologies include DESI and laser ablation electrospray ionization. These techniques utilize the substrate itself as a means of ionization and sampling. Specifically, DESI uses an electrospray of solvent droplets to dissolve and disperse charged molecules into an ionic cloud "sampling" of the surface. Operation and data acquisition occur in real-time. Since DESI's first appearance in 2004, ambient ionization MS technologies have become more prevalent today [116].

To create point-of-care assays with a high SNR and an ability to identify bioanalytes accurately, ambient ionization MS technologies are key candidates. These devices display good accuracy, possess the ability to detect low quantities of bioanalytes directly from samples, are simple to use, and provide rapid results. MS-driven point-of-care tests are well suited to the analysis of oral fluid from swabs [117,118] (Fig. 11.3B), samples contained on paper-like material, and direct injection of biofluids in, for example, in emergency or acute care situations [119,120].

11.5.4 Hand-held mass spectrometry for intraoperative tumor diagnostics

Surgical removal of solid tumors is often a first-line treatment followed by supportive or adjuvant therapy such as chemo or radiation. In curative surgery, the surgeon must remove all the primary tumor mass and any detectible metastases. Complete removal of the tumor is vital to outcomes. During operation, surgeons will apply a resection margin to aid incomplete excision of the tumor and minimal damage to healthy tissue [121–123]. Surgeons rely on preoperative and intraoperative techniques such as clinical/anatomic pathology and imaging techniques to differentiate tumor from healthy tissue and determine the extent of tumor resection. Preoperative procedures are often time-consuming and require technical expertise, while real-time intraoperative techniques offer advantages in terms of time-to-decision and accuracy in margin assessment [124].

The need to accurately assess tumor margins in real-time has fueled the development of novel intraoperative technologies, with the ultimate goal of providing a definitive intraoperative molecular diagnosis that can map tumor and nontumor tissues. In the context of the operating room, tools that enable diagnostic readout within the surgical time window and that positively impact surgical workflow are beneficial. Ambient MS direct-sampling technologies such as DESI-MS [125,126] and rapid evaporative ionization mass spectrometry (REIMS) [127] are powerful tools for studying the chemical constituents of tissue samples in vivo because they enable ionization of tissue samples without any further chemical modifications [128]. Recently, handheld electrosurgical devices coupled to MS (REIMS and DESI-MS) instruments may have the potential for rapid, robust, and easy-to-use mechanics necessary in clinical and surgical environments.

The *iKnife* [129], a handheld sampling device developed by scientists at Imperial College in London, uses REIMS. REIMS can be used to distinguish tumor tissue from benign tissue [130] intraoperatively [131]. Similarly, the *MasSpec Pen* [132] is a handheld instrument that allows gas and solvent transfer to the sampling surface and collects ionized droplets directly for MS analysis. Another handheld device, *Spider mass* [133], has been used to directly analyze human skin.

The application of MS technologies to intraoperative tumor diagnosis relies on detecting and distinguishing molecules with high sensitivity and specificity. Different classes of molecules can serve as qualified biomarkers. Both biased (targeted search for a particular m/z) and unbiased (untargeted search for a signature or set of m/z) MS approaches are used to distinguish cancer and healthy tissue. For example, a study conducted by Balog et al. identified a fingerprint of library-curated lipid changes using the *iKnife* with a sensitivity of 97.7% and a specificity of 96.5% (tumor vs metastasis) [129].

11.6 Mechanical and synthetic techniques

Atypical emerging assay methods are unique; spectrophotometric, magnetic, and electrical techniques are the oldest detection schemes. However, mechanical properties—size, viscosity, rigidity, stiffness—are well-suited to enable the isolation and detection of cancer-related targets and assess cancer cells' invasive characteristics. In addition, novel methodologies have generated artificial biomarkers that can be detected and are not dependent on the tumor type or expression levels of the endogenous target biomarkers. In this section, we will review emerging technologies in cancer detection that rely on the physical properties of the analyte and the generation of artificial biomarkers in situ.

11.6.1 Assays for invasiveness

The process of metastasis involves cancer cell invasion from the primary tumor and eventually seeding into distant organs. A cell's ability to invade is contained within sets of genes that promote the mechanical changes necessary for breaking through tissue matrix, motility, access to circulation (extravasation), dissemination, implanting at a metastatic site, invasion into a niche, and proliferation in the new tissue [134]. Thus the ability to quantify "invasiveness" can significantly impact diagnosis, prognosis, and clinical care.

Merkher et al. developed a mechanobiology polymer-based device to predict the metastatic potential of patient tumor samples [135]. While current assays used to determine the stage of malignancy and spread are based on evaluating the lymph-node status, tumor size, histopathology, and genetics, requiring days to weeks for prognosis, the mechanical invasiveness assay requires <2 hours. These researchers demonstrated their methodology with breast and pancreatic cancer cell lines and in vivo models. Briefly, their assay included rapid extraction of single cells from a tumor biopsy and then seeding the cells into a polymer-based matrix with embedded fluorescent nanoparticles. The assay monitored the degree of invasiveness through the determination of nanoparticle depth after the cells pushed them. The deeper the nanoparticles are moved, the greater the invasiveness of the cancer cells (Fig. 11.4A). The researchers confirmed the strong relationship between the degree of indentation depth and the degree of cellular invasiveness. This type of platform could have significant utility for treatment assignment and monitoring, diagnosis, prognosis, and positively impact survival rates.

11.6.2 Synthetic biopsy

Biomarkers, especially in oncology, play essential overlapping roles in diagnostics, fundamental research, and treatment. Despite considerable progress in discovering and validating new biomarkers that

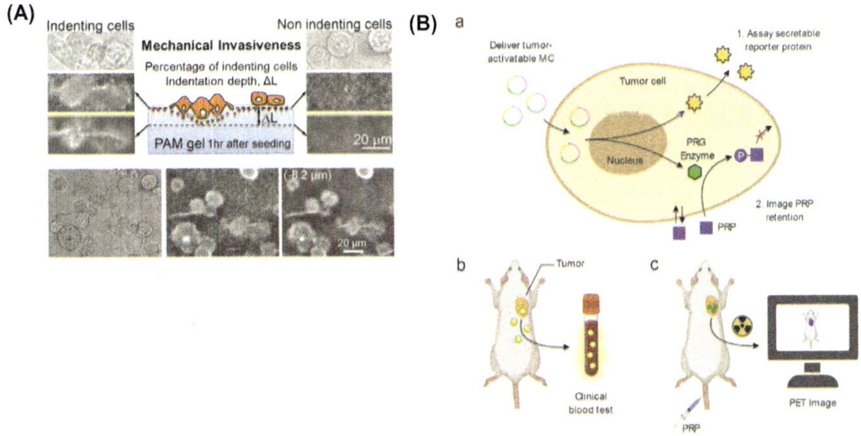

FIGURE 11.4 Emerging mechanical technologies for isolating, detecting, and reporting cancer markers.

(A) Mechanical invasiveness assay. (B) Schematic of a two-step synthetic biopsy blood biomarker and imaging-based cancer screening approach.

(A) Reprinted from Merkher Y, Horesh Y, Abramov Z, Shleifer G, Ben-Ishay O, Kluger Y, et al. Rapid cancer diagnosis and early prognosis of metastatic risk based on mechanical invasiveness of sampled cells. Ann Biomed Eng 2020;48(12):2846–58 [135]. (B) Reprinted from Robinson ER, Gowrishankar G, D'Souza AL, Kheirolomoom A, Haywood T, Hori SS, et al. Minicircles for a two-step blood biomarker and PET imaging early cancer detection strategy. J Controlled Rel 2021;335:281–9 [136].

stage cancers, the entire discovery pipeline requires a huge effort, high complexity-trained labor, and time. In addition, the prevalence, concentration, and analyte classes that characterize biomarkers vary between cancer types, stages, and patients. Some biomarkers are expressed in otherwise healthy patients who unfortunately limits clinical utility.

A novel strategy analogous to a biological Trojan horse was developed to introduce artificial biomarkers into cancer cells to overcome these challenges. Cells tricked to express these markers would become labeled and detectable by conventional assay. This technique was described as a "synthetic liquid biopsy." In principle, there is no targeting limitation to this technique. Prof. Sanjiv Sam Gambhir, Prof. Sangeeta Bhatia, and colleagues have pioneered this field.

In their initial proof of concept, Gambhir's group used DNA microcircles to transfect melanoma cells in vivo [137]. The DNA microcircles are circular plasmid constructs used to transfect cells in culture. Since DNA microcircles are much smaller than a typical plasmid (typically ~4 kb pairs), microcircles have superior durability, stability, and transfection efficiency and are integration-free and extra-chromosomal (Fig. 11.4B).

Gambhir's group then engineered DNA microcircles to contain a promoter for *survivin*, a gene believed to be activated only in cancer cells [138], and a reporter gene that can be assayed (embryonic alkaline phosphatase SEAP) [136]. Researchers then engineered the immune system to respond to a growing tumor and report this response using the same transfected reporter gene [139]. When transfected macrophages were injected in vivo into tumor-bearing mice (colorectal cancer), the macrophages naturally migrated to the tumor site, polarized, expressed, and secreted the reporter agent. By transforming immune cells into cancer sentinels that express readily detectable agents, continuous screening for cancer biomarkers is possible.

The group of Sangeeta developed a different approach to generate artificial biomarkers for both inflammation and cancer that can be measured using MS [140]. Kwong et al. synthesized synthetic biomarker library that contained mass-encoded tandem peptides conjugated onto polyethylene glycol–coated iron oxide ("nanoworm") nanoparticles. After injecting the nanoparticles intravenously into mice bearing either liver fibrosis or cancer, the nanoworm particles accumulate in the target disease tissues. Inside the diseased tissue, dysregulated proteases cleave peptides from nanoworms, allowing fragments to filter into the urine. The peptides were recovered from the urine samples and quantified by LC–MS/MS. Based on these measurements, the scientists could correlate both the type and the stage of the disease, such as colorectal cancer. This approach eliminates the need for invasive core biopsies and substantially improves early cancer detection compared with current clinically used blood biomarkers. This platform of generating synthetic biomarkers secreted from the body by urine is an excellent example of technology that should be used in pathophysiological processes and point-of-care diagnostics.

11.7 Data analytics

Current biomarker discovery efforts are facilitated by high-throughput techniques such as metabolomics, genetic sequencing, proteomics, and micro/nanoarrays, which generate large datasets for data-driven biomarker discovery. Although there has been gradual progress in processing and analyzing the data based on statistical models, it can be challenging to deconvolute disease-specific signatures within the complexity of growing dataset sizes. AI provides researchers with machines that can mimic human behavior, a suitable improvement for processing extensive amounts of data. AI systems are a valuable tool in biomarker discovery and sensor design.

AI is a conceptual departure from classical statistics because of the amount of data it deals with and the fewer assumptions about the underlying distributions and models (e.g., linearity vs normality). Large datasets and computational power replace these assumptions. ML (machine learning) algorithms learn from the data, they are exposed to and make predictions without explicit preprogramming or prompting. Deep learning is the technical term used when computers create deep neural networks consisting of many layers. The AI achieves unprecedented accuracy during tasks such as image recognition. In one sense, these networks go further than ML because they learn complex features embedded in the data without declaring the features beforehand.

11.7.1 Facilitating conventional biomarker discovery

ML uses various algorithms to iteratively learn, describe, and improve data to predict useful disease fingerprints, such as extracting panels of biomarkers from large-scale and complex datasets. An AI-based platform enables performance optimization and high precision in a high-dimensional setting such as in a situation where the number of features significantly larger than the sample size.

Genetic information used as a cancer biomarker is challenging because of genetic heterogeneity in malignancies and the often-low concentrations of tumor-derived molecules. ML algorithms can be implemented to identify disease-specific patterns in the genetic dataset. For instance, ML algorithms can be employed to identify overexpressed micro-RNA in cancer patients. In one case, ML algorithms were trained to recognize an ovarian cancer-associated micro-RNA in the blood by comparing a micro-RNA library of women with ovarian cancer and healthy participants [141]. ML guided panel design can improve the detection of tumor variants in cell-free DNAs [142] or identify disease-specific patterns in DNA methylation microarrays [143].

AI has been used in every step of proteomics and metabolomics workflows, including chromatographic separation, ionization, ion mobility separation, mass spectrometer optics, and postrun analyte identification and quantification. Several ML models and deep neural network architectures were employed to identify peptides and proteins, predict retention times, extrapolate mass spectra [144], quantify target analytes [145], and determine the best combination of biomarkers that would differentiate cancer patients from healthy individuals [146].

11.7.2 Artificial intelligence in cancer diagnosis

AI-based data analysis facilitates processing the multiplexed sensor array readouts for diagnostic interpretation. Biomarker-based diagnosis is sometimes ineffective because accurate detection of an analyte does not always confer high sensitivity and specificity for a disease. Known biomarkers may not represent the complete disease state and are present in many other disorders, attenuating specificity and preventing use in screening-based strategies due to an unmanageable number of false-positive results. AI-based approaches make decisions based on the prelearned dataset of multiple biomarkers and identify the most effective biomarker combinations to determine accurate disease states [147].

Macine Learning (ML)-based prediction algorithms are often employed to understand the design rule of sensors systematically. Conventional sensor development relies on a selection process based on molecular biological recognition. Synthetic molecular recognition can improve sensitivity and diversify the sensor responses to various types of analytes [148–150]. ML models can facilitate the design rule of such molecular recognition. Kelich et al. have shown that the fluorescence responses of ssDNA-wrapped carbon nanotube (DNA-CNT) to a small molecule, serotonin, can be correlated with the physical and chemical properties of the DNA-CNT, such as DNA length and DNA base content [151]. The dataset was used to train ML models to predict unique sensor designs with strong sensor responses to target analytes.

Yaari et al. have developed a perception-based optical nanosensor platform that detects multiple biomarkers in ovarian cancer patients' biofluids [152]. The team generated an array of DNA-wrapped SWCNTs that differ by the DNA sequence and the chirality of the SWCNT. The optical datasets were used to train and test ML models to predict the presence of several biomarkers in uterine washes samples from ovarian cancer patients. This platform has the modularity and flexibility to detect any type and number of targets in a given solution by applying specific training using ML.

Deep Learning (DL)-based imaging analysis benefits cancer pathology [153]. DL can simplify routine workflows that pathologists entirely perform during histopathology. Prominent examples include the detection of tumor tissue in biopsies and tumor subtyping on morphology. DL can be used to identify the latest information unnoticed by pathologists observing complex histology images. Further information identified by DL includes the prediction of mutation, survival, and biological response. For example, changes in the morphology of cancer cells, such as the nuclear and cytoplasmatic texture, size, and shape within a histological image suggest genetic mutation. Such identifications are important histology risk factors in certain diseases and positive response information in therapy.

11.7.3 Potential pitfalls in artificial intelligence–based sensor technology

Due to the limited size of clinical data typically available, including sample heterogeneity and complex assay output features, AI may risk overfitting. Overfitting produces an analysis that corresponds too closely to a limited set of data and may fail to fit additional data and predict future observations reliably. Multilevel validation is required to address such generalizable issues. For example, testing with an independent sample set that is not used in model development is beneficial for internal validation.

Statistical analyses, such as regression models where weights describe the coefficients of a parametrized curve, can be intuitively interpreted by experts. However, especially in deep learning, AI models often lack interpretability since their predictions and analyses often involve multilayer nodes. Thus it can be challenging to glean biological insights through direct correlations between input data and the AI's results. This issue can be circumvented with relatively simple models whose rules can be independently examined, logistic regression, and decision trees are classic examples. Interpretable algorithms are critical because they allow backtracking of the most influential features and decision mechanisms [154].

11.8 Summary Points

- New, promising technology platforms are entering the field, including microfluidics, magnetic sensors, and point-of-care mass spectrometry.
- Emerging technologies should be easy to use, rapid, and cost-effective.
- The development of point-of-care technologies is a priority for addressing patient needs.
- Artificial intelligence is emerging as an important component in diagnostic technologies and integration with clinical care.

11.9 Conclusions and future directions

This chapter highlighted state-of-the-art, robust, and tractable technologies that show qualities necessary in laboratory medicine. Technologies were chosen to emphasize utility in cancer care; however, many valuable tools unknown to oncology can transition well. Our position is that all helpful assay instrumentation should be mobile and accessible to realize future demand for operation in any environment. To meet this goal, instruments and devices should be stable, fast, easy to use, and cost-effective. Based on the technologies outlined above, together with AI and ML algorithms, clinical answers will become more accurate in the worlds of diagnostics, screenings, and predictions.

Profiling and quantifying biomolecules is a vital operation in the clinical laboratory. MS, magnetic, and electrochemical technologies can offer analytical measurements that acknowledge analyte versatility while retaining selectivity, accuracy, precision, and potential high-throughput functionality. MS technologies appear more frequently in research aimed at personalized therapy, biomarker validation, and surgical assistance. The development and commercialization of miniature MS, MRI, and μNMR that are cost-effective and capable of high-volume workflow should expedite the current movement of these instruments from academic laboratories and the clinical laboratorian's bench into point-of-care environments.

As for implantable devices for point-of-care uses, recent optical technologies can revolutionize the field of diagnostic. These technologies, such as SWCNTs-based platforms, have the high sensitivity, flexibility, modularity, and specificity to detect endless desired targets in situ. Combining AI algorithms allow these sensors to be limitless in terms of the number of cancer markers and the sensitivity for each target. We believe that the next generation of detective technologies is ready and waiting for the next breakthrough.

References

[1] Meites S. Otto folin: America's first clinical biochemist. American Association for Clinical Chemistry; 1989.

[2] Su J, Xu J, Chen Y, Xiang Y, Yuan R, Chai Y. Personal glucose sensor for point-of-care early cancer diagnosis. Chem Commun 2012;48 (55):6909–11.

[3] Amalfitano E, Karlikow M, Norouzi M, Jaenes K, Cicek S, Masum F, et al. A glucose meter interface for point-of-care gene circuit-based diagnostics. Nat Commun 2021;12(1):724.

[4] Huang X, Li J, Lu M, Zhang W, Xu Z, Yu B-Y, et al. Point-of-care testing of MicroRNA based on personal glucose meter and dual signal amplification to evaluate drug-induced kidney injury. Anal Chim Acta 2020;1112:72–9.

[5] Shi H, Zheng J, Wang Y, Zhu S, Xiang Y, Zhu X, et al. Point-of-care testing of protein biomarkers by integrating a personal glucose meter with a concatenated DNA amplifier. Sens Actuators B: Chem 2020;322:128659.
[6] Idili A, Parolo C, Ortega G, Plaxco KW. Calibration-free measurement of phenylalanine levels in the blood using an electrochemical aptamer-based sensor suitable for point-of-care applications. ACS Sens 2019;4(12):3227–33.
[7] Wang Y, Cui M, Jiao M, Luo X. Antifouling and ultrasensitive biosensing interface based on self-assembled peptide and aptamer on macroporous gold for electrochemical detection of immunoglobulin E in serum. Anal Bioanal Chem 2018;410(23):5871–8.
[8] Mayer MD, Lai RY. Effects of redox label location on the performance of an electrochemical aptamer-based tumor necrosis factor-alpha sensor. Talanta 2018;189:585–91.
[9] Zamay GS, Zamay TN, Kolovskii VA, Shabanov AV, Glazyrin YE, Veprintsev DV, et al. Electrochemical aptasensor for lung cancer-related protein detection in crude blood plasma samples. Sci Rep 2016;6(1):34350.
[10] Lai RY, Plaxco KW, Heeger AJ. Aptamer-based electrochemical detection of picomolar platelet-derived growth factor directly in blood serum. Anal Chem 2007;79(1):229–33.
[11] Xiong H, Yan J, Cai S, He Q, Peng D, Liu Z, et al. Cancer protein biomarker discovery based on nucleic acid aptamers. Int J Biol Macromol 2019;132:190–202.
[12] Tin A, Yu B, Ma J, Masushita K, Daya N, Hoogeveen RC, et al. Reproducibility and variability of protein analytes measured using a multiplexed modified aptamer assay. J Appl Lab Med 2019;4(1):30–9.
[13] Glavan AC, Christodouleas DC, Mosadegh B, Yu HD, Smith BS, Lessing J, et al. Folding analytical devices for electrochemical ELISA in hydrophobic RH paper. Anal Chem 2014;86(24):11999–2007.
[14] Su M, Ge L, Kong Q, Zheng X, Ge S, Li N, et al. Cyto-sensing in electrochemical lab-on-paper cyto-device for in-situ evaluation of multi-glycan expressions on cancer cells. Biosens Bioelectron 2015;63:232–9.
[15] Cao L, Han G-C, Xiao H, Chen Z, Fang C. A novel 3D paper-based microfluidic electrochemical glucose biosensor based on rGO-TEPA/PB sensitive film. Anal Chim Acta 2020;1096:34–43.

[16] Tu T, Liang B, Cao Q, Fang L, Zhu Q, Cai Y, et al. Fully transient electrochemical testing strips for eco-friendly point of care testing. RSC Adv 2020;10(12):7241–50.

[17] Cao Q, Liang B, Yu C, Fang L, Tu T, Wei J, et al. High accuracy determination of multi metabolite by an origami-based coulometric electrochemical biosensor. J Electroanal Chem 2020;873:114358.

[18] Zhang D, Lu Y, Zhang Q, Liu L, Li S, Yao Y, et al. Protein detecting with smartphone-controlled electrochemical impedance spectroscopy for point-of-care applications. Sens Actuators B: Chem 2016;222:994–1002.

[19] Martín A, Kim J, Kurniawan JF, Sempionatto JR, Moreto JR, Tang G, et al. Epidermal microfluidic electrochemical detection system: enhanced sweat sampling and metabolite detection. ACS Sens 2017;2(12):1860–8.

[20] Zheng L, Liu Y, Zhang C. A sample-to-answer, wearable cloth-based electrochemical sensor (WCECS) for point-of-care detection of glucose in sweat. Sens Actuators B: Chem 2021;130131.

[21] Bauer M, Wunderlich L, Weinzierl F, Lei Y, Duerkop A, Alshareef HN, et al. Electrochemical multi-analyte point-of-care perspiration sensors using on-chip three-dimensional graphene electrodes. Anal Bioanal Chem 2021;413(3):763–77.

[22] Si P, Honkala A, de la Zerda A, Smith BR. Optical microscopy and coherence tomography of cancer in living subjects. Trends Cancer 2020;6(3):205–22.

[23] Zysk AM, Nguyen FT, Oldenburg AL, Marks DL, Boppart SA. Optical coherence tomography: a review of clinical development from bench to bedside. J Biomed Opt 2007;12(5):21.

[24] Vakoc BJ, Fukumura D, Jain RK, Bouma BE. Cancer imaging by optical coherence tomography: preclinical progress and clinical potential. Nat Rev Cancer 2012;12(5):363–8.

[25] Nguyen FT, Zysk AM, Chaney EJ, Kotynek JG, Oliphant UJ, Bellafiore FJ, et al. Intraoperative evaluation of breast tumor margins with optical coherence tomography. Cancer Res 2009;69 (22):8790–6.

[26] Erickson-Bhatt SJ, Nolan RM, Shemonski ND, Adie SG, Putney J, Darga D, et al. Real-time imaging of the resection bed using a handheld probe to reduce incidence of microscopic positive margins in cancer surgery. Cancer Res 2015;75(18):3706–12.

[27] Wang J, Xu Y, Mesa KJ, South FA, Chaney EJ, Spillman Jr DR, et al. Complementary use of polarization-sensitive and standard OCT

metrics for enhanced intraoperative differentiation of breast cancer. Biomed Opt Express 2018;9(12):6519–28.

[28] Nguyen FT, Zysk AM, Chaney EJ, Adie SG, Kotynek JG, Oliphant UJ, et al. Optical coherence tomography the intraoperative assessment of lymph nodes in breast cancer. IEEE Eng Med Biol Mag 2010;29(2):63–70.

[29] Nolan RM, Adie SG, Marjanovic M, Chaney EJ, South FA, Monroy GL, et al. Intraoperative optical coherence tomography for assessing human lymph nodes for metastatic cancer. BMC Cancer 2016;16:144.

[30] De Carvalho N, Welzel J, Schuh S, Themstrup L, Ulrich M, Jemec GBE, et al. The vascular morphology of melanoma is related to Breslow index: an in vivo study with dynamic optical coherence tomography. Exp Dermatol 2018;27(11):1280–6.

[31] Pires L, Demidov V, Vitkin IA, Bagnato V, Kurachi C, Wilson BC. Optical clearing of melanoma in vivo: characterization by diffuse reflectance spectroscopy and optical coherence tomography. J Biomed Opt 2016;21(8):081210.

[32] Wang J, Xu Y, Boppart SA. Review of optical coherence tomography in oncology. J Biomed Opt 2017;22(12):1–23.

[33] Kut C, Chaichana KL, Xi J, Raza SM, Ye X, McVeigh ER, et al. Detection of human brain cancer infiltration ex vivo and in vivo using quantitative optical coherence tomography. Sci Transl Med 2015;7(292):292ra100.

[34] Walther J, Gaertner M, Cimalla P, Burkhardt A, Kirsten L, Meissner S, et al. Optical coherence tomography in biomedical research. Anal Bioanal Chem 2011;400(9):2721–43.

[35] Smith BR, Gambhir SS. Nanomaterials for in vivo imaging. Chem Rev 2017;117(3):901–86.

[36] Ralston TS, Adie SG, Marks DL, Boppart SA, Carney PS. Cross-validation of interferometric synthetic aperture microscopy and optical coherence tomography. Opt Lett 2010;35(10):1683–5.

[37] Xu Y, Liu YZ, Boppart SA, Carney PS. Automated interferometric synthetic aperture microscopy and computational adaptive optics for improved optical coherence tomography. Appl Opt 2016;55(8):2034–41.

[38] Ralston TS, Marks DL, Carney PS, Boppart SA. Interferometric synthetic aperture microscopy. Nat Phys 2007;3(2):129–34.

[39] Zhu D, Wang J, Marjanovic M, Chaney EJ, Cradock KA, Higham AM, et al. Differentiation of breast tissue types for surgical margin

assessment using machine learning and polarization-sensitive optical coherence tomography. Biomed Opt Express 2021;12(5):3021–36.

[40] Mojahed D, Ha RS, Chang P, Gan Y, Yao X, Angelini B, et al. Fully automated postlumpectomy breast margin assessment utilizing convolutional neural network based optical coherence tomography image classification method. Acad Radiol 2020;27(5):e81–6.

[41] You S, Sun Y, Yang L, Park J, Tu H, Marjanovic M, et al. Real-time intraoperative diagnosis by deep neural network driven multiphoton virtual histology. NPJ Precis Oncol 2019;3:33.

[42] Mayinger B, Horner P, Jordan M, Gerlach C, Horbach T, Hohenberger W, et al. Endoscopic fluorescence spectroscopy in the upper GI tract for the detection of GI cancer: initial experience. Am J Gastroenterol 2001;96(9):2616–21.

[43] Chang SK, Pavlova I, Marin NM, Follen M, Richards-Kortum R. Fluorescence spectroscopy as a diagnostic tool for detecting cervical pre-cancer. Gynecol Oncol 2005;99(3 Suppl. 1):S61–3.

[44] Brancaleon L, Durkin AJ, Tu JH, Menaker G, Fallon JD, Kollias N. In vivo fluorescence spectroscopy of nonmelanoma skin cancer. Photochem Photobiol 2001;73(2):178–83.

[45] Uspenskii LV, Kuzin MI, Ablitsov IA, Rybin VK, Loshchenov VB. Laser fluorescence spectroscopy in intraoperative diagnosis and staging of lung cancer. Khirurgiia (Mosk) 1996;(3):31–3.

[46] Chidananda SM, Satyamoorthy K, Rai L, Manjunath AP, Kartha VB. Optical diagnosis of cervical cancer by fluorescence spectroscopy technique. Int J Cancer 2006;119(1):139–45.

[47] Scepanovic OR, Volynskaya Z, Kong CR, Galindo LH, Dasari RR, Feld MS. A multimodal spectroscopy system for real-time disease diagnosis. Rev Sci Instrum 2009;80(4):043103.

[48] Volynskaya Z, Haka AS, Bechtel KL, Fitzmaurice M, Shenk R, Wang N, et al. Diagnosing breast cancer using diffuse reflectance spectroscopy and intrinsic fluorescence spectroscopy. J Biomed Opt 2008;13(2):024012.

[49] Tunnell JW, Desjardins AE, Galindo L, Georgakoudi I, McGee SA, Mirkovic J, et al. Instrumentation for multi-modal spectroscopic diagnosis of epithelial dysplasia. Technol Cancer Res Treat 2003;2 (6):505–14.

[50] Muller MG, Valdez TA, Georgakoudi I, Backman V, Fuentes C, Kabani S, et al. Spectroscopic detection and evaluation of morphologic and biochemical changes in early human oral carcinoma. Cancer 2003;97(7):1681–92.

[51] Georgakoudi I, Sheets EE, Muller MG, Backman V, Crum CP, Badizadegan K, et al. Trimodal spectroscopy for the detection and characterization of cervical precancers in vivo. Am J Obstet Gynecol 2002;186(3):374–82.

[52] Sun Y, You S, Tu H, Spillman Jr DR, Chaney EJ, Marjanovic M, et al. Intraoperative visualization of the tumor microenvironment and quantification of extracellular vesicles by label-free nonlinear imaging. Sci Adv 2018;4(12):eaau5603.

[53] You S, Barkalifa R, Chaney EJ, Tu H, Park J, Sorrells JE, et al. Label-free visualization and characterization of extracellular vesicles in breast cancer. Proc Natl Acad Sci U S A 2019;116(48):24012–18.

[54] Sorrells JE, Martin EM, Aksamitiene E, Mukherjee P, Alex A, Chaney EJ, et al. Label-free characterization of single extracellular vesicles using two-photon fluorescence lifetime imaging microscopy of NAD(P)H. Sci Rep 2021;11(1):3308.

[55] Iverson NM, Barone PW, Shandell M, Trudel LJ, Sen S, Sen F, et al. In vivo biosensing via tissue-localizable near-infrared-fluorescent single-walled carbon nanotubes. Nat Nanotechnol 2013;8(11):873–80.

[56] Graff RA, Swanson JP, Barone PW, Baik S, Heller DA, Strano MS. Achieving individual-nanotube dispersion at high loading in single-walled carbon nanotube composites. Adv Mater 2005;17(8):980.

[57] Liu Z, Tabakman S, Welsher K, Dai HJ. Carbon nanotubes in biology and medicine: in vitro and in vivo detection, imaging and drug delivery. Nano Res 2009;2(2):85–120.

[58] Cherukuri P, Bachilo SM, Litovsky SH, Weisman RB. Near-infrared fluorescence microscopy of single-walled carbon nanotubes in phagocytic cells. J Am Chem Soc 2004;126(48):15638–9.

[59] Yi H, Ghosh D, Ham MH, Qi J, Barone PW, Strano MS, et al. M13 phage-functionalized single-walled carbon nanotubes as nanoprobes for second near-infrared window fluorescence imaging of targeted tumors. Nano Lett 2012;12(3):1176–83.

[60] Mu B, Ahn J, McNicholas TP, Strano MS. Generating selective saccharide binding affinity of phenyl boronic acids by using single-walled carbon nanotube corona phases. Chem-Eur J 2015;21(12):4523–8.

[61] Barone PW, Parker RS, Strano MS. In vivo fluorescence detection of glucose using a single-walled carbon nanotube optical sensor: design, fluorophore properties, advantages, and disadvantages. Anal Chem 2005;77(23):7556–62.

[62] Landry MP, Vuković L, Kruss S, Bisker G, Landry AM, Islam S, et al. Comparative dynamics and sequence dependence of DNA and RNA binding to single walled carbon nanotubes. J Phys Chem C 2015;119(18):10048–58.

[63] Ulissi ZW, Sen F, Gong X, Sen S, Iverson N, Boghossian AA, et al. Spatiotemporal intracellular nitric oxide signaling captured using internalized, near-infrared fluorescent carbon nanotube nanosensors. Nano Lett 2014;14(8):4887–94.

[64] Jin H, Heller DA, Kalbacova M, Kim J-H, Zhang J, Boghossian AA, et al. Detection of single-molecule H2O2 signalling from epidermal growth factor receptor using fluorescent single-walled carbon nanotubes. Nat Nanotechnol 2010;5(4):302–9.

[65] Harvey JD, Jena PV, Baker HA, Zerze GH, Williams RM, Galassi TV, et al. A carbon nanotube reporter of microRNA hybridization events in vivo. Nat Biomed Eng 2017;1(4):0041.

[66] Williams RM, Lee C, Galassi TV, Harvey JD, Leicher R, Sirenko M, et al. Noninvasive ovarian cancer biomarker detection via an optical nanosensor implant. Sci Adv 2018;4(4):eaaq1090.

[67] Bisker G, Dong J, Park HD, Iverson NM, Ahn J, Nelson JT, et al. Protein-targeted corona phase molecular recognition. Nat Commun 2016;7.

[68] Yaari Z, Cheung JM, Baker HA, Frederiksen RS, Jena PV, Horoszko CP, et al. Nanoreporter of an enzymatic suicide inactivation pathway. Nano Lett 2020;20(11):7819–27.

[69] Giraldo JP, Landry MP, Kwak S-Y, Jain RM, Wong MH, Iverson NM, et al. A ratiometric sensor using single chirality near-infrared fluorescent carbon nanotubes: application to in vivo monitoring. Small 2015;11(32):3973–84.

[70] Zhang J, Landry MP, Barone PW, Kim J-H, Lin S, Ulissi ZW, et al. Molecular recognition using corona phase complexes made of synthetic polymers adsorbed on carbon nanotubes. Nat Nano 2013;8(12):959–68.

[71] Zhang JQ, Boghossian AA, Barone PW, Rwei A, Kim JH, Lin DH, et al. Single molecule detection of nitric oxide enabled by d(AT)(15) DNA adsorbed to near infrared fluorescent single-walled carbon nanotubes. J Am Chem Soc 2011;133(3):567–81.

[72] Kim JH, Patra CR, Arkalgud JR, Boghossian AA, Zhang JQ, Han JH, et al. Single-molecule detection of H(2)O(2) mediating angiogenic redox signaling on fluorescent single-walled carbon nanotube array. ACS Nano 2011;5(10):7848–57.

[73] Boghossian AA, Zhang J, Barone PW, Reuel NF, Kim JH, Heller DA, et al. Near-infrared fluorescent sensors based on single-walled carbon nanotubes for life sciences applications. ChemSusChem 2011;4(7):848–63.

[74] Kim JH, Ahn JH, Barone PW, Jin H, Zhang JQ, Heller DA, et al. A luciferase/single-walled carbon nanotube conjugate for near-infrared fluorescent detection of cellular ATP. Angew Chem Int Ed 2010;49 (8):1456–9.

[75] Jeng ES, Nelson JD, Prather KLJ, Strano MS. Detection of a single nucleotide polymorphism using single-walled carbon-nanotube near-infrared fluorescence. Small 2010;6(1):40–3.

[76] Kim JH, Heller DA, Jin H, Barone PW, Song C, Zhang J, et al. The rational design of nitric oxide selectivity in single-walled carbon nanotube near-infrared fluorescence sensors for biological detection. Nat Chem 2009;1(6):473–81.

[77] Heller DA, Jin H, Martinez BM, Patel D, Miller BM, Yeung TK, et al. Multimodal optical sensing and analyte specificity using single-walled carbon nanotubes. Nat Nanotechnol 2009;4(2):114–20.

[78] Barone PW, Yoon H, Ortiz-Garcia R, Zhang JQ, Ahn JH, Kim JH, et al. Modulation of single-walled carbon nanotube photoluminescence by hydrogel swelling. ACS Nano 2009;3(12):3869–77.

[79] Heller DA, Pratt GW, Zhang J, Nair N, Hansborough AJ, Boghossian AA, et al. Peptide secondary structure modulates single-walled carbon nanotube fluorescence as a chaperone sensor for nitroaromatics. Proc Natl Acad Sci U S A 2011;108(21):8544.

[80] Heller Daniel A, Jeng Esther S, Yeung T-K, Martinez Brittany M, Moll Anthonie E, Gastala Joseph B, et al. Optical detection of DNA conformational polymorphism on single-walled carbon nanotubes. Science 2006;311(5760):508–11.

[81] Roxbury D, Jena PV, Shamay Y, Horoszko CP, Heller DA. Cell membrane proteins modulate the carbon nanotube optical bandgap via surface charge accumulation. ACS Nano 2016;10(1):499–506.

[82] Williams RM, Lee C, Heller DA. A fluorescent carbon nanotube sensor detects the metastatic prostate cancer biomarker uPA. ACS Sens 2018;3(9):1838–45.

[83] Kong K, Kendall C, Stone N, Notingher I. Raman spectroscopy for medical diagnostics—from in-vitro biofluid assays to in-vivo cancer detection. Adv Drug Deliv Rev 2015;89:121–34.

[84] Jermyn M, Desroches J, Aubertin K, St-Arnaud K, Madore WJ, De Montigny E, et al. A review of Raman spectroscopy advances with

an emphasis on clinical translation challenges in oncology. Phys Med Biol 2016;61(23):R370–400.

[85] Auner GW, Koya SK, Huang C, Broadbent B, Trexler M, Auner Z, et al. Applications of Raman spectroscopy in cancer diagnosis. Cancer Metastasis Rev 2018;37(4):691–717.

[86] Maheedhar K, Bhat RA, Malini R, Prathima NB, Keerthi P, Kushtagi P, et al. Diagnosis of ovarian cancer by Raman spectroscopy: a pilot study. Photomed Laser Surg 2008;26(2):83–90.

[87] Chin LK, Son T, Hong JS, Liu AQ, Skog J, Castro CM, et al. Plasmonic sensors for extracellular vesicle analysis: from scientific development to translational research. ACS Nano 2020;14(11):14528–48.

[88] Mehrmohammadi M, Yoon SJ, Yeager D, Emelianov SY. Photoacoustic imaging for cancer detection and staging. Curr Mol Imaging 2013;2(1):89–105.

[89] Steinberg I, Huland DM, Vermesh O, Frostig HE, Tummers WS, Gambhir SS. Photoacoustic clinical imaging. Photoacoustics 2019;14:77–98.

[90] Beckonert O, Keun HC, Ebbels TMD, Bundy J, Holmes E, Lindon JC, et al. Metabolic profiling, metabolomic and metabonomic procedures for NMR spectroscopy of urine, plasma, serum and tissue extracts. Nat Protoc 2007;2(11):2692–703.

[91] Gouilleux B, Charrier B, Akoka S, Giraudeau P. Gradient-based solvent suppression methods on a benchtop spectrometer. Magnetic Reson Chem 2017;55(2):91–8.

[92] Danieli E, Perlo J, Blümich B, Casanova F. Small magnets for portable NMR spectrometers. Angew Chem Int Ed 2010;49 (24):4133–5.

[93] Percival BC;, Grootveld M, Gibson M, Osman Y, Molinari M, Jafari F, et al. Low-field, benchtop NMR spectroscopy as a potential tool for point-of-care diagnostics of metabolic conditions: validation, protocols and computational models. High-Throughput 2018;8(1):2.

[94] Haun Jered B, Castro Cesar M, Wang R, Peterson Vanessa M, Marinelli Brett S, Lee H, et al. Micro-NMR for rapid molecular analysis of human tumor samples. Sci Transl Med 2011;3(71) 71ra16.

[95] Daniel KD, Kim GY, Vassiliou CC, Jalali-Yazdi F, Langer R, Cima MJ. Multi-reservoir device for detecting a soluble cancer biomarker. Lab Chip 2007;7(10):1288–93.

[96] Gao Y, Huo W, Zhang L, Lian J, Tao W, Song C, et al. Multiplex measurement of twelve tumor markers using a GMR multi-biomarker immunoassay biosensor. Biosens Bioelectron 2019;123:204–10.

[97] Weissleder R, Pittet MJ. Imaging in the era of molecular oncology. Nature 2008;452(7187):580–9.
[98] Lee H, Sun E, Ham D, Weissleder R. Chip–NMR biosensor for detection and molecular analysis of cells. Nat Med 2008;14(8):869–74.
[99] Haun JB, Devaraj NK, Hilderbrand SA, Lee H, Weissleder R. Bioorthogonal chemistry amplifies nanoparticle binding and enhances the sensitivity of cell detection. Nat Nanotechnol 2010;5(9):660–5.
[100] Sun EY, Weissleder R, Josephson L. Continuous analyte sensing with magnetic nanoswitches. Small 2006;2(10):1144–7.
[101] Wang W, Wang Y, Tu L, Feng Y, Klein T, Wang J-P. Magnetoresistive performance and comparison of supermagnetic nanoparticles on giant magnetoresistive sensor-based detection system. Sci Rep 2014;4(1):5716.
[102] McDonnell LA, Angel PM, Lou S, Drake RR. Mass spectrometry imaging in cancer research. Adv Cancer Res 2017;283–90.
[103] Arentz G, Mittal P, Zhang C, Ho YY, Briggs M, Winderbaum L, et al. Applications of mass spectrometry imaging to cancer. Adv Cancer Res 2017;27–66.
[104] Lin L-E, Chen C-L, Huang Y-C, Chung H-H, Lin C-W, Chen K-C, et al. Precision biomarker discovery powered by microscopy image fusion-assisted high spatial resolution ambient ionization mass spectrometry imaging. Anal Chim Acta 2020;1100:75–87.
[105] Zhang Z, Bast Jr RC, Yu Y, Li J, Sokoll LJ, Rai AJ, et al. Three biomarkers identified from serum proteomic analysis for the detection of early stage ovarian cancer. Cancer Res 2004;64(16):5882–90.
[106] Losman JA, Kaelin WG. What a difference a hydroxyl makes: mutant IDH, (R)-2-hydroxyglutarate, and cancer. Genes & Dev 2013;27(8):836–52.
[107] Keshishian H, Burgess MW, Specht H, Wallace L, Clauser KR, Gillette MA, et al. Quantitative, multiplexed workflow for deep analysis of human blood plasma and biomarker discovery by mass spectrometry. Nat Protoc 2017;12(8):1683–701.
[108] Le Tourneau C, Borcoman E, Kamal M. Molecular profiling in precision medicine oncology. Nat Med 2019;25(5):711–12.
[109] Garg U, Zhang YV. Mass spectrometry in clinical laboratory: applications in therapeutic drug monitoring and toxicology. Methods Mol Biol 2016;1383:1–10.
[110] Clarke NJ. Mass spectrometry in precision medicine: phenotypic measurements alongside pharmacogenomics. Clin Chem 2016;62(1):70–6.

[111] Decosterd LA, Widmer N, André P, Aouri M, Buclin T. The emerging role of multiplex tandem mass spectrometry analysis for therapeutic drug monitoring and personalized medicine. Trends Anal Chem 2016;84:5 13.

[112] Zhang W, Hankemeier T, Ramautar R. Next-generation capillary electrophoresis−mass spectrometry approaches in metabolomics. Curr Opin Biotechnol 2017;43:1−7.

[113] Prasad B, Vrana M, Mehrotra A, Johnson K, Bhatt DK. The promises of quantitative proteomics in precision medicine. J Pharm Sci 2017;106(3):738−44.

[114] Kowalczyk T, Ciborowski M, Kisluk J, Kretowski A, Barbas C. Mass spectrometry based proteomics and metabolomics in personalized oncology. Biochim Biophys Acta Mol Basis Dis 2020;1866 (5):165690.

[115] Zou R, Cao W, Chong L, Hua W, Xu H, Mao Y, et al. Point-of-care tissue analysis using miniature mass spectrometer. Anal Chem 2019;91(1):1157−63.

[116] Feider CL, Krieger A, DeHoog RJ, Eberlin LS. Ambient ionization mass spectrometry: recent developments and applications. Anal Chem 2019;91(7):4266−90.

[117] Pirro V, Jarmusch AK, Vincenti M, Cooks RG. Direct drug analysis from oral fluid using medical swab touch spray mass spectrometry. Anal Chim Acta 2015;861:47−54.

[118] Garza KY, Silva AAR, Rosa JR, Keating MF, Povilaitis SC, Spradlin M, et al. Rapid screening of COVID-19 directly from clinical nasopharyngeal swabs using the MasSpec pen. Anal Chem 2021;93(37):12582−93.

[119] Shi R-Z, El Gierari ETM, Manicke NE, Faix JD. Rapid measurement of tacrolimus in whole blood by paper spray-tandem mass spectrometry (PS-MS/MS). Clin Chim Acta 2015;441:99−104.

[120] Lee C-W, Su H, Lee R-H, Lin Y-P, Tsai Y-D, Wu D-C, et al. Point-of-care identification of organophosphates in gastric juice by ambient mass spectrometry in emergency settings. Clin Chim Acta 2018;485:288−97.

[121] Sanai N, Berger MS. Surgical oncology for gliomas: the state of the art. Nat Rev Clin Oncol 2018;15(2):112−25.

[122] Shin D, Park S-S. Clinical importance and surgical decision-making regarding proximal resection margin for gastric cancer. World J Gastrointest Oncol 2013;5(1):4−11.

[123] Schwarz J, Schmidt H. Correction to: Technology for intraoperative margin assessment in breast cancer. Ann Surg Oncol 2020;27 (Suppl. 3):972–5.

[124] Pradipta AR, Tanei T, Morimoto K, Shimazu K, Noguchi S, Tanaka K. Emerging technologies for real-time intraoperative margin assessment in future breast-conserving surgery. Adv Sci 2020;7 (9):1901519.

[125] Takáts Z, Wiseman JM, Gologan B, Cooks RG. Mass spectrometry sampling under ambient conditions with desorption electrospray ionization. Science 2004;306(5695):471–3.

[126] Calligaris D, Caragacianu D, Liu X, Norton I, Thompson CJ, Richardson AL, et al. Application of desorption electrospray ionization mass spectrometry imaging in breast cancer margin analysis. Proc Natl Acad Sci U S A 2014;111(42):15184–9.

[127] Schäfer K-C, Dénes J, Albrecht K, Szaniszló T, Balog J, Skoumal R, et al. In vivo, in situ tissue analysis using rapid evaporative ionization mass spectrometry. Angew Chem Int Ed Engl 2009;48(44):8240–2.

[128] Takats Z, Strittmatter N, McKenzie JS. Ambient mass spectrometry in cancer research. Adv Cancer Res 2017;134:231–56.

[129] Balog J, Sasi-Szabó L, Kinross J, Lewis MR, Muirhead LJ, Veselkov K, et al. Intraoperative tissue identification using rapid evaporative ionization mass spectrometry. Sci Transl Med 2013;5 (194):194ra93.

[130] Phelps DL, Balog J, Gildea LF, Bodai Z, Savage A, El-Bahrawy MA, et al. The surgical intelligent knife distinguishes normal, borderline and malignant gynaecological tissues using rapid evaporative ionisation mass spectrometry (REIMS). Br J Cancer 2018;118 (10):1349–58.

[131] Alexander J, Gildea L, Balog J, Speller A, McKenzie J, Muirhead L, et al. A novel methodology for in vivo endoscopic phenotyping of colorectal cancer based on real-time analysis of the mucosal lipidome: a prospective observational study of the iKnife. Surg Endosc 2017;31(3):1361–70.

[132] Zhang J, Rector J, Lin JQ, Young JH, Sans M, Katta N, et al. Nondestructive tissue analysis for ex vivo and in vivo cancer diagnosis using a handheld mass spectrometry system. Sci Transl Med 2017;9(406):eaan3968.

[133] Fatou B, Saudemont P, Leblanc E, Vinatier D, Mesdag V, Wisztorski M, et al. In vivo real-time mass spectrometry for guided surgery application. Sci Rep 2016;6:25919.

[134] Fidler IJ. The pathogenesis of cancer metastasis: the 'seed and soil' hypothesis revisited. Nat Rev Cancer 2003;3(6):453–8.

[135] Merkher Y, Horesh Y, Abramov Z, Shleifer G, Ben-Ishay O, Kluger Y, et al. Rapid cancer diagnosis and early prognosis of metastatic risk based on mechanical invasiveness of sampled cells. Ann Biomed Eng 2020;48(12):2846–58.

[136] Robinson ER, Gowrishankar G, D'Souza AL, Kheirolomoom A, Haywood T, Hori SS, et al. Minicircles for a two-step blood biomarker and PET imaging early cancer detection strategy. J Controlled Rel 2021;335:281–9.

[137] Ronald JA, Chuang H-Y, Dragulescu-Andrasi A, Hori SS, Gambhir SS. Detecting cancers through tumor-activatable minicircles that lead to a detectable blood biomarker. Proc Natl Acad Sci U S A 2015;112(10):3068–73.

[138] Jaiswal P, Goel A, Mittal R. Survivin: a molecular biomarker in cancer. Indian J Med Res 2015;141(4):389–97.

[139] Aalipour A, Chuang H-Y, Murty S, D'Souza AL, Park S-M, Gulati GS, et al. Engineered immune cells as highly sensitive cancer diagnostics. Nat Biotechnol 2019;37(5):531–9.

[140] Kwong GA, von Maltzahn G, Murugappan G, Abudayyeh O, Mo S, Papayannopoulos IA, et al. Mass-encoded synthetic biomarkers for multiplexed urinary monitoring of disease. Nat Biotechnol 2013;31(1):63–70.

[141] Elias KM, Fendler W, Stawiski K, Fiascone SJ, Vitonis AF, Berkowitz RS, et al. Diagnostic potential for a serum miRNA neural network for detection of ovarian cancer. eLife 2017;6.

[142] Tao K, Bian Z, Zhang Q, Guo X, Yin C, Wang Y, et al. Machine learning-based genome-wide interrogation of somatic copy number aberrations in circulating tumor DNA for early detection of hepatocellular carcinoma. EBioMedicine 2020;56:102811.

[143] Maros ME, Capper D, Jones DTW, Hovestadt V, Von Deimling A, Pfister SM, et al. Machine learning workflows to estimate class probabilities for precision cancer diagnostics on DNA methylation microarray data. Nat Protoc 2020;15(2):479–512.

[144] Gessulat S, Schmidt T, Zolg DP, Samaras P, Schnatbaum K, Zerweck J, et al. Prosit: proteome-wide prediction of peptide tandem mass spectra by deep learning. Nat Methods 2019;16(6):509–18.

[145] The M, Käll L. Focus on the spectra that matter by clustering of quantification data in shotgun proteomics. Nat Commun 2020;11(1).

[146] Xie Y, Meng W-Y, Li R-Z, Wang Y-W, Qian X, Chan C, et al. Early lung cancer diagnostic biomarker discovery by machine learning methods. Transl Oncol 2021;14(1):100907.

[147] Kawakami E, Tabata J, Yanaihara N, Ishikawa T, Koseki K, Iida Y, et al. Application of artificial intelligence for preoperative diagnostic and prognostic prediction in epithelial ovarian cancer based on blood biomarkers. Clin Cancer Res 2019;25(10):3006.

[148] Lee MA, Wang S, Jin X, Bakh NA, Nguyen FT, Dong J, et al. Implantable nanosensors for human steroid hormone sensing in vivo using a self-templating corona phase molecular recognition. Adv Healthc Mater 2020;9(21):2000429.

[149] Stojanovic MN, Landry DW. Aptamer-based colorimetric probe for cocaine. J Am Chem Soc 2002;124(33):9678–9.

[150] Jin H, Jeng ES, Heller DA, Jena PV, Kirmse R, Langowski J, et al. Divalent ion and thermally induced DNA conformational polymorphism on single-walled carbon nanotubes. Macromolecules 2007;40 (18):6731–9.

[151] Kelich P, Jeong S, Navarro N, Adams J, Sun X, Zhao H, et al. Machine learning enables discovery of DNA-carbon nanotube sensors for serotonin. bioRxiv 2021. Available from: https://doi.org/10.1101/2021.08.20.457145.

[152] Yaari Z, Yang Y, Apfelbaum E, Cupo C, Settle Alex H, Cullen Q, et al. A perception-based nanosensor platform to detect cancer biomarkers. Sci Adv 2021;7(47):eabj0852.

[153] Echle A, Rindtorff NT, Brinker TJ, Luedde T, Pearson AT, Kather JN. Deep learning in cancer pathology: a new generation of clinical biomarkers. Br J Cancer 2021;124(4):686–96.

[154] Zednik C. Solving the black box problem: a normative framework for explainable artificial intelligence. Philosophy Technol 2021;34 (2):265–88.

CHAPTER

Oncometabolites and their role in cancer

12

Emily L. Gill[1,2], Khushbu Patel[1,2] and Dinesh Rakheja[3,4]

[1]*Department of Pathology and Laboratory Medicine, Children's Hospital of Philadelphia, Philadelphia, PA, United States*

[2]*Department of Pathology and Laboratory Medicine, University of Pennsylvania, Philadelphia, PA, United States*

[3]*Department of Pathology and Laboratory Medicine, Children's Health, Dallas, TX, United States*

[4]*Department of Pathology, University of Texas Southwestern Medical Center, Dallas, TX, United States*

Objectives

1. Discuss the origin of oncometabolites and their role in cancer pathogenesis.
2. Discuss the role of oncometabolites as cancer biomarkers.
3. List laboratory techniques used for the measurement of oncometabolites.

12.1 Introduction

"Oncometabolites" are metabolic intermediates with oncogenic properties. They aberrantly accumulate in cancer cells either upstream or downstream of metabolic defects, driving cancer progression and therapeutic resistance. Since the recognition of oncometabolites and the discovery of the underlying metabolic aberrancies, it has been recognized that they can be utilized as biomarkers for cancer diagnostics, prognostics, and therapeutics.

Reprogramming of energy metabolism was recognized as one of the hallmarks of cancer over a decade ago [1]. Changes in energy

Cancer Biomarkers: Clinical Aspects and Laboratory Determination. DOI: https://doi.org/10.1016/B978-0-12-824302-2.00003-5

© 2022 Elsevier Inc. All rights reserved.

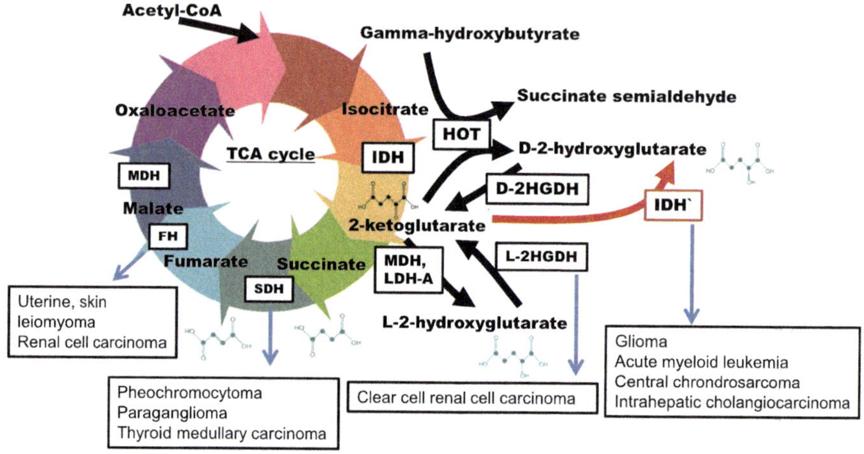

FIGURE 12.1

Location of oncometabolites D-2-hydroxyglutarate, L-2-hydroxyglutarate, succinate, and fumarate in the tricarboxylic acid cycle (TCA) cycle.

metabolism are necessary to fuel cancer cell growth and division. This was first described by Otto Warburg in 1930 who observed that cancer cells can reprogram glucose metabolism by restricting themselves strictly to glycolysis even under aerobic conditions [2]. This contrasts with normal cells that under aerobic conditions process glucose to carbon dioxide by routing pyruvate metabolism to the mitochondria. Cancer cells that undergo this type of "aerobic glycolysis" compensate for the lowered adenosine triphosphate (ATP) production and can thrive on glucose as an energy source by upregulating glucose transporters and increasing glucose import into the cytoplasm. Glycolytic fueling of cancer cells enables diversion of glycolytic intermediates to various biosynthetic pathways for synthesis of macromolecules and organelles required for cell proliferation.

More recently, specific cancer types with aberrances in or proximate to the citric acid/tricarboxylic acid (TCA) cycle have been described [3]. These cancer-related metabolic aberrances affect isocitrate dehydrogenase 1 (IDH1) or isocitrate dehydrogenase 2 (IDH2), L-2-hydroxyglutarate dehydrogenase (L2HGDH), succinate dehydrogenase (SDH), and fumarate hydratase (FH) (Fig. 12.1). These enzymes share metabolic proximity within the TCA cycle and either gain- or loss-of-function mutations can result in overproduction of the four currently recognized oncometabolites: D-2-hydroxyglutarate (D2HG), L-2-hydroxyglutarate (L2HG), succinate, and fumarate [4,5].

12.2 Oncometabolites and their associated cancers

D2HG and L2HG are enantiomers with identical physical properties, but because of different spatial orientation they are recognized by different enzymes and thus produced through distinct biochemical reactions. They both are by-products of side reactions that utilize 2-ketoglutarate, a TCA cycle intermediate. Physiologically, D2HG is produced from 2-ketoglutarate in a reaction catalyzed by hydroxy acid-oxoacid transhydrogenase, while L2HG is produced from 2-ketoglutarate by the action of the enzymes malate dehydrogenase and lactate dehydrogenase A [6].

The overproduction of D2HG in cancer results from gain-of-function mutations in the genes encoding IDH1 or IDH2. Wild-type IDHs convert isocitrate to 2-ketoglutarate via reversible oxidative decarboxylation. In many cancers such as gliomas, acute myeloid leukemia, intrahepatic cholangiocarcinoma, and central chondrosarcoma, somatic gain-of-function hotspot mutations in *IDH1* (single amino acid substitutions at R132) or *IDH2* (single amino acid substitutions at R140 or R172) introduce a neomorphic activity in IDHs catalyzing the reduction of 2-ketoglutarate to D2HG that leads to markedly increased levels of D2HG in these cancer cells [7–9]. The autosomal recessive inborn error of metabolism L-2-hydroxyglutaric aciduria, caused by loss-of-function mutations of *L2HGDH*, is associated with brain and extracranial tumors [10]. Recently, increased L2HG levels in a subset of clear cell renal cell carcinomas have been attributed to reduced expression of L2HGDH in the cancer cells [11].

Succinate dehydrogenase is a heterotetrameric complex composed of four subunits, SDHA, SDHB, SDHC, and SDHD, and two assembly factors, SDHAF1 and SDHAF2. Mutations in the *SDH* genes have been described in autosomal dominant hereditary paraganglioma, pheochromocytoma, renal carcinoma, thyroid cancer, neuroblastoma, T-cell leukemia, and gastrointestinal stromal tumor. Heterozygous germline mutations of *FH* followed by a "second hit" loss of heterozygosity cause hereditary leiomyomatosis and renal cell cancer (HLRCC) syndrome. *FH* mutations are also found in a subset of paragangliomas/pheochromocytomas and *FH* deletions have been described in neuroblastoma [6,12–15].

All four of these oncometabolites share structural similarity and exhibit oncogenic properties by inhibiting a superfamily of enzymes known as 2-ketoglutarate-dependent dioxygenases. This superfamily comprises a variety of enzymes exhibiting a wide range of biological functions. Notably, inhibition of prolyl hydroxylase domain proteins (PDHs), a member of the 2-ketoglutarate-dependent dioxygenase family, causes stabilization of hypoxia-inducible factors (HIFs), which promotes angiogenesis and cancer cell growth. Inhibition of ten-eleven translocation (TET) enzymes, another family member, leads to aberrant DNA hypermethylation and epigenetic changes in cancer cells. Other downstream effects of these oncometabolites include production of argininosuccinate by fumarate leading to diversion of urea cycle metabolites to urine and pyrimidine biosynthesis. Fumarate accumulation also leads to a posttranslational protein modification called succination, in which fumarate reacts with the thiol groups of cysteine residues ultimately leading to oxidative stress. The biological mechanisms by which oncometabolites exhibit oncogenic properties is a growing area of research. Furthermore, since the discovery of the genetic molecular mechanisms that drive their overproduction, oncometabolites are ideal biomarkers that give insight into cancer diagnosis, progression, and prognosis. The remainder of the chapter will focus on the biochemical methods used for detecting and measuring oncometabolites as well as implications for the diagnosis and management of cancers.

12.3 In vitro profiling and measurement of oncometabolites

Metabolomic workflows are either global (untargeted) or targeted, depending on the analytical goals(s) [16]. Clinically validated assays use targeted methods optimized for the detection of a single or, less commonly, a panel of therapeutic markers and/or diagnostic markers.

The common methodologies currently used for targeted identification and measurement of oncometabolites include gas chromatography–mass spectrometry (GC–MS) and liquid chromatography–tandem mass spectrometry (LC–MS/MS). GC–MS relies on the analytes of interest being volatile; derivatization steps are often required during sample preparation to increase analyte volatility. Once in the gas phase, an inert carrier gas pulls the volatile mixture through the column, where each individual component interacts with a stationary phase with varying affinities. This generates a wide range of retention times (RT) because each component elutes from the "column" based on its interaction with the stationary phase. Once separated, volatile molecules are ionized and fragmented (e.g., by being bombarded with electrons) and the fragment ions are then passed into a quadrupole prior to mass/charge ratio detection.

Yost and colleagues introduced the tandem-in-space triple quadrupole mass spectrometer (TQ-MS) in 1979, extending the analytical capabilities of MS [17]. The tandem-in-space mass spectrometer offers greater specificity and sensitivity. Select reaction monitoring (SRM) allows ions of interest to be preferentially selected within a 1 atomic mass unit (amu) window. The ions are then passed into a higher-pressure cell where they collide with neutral gas molecules and are fragmented by collision-induced dissociation (CID). Then, the corresponding fragment ions are selected, generating a mass spectrum of fragment ions specific to the preselected ions of interest. Coupling TQ-MS with liquid chromatography (LC), known as liquid chromatography–tandem mass spectrometry (LC–MS/MS), is used to separate the analytes of interest based on their interaction with a stationary phase and liquid mobile phase within a complex mixture. LC does not require derivatization to increase volatility, unlike gas chromatography (GC), thus simplifying sample preparation prior to analysis.

12.4 Omics approaches to metabolite profiling in cancers

Global oncometabolomics are research approaches and can be defined as the comprehensive study of the metabolome of malignant neoplastic disease using high-resolution mass spectrometry (HRMS) platforms. The goal is to identify metabolic signatures of disease by comparing biological samples with healthy control subjects. Metabolomic studies can be carried out using cells, biofluids, or tissues. Biofluids can include serum, plasma, urine, ascitic fluid, saliva, among others. Global workflows require high mass accuracy; therefore, liquid chromatography-high-resolution mass spectrometry (LC-HRMS) platforms are the instruments of choice. LC-HRMS is predominantly a research platform and has not yet been sufficiently validated for clinical use.

Orbitrap is the most common mass analyzer, which offers high-resolution, accurate mass capabilities (<5 ppm) and a wide dynamic range. Multiplexing HRMS with LC at μL/minute flow rates enables the identification of polar and nonpolar species in a single extraction/injection. Nonetheless, comprehensive assays are significantly more challenging than targeted assays, requiring more sophisticated workflows. Global analysis generates a feature list (relating to specific m/z and RTs) that is identified using in-house libraries and proprietary software. Thousands of features are detected per injection, including features related to sample degradation due to poor sample handling, adducts, clusters, contamination, and artifacts. Many endogenous features remain unidentified because feature matching libraries are not exhaustive. That said, within these identified and unknown feature lists there may be distinct metabolic signatures of malignancy.

Another challenge of global metabolomics is that lower abundance species are often neglected during fragmentation. Koelmel et al. introduced a data-dependent iterative exclusion (ddMS2-IE) methodology known as IE-omics using LC-HRMS for lipidomics analysis [18]. Lipidomics is a subdiscipline of metabolomics; synthesis or uptake of lipids contributes to rapid cancer cell growth and tumor formation [19]. IE-omics data acquisition involves creating an exclusion list between traditional data-dependent (ddMS2) runs. Traditional ddMS2 involves selecting the most abundant ions for fragmentation per scan. Applying an exclusion list ensures that lower abundance species are not neglected in subsequent runs; this approach increases the potential for finding and identifying biomarkers. This method has also been applied to proteomics [20].

Fluxomics is another approach to studying the metabolome. This approach involves tracing the pathway(s) of an isotopically labeled carbon- or nitrogen-containing compound. Typically, cells are cultured in vitro in isotopically labeled media; 13C-glucose is an example. The resulting data using labeled media, following extraction and LC-HRMS analysis, provides enrichment pathway mapping rather than a single quantified value relating solely to glucose. As described earlier, cancer cells consume glucose at a much faster rate compared to normal cells. Thus fluxomics provides a unique approach to further understanding and elucidating cancer metabolism [21].

An alternative analytical approach to MS is nuclear magnetic resonance (NMR), which exploits the behavior of molecules when placed in a constant magnetic field [22]. Molecules are subsequently identified based on the precise resonant frequency of their nuclei (such as 1 H, 13 C, 15 N, or 31 P). For example, 31P-NMR can be used to identify products of energy or phospholipid metabolism, whereas 13C-NMR measures dynamic carbon fluxes, such as glucose metabolism. Because all nuclei are electrically charged and most have spin (i.e., they behave like magnets when a strong external magnetic field is applied), they are transiently excited, causing them to flip their spin state. As these nuclei relax, they produce a characteristic spectroscopic pattern (chemical shift). The field is affected by electron shielding, which in itself is dependent on the chemical environment. Therefore the chemical shift reflects the type, location, and electromagnetic environment of excited atoms in a given molecule. The major limitation of NMR is that it is generally less sensitive than MS, making the detection of low abundance metabolites challenging. That said, 1H-NMR can be used for liquid or solid samples, using magic angle spinning (MAS) techniques [23], with minimal sample preparation. By spinning the sample at the magic angle (θm) 54.74°, with respect to the direction of the magnetic field, the normally broad lines become narrower, therefore increasing the resolution.

12.5 Imaging for oncometabolites

Based on the operating principles of NMR, magnetic resonance spectral imaging (MRSI) is an imaging technique which generates parallel spectroscopic information, providing real-time oncometabolic activity. For example, in the context of oncology, a T1- or T2-weighted MRI image provides information regarding the shape and size of a tumor. MRSI, on the other hand, provides additional information about the metabolic activity occurring in the tumor. Spatially resolved oncometabolic activity can be achieved using a standard MRI instrument providing multimodal imaging and metabolic information.

MS imaging (MSI) techniques, such as matrix-assisted laser desorption ionization MS (MALDI-MS), are also useful for studying the metabolic distribution of oncometabolites in thin tissue sections (5–25 μm) [24,25]. Unlike MRSI, studies are carried out in vitro using cryosectioned tissue using MALDI-MS[26]. The tissue section(s) are mounted onto a glass slide and coated with a matrix using a nebulizer. The matrix is an organic acid, base, or metal composite that contains additives to increase the signal-to-noise ratio. The matrix forms a fine crystalline solid; the smaller the crystal size, the greater the homogeneity, which maximizes uniformity of matrix-analyte cocrystals, preventing analyte migration while also reducing spectral complexity. The glass slide is then secured onto a moving sample plate and inserted into a vacuum chamber, where a stationary laser raster across the tissue section(s), collecting a mass spectrum at each pixel. The resulting spatial resolution is dependent on the laser spot size and step size. The intensities of each molecular ion can be extracted to produce pixel spectra. These spectra combine to produce pseudo-colored images. The color-intensity is proportional to the molecular-ion intensity, highlighting the distribution of the user-determined *m/z* value.

12.6 Recent innovations in oncometabolomics

MS is an evolving field with instrumentation continually being developed and optimized, thus expanding the list of applications. In oncometabolomics, the MasSpec Pen, developed by Zhang et al. is one recent innovation [27]. The MasSpec Pen facilitates intraoperative screening of cancer and identification of cancer margins. The handheld pen-like device rapidly identifies the molecular profile of tissues using a small volume water droplet venturi-style mechanism to draw in analytes for MS analysis. Keating et al. integrated a laparoscopic version of the MasSpec Pen with the da Vinci Surgical System [28]. This instrument was used during robotic-assisted porcine surgery by introducing it into the animal's upper gastrointestinal system and performing in vivo analyses of the stomach and liver. Balog et al. developed similar technology for intraoperative tissue identification, known as the intelligent knife (iKnife), which uses rapid evaporative ionization mass spectrometry (REIMS) [29]. It connects an electrosurgical knife to a MS. Electrosurgical knives use an electric current to rapidly heat tissue, causing vaporization of the tissue to occur. The iKnife samples the smoke generated by this process that is normally removed by an extraction system. This smoke may provide real-time information about the neoplastic growth in terms of its malignancy. The iKnife has been used to assess tumor margins in surgical resections of brain, liver, lung, breast, or colorectal tumors [29].

Cancer has a myriad of effects on metabolism; rewiring of intracellular metabolism may produce metabolic phenotypes that can be targeted for early diagnosis, therapeutic response, and clinical trials. This phenotypic change enables cancer cells to proliferate inappropriately (e.g., Warburg effect) and adapt to the tumor microenvironment [30]. Alterations in metabolism can affect both cancer cells and normal tissue, which, in turn, interact with intrinsic and extrinsic factors to create an oncometabolic network. Intrinsic and extrinsic factors can include genetics (germline and somatic), diet, and physical activity, among others. Global metabolomics strives to identify oncometabolic changes and therefore potential candidates for targeted clinical assays. Furthermore, real-time measurements are promising for rapid diagnosis and surgical removal of oncological tissue. While most omics approaches are still in their infancy and some have or are starting to translate into the clinics and operating rooms, the technical complexity of these methods may be a limiting factor for large-scale adoption. On the other hand, many clinical laboratories that perform mass spectrometry testing are well-situated for targeted metabolite measurement. In the next section we describe methods for clinical measurement of D- and L-2-hydroxylgutarate.

12.7 Measuring oncometabolites D-2-hydroxyglutarate and L-2-hydroxyglutarate in the clinical laboratory

Overall, both LC–MS/MS and GC–MS instruments allow for rapid detection and/or quantitation of total 2-hydroxyglutarte and separated D2HG and L2HG. Urine is the sample of choice because concentrations of both enantiomers are significantly higher in urine compared with plasma and cerebrospinal fluid (CSF). The most common method for the screening urine metabolites by GC–MS begins with an ethyl acetate extraction and trimethylsilyl (TMS) derivatization. These assays are easily adaptable to targeting oncometabolic markers such as 2HG, fumarate, and succinate in a variety of body fluids and cell cultures. It is important to note, however, that common GC–MS assays do not distinguish between the enantiomers of 2HG and measure total 2HG [6]. See Fig. 12.2.

Chiral differentiation can be achieved using LC–MS/MS as well as GC–MS. Gibson et al. used D-2-butanol to achieve chiral separation via GC–MS [31]. This method has been extensively validated using clinical samples. However, less time-consuming alternatives using LC–MS/MS methods have since been described. Approaches to separating stereoisomers include the addition of chiral additives in the mobile phase or by using a chiral column. Chiral stationary phases (CSP) are prepared using a single optical isomer, either the R or the S isomer. Another method using ristocetin A glycopeptide antibiotic silica gel bonded column coupled with tandem-MS to achieve enantiomeric separation of 2HG has also been described [32].

Jones et al.[33] described a clinical validation of LC–MS/MS-based assay that is currently being used as a laboratory developed test for following patients with IDH-mutated myeloid neoplasms.

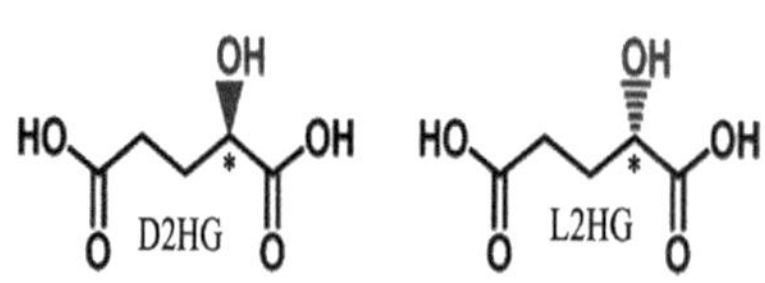

FIGURE 12.2

D-2-Hydroxyglutarate (D2HG) and L-2-hydroxyglutarate (L2HG).

Previously, the circulating levels of D2HG, total 2HG, and D2HG/L2HG ratio were shown to be highly predictive of IDH1/IDH2 mutation in acute myeloid leukemia in a preliminary analysis [34]. The assay design includes a five-point calibration containing native D2HG and L2HG and a high- and a low-quality control (QC). In clinical samples collected from patients, for example, serum or plasma are spiked with a stable isotope-labeled internal standard, D_4- D, L-2-hydroxyglutaric acid (synthesized as a zinc salt from D_6- D, L-2-ketopentanedioic acid) (Fig. 12.3).

A mixture of ice-cold 80% methanol and 20% water is then used to extract 2HG. The samples are vortexed and incubated at −80°C and then centrifuged before being dried under nitrogen and derivatized with (+)-Di-O-acetyl-l-tartaric anhydride (DATAN) (Fig. 12.4).

Derivatization converts the enantiomers into diastereoisomers to facilitate column separation. The LC–MS/MS method follows a 30-minute gradient of 125 mg/L ammonium formate in water and acetonitrile. Electrospray ionization (ESI) is used for ionization and introduction into the MS for SRM. Quantifier transitions are monitored at m/z 363.2 → m/z 147.1 and m/z 367.2 → m/z 151.1 for the native D2HG/L2HG and for the stable isotope labeled internal

FIGURE 12.3

D_6-D, L-2-ketopentandioic acid to D_4- D, L-2-hydroxyglutaric acid; zinc salt.

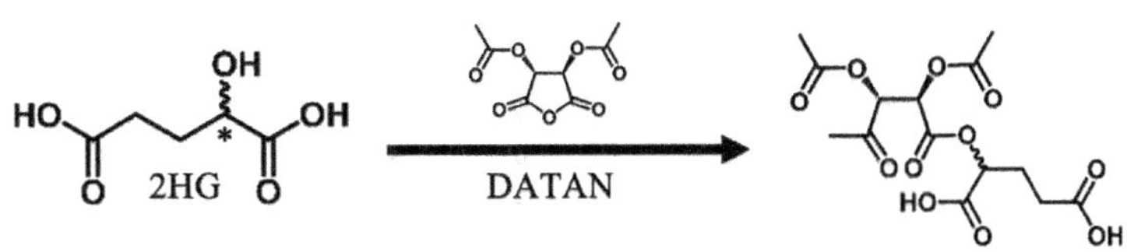

FIGURE 12.4

Derivatized with (+)-Di-O-acetyl-l-tartaric anhydride (DATAN).

standard, respectively. The product ions of these transitions are used as the quantifier ions, generating a standard curve by plotting the ratio of the native D2HG and L2HG to the stable isotope-labeled internal standard. Similarly, the ratio of the native D2HG and L2HG to stable isotope-labeled internal standard is calculated to derive the final concentration of endogenous D2HG and L2HG from the standard curve equation. A qualifier transition is monitored at m/z 363.2 → m/z 129.1 for the native 2HG enantiomers. The reference intervals reported for peripheral blood plasma levelsare:L2HG (6–147 ng/mL), D2HG (18–263 ng/mL), total 2HG (24–410 ng/mL), and D2HG/L2HG ratio (0.2–3.8) [35].

The clinical assay is used in patients with myeloid neoplasms as a preliminary screen while awaiting molecular results. The assay is highly sensitive and therefore especially useful in patients with low blast counts when molecular assays may be falsely negative [35]. The sensitivity of the D2HG/L2HG ratio is similar to the sensitivity of quantitative allele-specific digital PCR for detection of IDH1/IDH2 mutations in bone marrow samples at mutant allele burden of >1% [36]. Thus the biochemical assay is helpful for following treatment in patients with IDH-mutated myeloid neoplasms and can predict early relapse of patients with IDH-mutated myeloid neoplasm on treatment or after bone marrow transplantation. Moreover, the assay can be used to study the impact of treatment using IDH inhibitors, for example, if a patient being treated with IDH inhibitor initially responds but then relapses, the high circulating levels of D2HG (and D2HG/L2HG ratio) would indicate that the IDH mutated clone has relapsed, while normal levels would suggest that a different clone has relapsed, thereby indicating the need for alternative therapeutic strategy.

12.8 Conclusion and future directions

Understanding the metabolic pathways that lead to the production of oncometabolites D2HG, L2HG, succinate, and fumarate along with the downstream signaling pathways that are activated by these metabolites is crucial for identifying new therapeutic targets. TCA cycle oncometabolites discussed in this chapter show promise for translation into the clinical laboratory as biomarkers of cancer diagnosis, progression, recurrence, and treatment monitoring. Ongoing and future work will add to the number of oncometabolites as omics technologies and data analytics mature. Current clinical assays can be used as screening tools and to measure response to treatment, and remission. In the future, the highly sensitive oncometabolite assays maybe used as screening tools for early detection of cancers, real-time sampling of tumor margins will gain a foothold in oncological operative care, and technological advances will lower the cost and bring HRMS approaches to the clinical laboratory.

12.9 Summary points

- Succinate, fumarate, D- and L-2-hydroxygultarate are canonical TCA cycle oncometabolites that are aberrantly elevated in several types of cancers.
- Oncometabolites exhibit oncogenic properties through a variety of biochemical mechanisms.
- Notable downstream mechanisms of oncometabolites include induction of oxidative stress and epigenetic changes.
- Metabolic profiling of cancers can be achieved through multiple omics approaches using a variety of MS modalities and NMR.
- Targeted quantification of D- and L-2-hydroxyglutarate by tandem mass spectrometry has been used clinically for diagnosis and management of myeloid neoplasms.

References

[1] Hanahan D, Weinberg RA. Hallmarks of cancer: the next generation. Cell Elsevier 2011;144:646–74.

[2] Warburg O, Wind F, Negelein E. The metabolism of tumors in the body. J Gen Physiol 1927;8:519–30.

[3] DeBerardinis RJ, Lum JJ, Hatzivassiliou G, Thompson CB. The biology of cancer: metabolic reprogramming fuels cell growth and proliferation. Cell Metab 2008;7:11–20.

[4] Yong C, Stewart GD, Frezza C. Oncometabolites in renal cancer. Nat Rev Nephrol 2020;16:156–72.

[5] Yang M, Soga T, Pollard PJ. Oncometabolites: linking altered metabolism with cancer. J Clin Invest 2013;123:3652–8.

[6] Collins RRJ, Patel K, Putnam WC, Kapur P, Rakheja D. Oncometabolites: a new paradigm for oncology, metabolism, and the clinical laboratory. Clin Chem 2017;63:1812–20.

[7] Ward PS, Patel J, Wise DR, Abdel-Wahab O, Bennett BD, Coller HA, et al. The common feature of leukemia-associated IDH1 and IDH2 mutations is a neomorphic enzyme activity converting alpha-ketoglutarate to 2-hydroxyglutarate. Cancer Cell 2010;17:225–34.

[8] Dang L, White DW, Gross S, Bennett BD, Bittinger MA, Driggers EM, et al. Cancer-associated IDH1 mutations produce 2-hydroxyglutarate. Nature 2009;462:739–44.

[9] Borger DR, Goyal L, Yau T, Poon RT, Ancukiewicz M, Deshpande V, et al. Circulating oncometabolite 2-hydroxyglutarate is a potential surrogate biomarker in patients with isocitrate dehydrogenase-mutant intrahepatic cholangiocarcinoma. Clin Cancer Res 2014;20:1884–90.

[10] Aghili M, Zahedi F, Rafiee E. Hydroxyglutaric aciduria and malignant brain tumor: a case report and literature review. J Neurooncol 2009;91:233–6.

[11] Shim E-H, Livi CB, Rakheja D, Tan J, Benson D, Parekh V, et al. L-2-Hydroxyglutarate: an epigenetic modifier and putative oncometabolite in renal cancer. Cancer Discov 2014;4:1290–8.

[12] Niemann S, Müller U. Mutations in SDHC cause autosomal dominant paraganglioma, type 3. Nat Genet 2000;26:268–70.

[13] Baysal BE, Ferrell RE, Willett-Brozick JE, Lawrence EC, Myssiorek D, Bosch A, et al. Mutations in SDHD, a mitochondrial complex II gene, in hereditary paraganglioma. Science 2000;287:848–51.

[14] Hao H-X, Khalimonchuk O, Schraders M, Dephoure N, Bayley J-P, Kunst H, et al. SDH5, a gene required for flavination of succinate dehydrogenase, is mutated in paraganglioma. Science 2009;325:1139–42.

[15] Williamson SR, Eble JN, Amin MB, Gupta NS, Smith SC, Sholl LM, et al. Succinate dehydrogenase-deficient renal cell carcinoma: detailed characterization of 11 tumors defining a unique subtype of renal cell carcinoma. Mod Pathol 2015;28:80–94.

[16] Armitage EG, Barbas C. Metabolomics in cancer biomarker discovery: current trends and future perspectives. J Pharm Biomed Anal 2014;87:1–11.

[17] Yost RA, Enke CG. Triple quadrupole mass spectrometry for direct mixture analysis and structure elucidation. Anal Chem 1979;51:1251–64.

[18] Koelmel JP, Kroeger NM, Gill EL, Ulmer CZ, Bowden JA, Patterson RE, et al. Expanding lipidome coverage using LC-MS/MS data-dependent acquisition with automated exclusion list generation. J Am Soc Mass Spectrom 2017;28:908–17.

[19] Yan F, Zhao H, Zeng Y. Lipidomics: a promising cancer biomarker. Clin Transl Med 2018;7:21.

[20] Tunyasuvunakool K, Adler J, Wu Z, Green T, Zielinski M, Žídek A, et al. Highly accurate protein structure prediction for the human proteome. Nature 2021;596:590–6.

[21] Antoniewicz MR. A guide to 13C metabolic flux analysis for the cancer biologist. Exp Mol Med 2018;5:1–13.

[22] Spratlin JL, Serkova NJ, Eckhardt SG. Clinical applications of metabolomics in oncology: a review. Clin Cancer Res 2009;15:431–40.

[23] Polenova T, Gupta R, Goldbourt A. Magic angle spinning NMR spectroscopy: a versatile technique for structural and dynamic analysis of solid-phase systems. Anal Chem 2015;87:5458–69.

[24] Gill EL, Yost RA, Vedam-Mai V, Garrett TJ. Precast gelatin-based molds for tissue embedding compatible with mass spectrometry imaging. Anal Chem 2017;89:576–80.

[25] Kriegsmann J, Kriegsmann M, Casadonte R. MALDI TOF imaging mass spectrometry in clinical pathology: a valuable tool for cancer diagnostics (review). Int J Oncol 2015;46:893–906.

[26] Rodrigo MAM, Zitka O, Krizkova S, Moulick A, Adam V, Kizek R. MALDI-TOF MS as evolving cancer diagnostic tool: a review. J Pharm Biomed Anal 2014;95:245–55.

[27] Zhang J, Rector J, Lin JQ, Young JH, Sans M, Katta N, et al. Nondestructive tissue analysis for ex vivo and in vivo cancer diagnosis using a handheld mass spectrometry system. Sci Transl Med 2017;9:3968.

[28] Keating MF, Zhang J, Feider CL, Retailleau S, Reid R, Antaris A, et al. Integrating the masspec pen to the da Vinci surgical system for in vivo tissue analysis during a robotic assisted porcine surgery. Anal Chem 2020;92:11535–42.

[29] Balog J, Sasi-Szabó L, Kinross J, Lewis MR, Muirhead LJ, Veselkov K, et al. Intraoperative tissue identification using rapid evaporative ionization mass spectrometry. Sci Transl Med 2013;5:194ra93.

[30] Schmidt DR, Patel R, Kirsch DG, Lewis CA, Vander Heiden MG, Locasale JW. Metabolomics in cancer research and emerging applications in clinical oncology. CA Cancer J Clin 2021;71:333–58.

[31] Gibson KM, ten Brink HJ, Schor DS, Kok RM, Bootsma AH, Hoffmann GF, et al. Stable-isotope dilution analysis of D- and L-2-hydroxyglutaric acid: application to the detection and prenatal diagnosis of D- and L-2-hydroxyglutaric acidemias. Pediatr Res 1993;34:277–80.

[32] Rashed MS, AlAmoudi M, Aboul-Enein HY. Chiral liquid chromatography tandem mass spectrometry in the determination of the configuration of 2-hydroxyglutaric acid in urine. Biomed Chromatogr 2000;14:317–20.

[33] Jones PM, Boriack R, Struys EA, Rakheja D. Measurement of Oncometabolites D-2-Hydroxyglutaric Acid and L-2-Hydroxyglutaric Acid. Methods Mol Biol. 2017;1633:219-234. https://doi.org/10.1007/978-1-4939-7142-8_14. PMID: 28735490.

[34] Chu A, Rakheja D, Boriack R, Firan M, Collins R, Frankel AE, et al. Circulating D-2-hydroxyglutarate and the D-2HG/L-2HG ratio predict IDH mutation in acute myeloid leukemia. Annual meeting of the United States and Canadian Academy of Pathology. Seattle, WA, March 2016.

[35] McGehee E, Rakheja D, Oliver D, Chen W, Boriack R, Collins RH. The importance of plasma D-2HG measurement in screening for IDH mutations in acute myeloid leukaemia. Br J Haematol 2016;173:323–6.

[36] Wiseman DH, Struys EA, Wilks DP, Clark CI, Dennis MW, Jansen EEW, et al. Direct comparison of quantitative digital PCR and 2-hydroxyglutarate enantiomeric ratio for IDH mutant allele frequency assessment in myeloid malignancy. Leukemia 2015;29:2421–3.

CHAPTER

13 Circulating cancer biomarkers: current status and future prospects

Michael J. Duffy[1,2]

[1]UCD School of Medicine, Conway Institute of Biomolecular and Biomedical Research, University College Dublin, Dublin, Ireland

[2]UCD Clinical Research Centre, St. Vincent's University Hospital, Dublin, Ireland

Objectives

- Review the clinical uses of established serum protein biomarkers in cancer detection, assessing prognosis, postoperative surveillance, and monitoring response to therapy.
- Discuss the benefits and limitations of PSA in screening for prostate cancer, CA125 in screening for ovarian cancer, and AFP in screening high-risk individuals for hepatocellular cancer.
- Present the advantages and limitations of biomarkers in surveillance following curative-intent surgery and in monitoring therapy in advanced cancer.
- Review the current status of emerging biomarkers such as ctDNA, CTCs, miRs, and extracellular vesicles.
- Discuss the use of ctDNA in selecting the most appropriate therapy for patients with advanced cancers.

13.1 Introduction

Circulating biomarkers are indispensable for the optimum management of patients with several different cancer types. Although of

Cancer Biomarkers: Clinical Aspects and Laboratory Determination. DOI: https://doi.org/10.1016/B978-0-12-824302-2.00009-6
© 2022 Elsevier Inc. All rights reserved.

limited value in screening or early diagnosis, circulating biomarkers are widely used in evaluating prognosis, surveillance following curative-intent surgery, and monitoring response to therapy in advanced disease (Table 13.1) [1,2]. The aim of this chapter is to discuss the use of circulating biomarkers in these multiple settings. The first part of this chapter focuses on the standard protein biomarkers while the second part discusses emerging biomarkers and their likely clinical utility. See also, Chapter 1 for a general overview on tumor markers.

13.2 Clinical uses of established circulating protein biomarkers

13.2.1 Screening for early cancer

The aim of cancer screening is to detect early-stage disease or high-risk preinvasive lesions that can be resected by surgery. Screening is widely regarded as being one of the most potentially effective strategies for reducing mortality from cancer. To have clinical utility, screening should ideally detect cancer early and result in an efficacious treatment. The earlier use of treatment should lead to a better outcome compared to receiving treatment at the time of developing symptoms [3]. Screening for cancer is widely promoted, especially in the lay media. Most of this promotion focuses on the potential benefits of the practice, with little references to its limitations or its potential for harm. Thus, while screening can potentially reduce mortality from cancer, it can also result in false-positive findings leading to anxiety and stress as well as overdiagnosis. Overdiagnosis, which is the detections of indolent disease that may never cause problems in a persons' life, can lead to overtreatment and the potential for harm. Overall, it has been estimated that approximately 20% of patients who have cancer detected using the current screening strategies are overdiagnosed [4].

The gold standard methodology for establishing clinical utility of a cancer screening test is demonstration of reduced mortality in a larger prospective randomized trial. Cancer screening tests validated for reducing mortality using this high-level methodology include

Table 13.1 Clinical uses of the established circulating protein biomarkers.

Biomarker	Cancer	Screening	Aiding diagnosis	Prognosis/ staging	Surveillance	Therapy monitoring
AFP	HCC	√*	√	√	√	√
AFP	Testicular (nonseminoma)		√	√	√	√
CA19-9	Pancreatic		√		√	√
CA125	Ovarian		√	√	√	√
CA15-3	Breast				?	√
PSA	Prostate	√	√	√	√	√
Thyrogloublin	Thyroid (differentiated)		√		√	√
Calcitonin	Thyroid (medullary)		√		√	√
hCG	GTD		√	√	√	√
hCG	Germ cell (testicular)		√	√	√	√

Notes: "√", *yes should be used or considered for use;* "?", *evidence unclear;* "*"*in high-risk subjects.* GTD, *gestational trophoblastic disease;* HCC, *hepatocellular cancer.*

mammography for breast cancer (especially in women >50 years of age), fecal occult blood testing or flexible sigmoidoscopy for colorectal cancer (CRC) and low-dose computed tomography for subjects at high-risk for lung cancer [5]. Although the Papanicalaou (PAP) test is widely used in screening for cervical cancer, this test never underwent validation in a prospective trial. Large population-based studies, however, have convincingly shown that PAP testing reduces mortality from cervical cancer.

Because of their ease of measurement, minimal inconvenience for those undergoing screening, and relatively low costs, measurement of circulating protein biomarker might be expected to be an attractive strategy for cancer screening. As screening tests, however, the available circulating biomarkers have several problems, especially lack of sensitivity for early lesions and lack of specificity for malignancy [6]. These limitations when combined with the low prevalence of specific cancer types in the general population at a specific time, limit the use of most standard biomarkers in screening asymptomatic populations for cancer. Despite these limitations, several circulating biomarkers, either alone or in combination with imaging, have been evaluated for cancer screening in large randomized prospective trials. These include prostate-specific antigen (PSA) for prostate cancer, CA125 [in combination with transvaginal ultrasound (TVU)] for ovarian cancer, and alpha-fetoprotein (AFP) [in combination with ultrasound (US)] for hepatocellular cancer (HCC).

13.2.1.1 Prostate-specific antigen in screening for prostate cancer

Prostate cancer is the most frequently diagnosed cancer and the second leading cause of cancer-related death in American men. Although controversial, PSA is widely used in opportunistic screening for prostate cancer (see also Chapter 5). This practice began in the 1990s, several years prior to initiating clinical trials to assess its clinical utility in this setting. Within the last decade, however, the potential value of PSA as a screening test for prostate cancer in asymptomatic men has been evaluated in three large prospective randomized trials, that is, in the Prostate, Lung, Colorectal and

Ovarian (PLCO) trial which was carried out at 10 sites in the United States [7], the European Randomised Study for Screening of Prostate Cancer (ERSPC) trial which was performed across eight European countries [8], and the Cluster Randomised Trial of PSA Testing for Prostate Cancer (CAP) which was conducted at 785 general practitioner sites in England and Wales [9]. Both the PLCO and the CAP trials concluded that screening with PSA had no significant impact on reducing mortality from prostate cancer. In contrast, the ERSPC trial found that PSA screening led to a 20% reduction in prostate cancer–specific mortality in men aged 55–69 years. The decrease in prostate cancer mortality observed in the ERSPC trial is consistent with population-based studies showing a decrease in men initially presenting with metastatic prostatic disease as well as a substantial decline in death rates from prostate cancer, since the introduction of screening for this disease [10]. Importantly, in none of the above trials was screening associated with a significant reduced all-cause mortality, possibly because the studies were underpowered to address this endpoint.

An intriguing question is: why did the screening trials produce conflicting results with respect to their impact on reducing prostate cancer mortality. Looking at the design of each trial, it was clear that all had limitations, the most serious occurring in PLCO and the CAP trials [7,9]. Thus the PLCO trial did not rigorously compare PSA screening with a no screening arm, as over 80% of men in the "control arm" were found to have undergone PSA testing either prior to the start of the trial or during the trial proper [11]. This high rate of screening in the control group may have diluted a potential benefit in the screened arm. The PLCO trial has therefore been described as a comparison between regular and sporadic screening or between organized and opportunistic screening [12]. A major limitation of the English-Welsh trial was that the men in the screening arm had only one PSA test performed during the study. A single PSA determination may not have been sufficient to have a significant impact on patient outcome.

Although the ERSPC trial [8] had minor differences in design across the various participating countries (frequency of PSA testing, age of men participating in trial, follow-up tests, and cut-off concentration for defining PSA positivity), most expert panels accept that it provides the best direct evidence of the potential benefits and harms of regular (every 2–4 years) PSA screening. Indeed, based primarily on the results of this trial, many professional organizations currently recommend regular measurement of PSA for prostate cancer screening in men 55–69 years of age following a process of shared decision-making and informed consent, for review, see ref. [12]. Finally, in the context of PSA screening, it is important to state that while the practice appears to reduce mortality in men aged 55–69 years of age, it also leads to substantial overdiagnosis. Although it is difficult to determine the exact extent of overdiagnosis, it was estimated to occur in 16%–50% of cases in the above-mentioned trials [13].

13.2.1.2 CA125 in screening for ovarian cancer

Compared with prostate cancer, ovarian cancer is relatively uncommon, affecting 1 in approximately 78 women [14]. It is, however, the fourth most common cause of cancer-related death in women and is one of the most lethal malignancies affecting women [14]. Overall, about two-thirds of women diagnosed with ovarian cancer die from the disease. This high rate of lethality relates to the fact that 80%–90% of women have advanced disease at the time of diagnosis. In contrast, the approximate 15% of women diagnosed with early-stage disease have a 5-year survival of about 90%. This poor outcome for most women diagnosed with ovarian cancer emphasizes the urgency to develop screening/early diagnostic biomarkers for the disease.

Unlike prostate cancer in which a core needle biopsy can be used to confirm malignancy, ovarian cancer screening requires invasive surgery to establish a definitive diagnosis. It is, therefore, important than any screening strategy has a low rate of false-positive finding to limit the number of surgeries performed. Adding further complexity to the development of an effective screening strategy for ovarian cancer is the low prevalence of the disease in asymptomatic women (approximately, 1 in 2500). Due to its low prevalence, Bast et al. [15] suggested that a clinically useful screening strategy for ovarian cancer would require a positive predictive value (PPV) of at least

10% [15]. To obtain this PPV, Bast et al. [15] also suggested that a screening strategy should have a sensitivity of >75% for early malignancy and a specificity of >99.6%. This high specificity can be achieved by a dual strategy involving a combination of CA125 testing and TVU [16].

Two large prospective randomized trials have evaluated this dual strategy in screening asymptomatic women for ovarian cancer. One of the trials was a component of the PLCO study mentioned above for PSA screening. The PLCO ovarian cancer trial randomized 78,216 asymptomatic women, aged 55–74 years, to either annual screening with CA125 for 6 years and TVU for 4 years versus standard care. After a median follow-up of 12.4 years, rates of mortality were similar in the screened and control groups [17]. Surprisingly, there was no increase in the number of women diagnosed with early-stage disease in the screened group compared to the control group. Even more concerning was that screening resulted in a relatively high number of false-positive findings, that is, 3285, which led to unnecessary treatment [17]. Indeed, resulting from the false-positive results, 1080 (33%) women underwent unnecessary surgery, of whom, 163 (15%) were reported to develop at least one serious complication.

The other large randomized trial to have evaluated CA125 and TVU for ovarian cancer, known as the UK Collaborative Trial of Ovarian Cancer Screening (UKCTOCS) trial, randomized 202,638 asymptomatic average-risk postmenopausal women, aged 50–74 years to annual TVU, annual CA125 measurement or a control group [18]. In contrast to the PLCO trial which relied on static concentrations of CA125, the UK trial used an algorithm known as the risk of ovarian cancer algorithm (ROCA). ROCA includes woman's age, the rate of change in CA125 concentration over time, and the absolute concentration of CA125 [18].

After a median follow-up of 16.3 years, similar rates of death were found in all arms of the study, that is, as in the PLCO trial, screening failed to significantly reduce mortality from ovarian cancer [18]. However, in contrast to the PLCO trial, compared with no screening, the ROCA strategy resulted in a greater proportion of cancers identified at a localized stage. Clearly, based on available evidence, screening asymptomatic average-risk women for ovarian cancer with CA125 and TVU cannot currently be recommended.

13.2.1.3 Alpha-fetoprotein and ultrasound in screening for hepatocellular cancer

HCC is the fourth most frequent cause of cancer-related death, worldwide and one of the most rapidly increasing causes of cancer death in some Western countries [19]. Most cases of HCC occur in subjects with risk factors such as infection with hepatitis B virus (HBV), infection with hepatitis C virus (HCV), nonalcoholic fatty liver disease (NAFLD), alcoholic cirrhosis, primary biliary cirrhosis, obesity, or genetic hemochromatosis [20]. These risk factors can lead to cirrhosis that in turn can lead to HCC.

Since subjects at high risk of developing HCC can be identified, screening using regular surveillance can be performed with the aim of detecting early HCC. Two tests are widely used to screen high-risk subjects for HCC, that is, AFP and liver US [20]. A combination of these tests has been investigated for screening subjects with HBV infection in two large randomized prospective trials conducted in China. Both these trials concluded that screening HBV-infected subjects with AFP and US significantly reduced mortality from HCC [21,22].

The relevance of these Chinese trials involving subjects with chronic HBV infection in screening high-risk subjects for HCC in the West is unclear, as most high-risk subjects in Western countries have cirrhosis due to HCV infection, alcohol consumption, or NAFLD [20]. Although there has never been a randomized trial to address HCC screening in patients with cirrhosis, multiple cohort studies concluded that the process resulted in increased early detection and improved overall survival [23,24]. Thus, following a systematic review and *meta*-analysis involving 47 studies with a total of 15,158 patients, Singal et al. [23] reported that screening using US with or without AFP was associated with improved early-stage detection [odds ratio (OR) 2.08, 95% confidence interval (CI), 1.80–2.37], curative treatment rates (OR 2.24, 95% CI, 1.99–2.52), and significantly prolonged 3-year overall survival (OR 1.90, 95% CI, 1.67–2.17). A further systematic review plus *meta*-analysis involving 32 studies containing 13,367 patients, concluded that the addition of AFP measurement to US increased sensitivity for detecting early HCC but resulted in decreased specificity [24].

Despite lack of evidence from a large prospective randomized trial, several expert panels recommend HCC screening for all subjects with cirrhosis (irrespective of the cause and provided liver function and comorbidities allow curative or palliative treatment) and specific subgroups of subjects with noncirrhotic HBV infection (e.g., Asian males $\geq$40 years, Asian females $\geq$50 years, subjects with viral load > 10,000 copies/mL) [20]. Although most organizations recommend screening using both AFP and US, some advocate only the use of US [20].

13.2.2 Evaluating prognosis following diagnosis of a primary cancer

Following diagnosis and surgical removal of a primary cancer, the key clinical questions that require addressing are: is the tumor aggressive or life threatening and is administration of adjuvant systemic therapy necessary? Thus, if a tumor is deemed to be indolent, the patient may be able to avoid having to receive adjuvant systemic treatment. On the other hand, if the tumor is identified as being potentially aggressive and life threatening, the patient would normally be advised to have additional therapy, such as adjuvant chemotherapy.

Historically, histological (tumor size, tumor grade, and status of local lymph nodes) and clinical (patient age and comorbidity) criteria were used as aids for evaluating prognosis in patients with cancer. Although these factors continue to be universally used, they are now increasingly being supplemented with biomarkers in several different types of cancer. Although many of the prognostic biomarkers that have established clinical utility are measured in tumor tissue, some circulating biomarkers are also used for this purpose. These include AFP, human chorionic gonadatrophin hCG), and lactate dehydrogenase (LDH) in patients with nonseminomatous testicular cancer [25], hCG and LDH in patients with seminomatous-type testicular cancer [25], and PSA in patients with prostate cancer [26].

In all these situations, the circulating biomarker levels are not used alone but in combination with the standard clinical and pathological prognostic factors for the respective cancer types.

It is important to mention that the cut-off levels used for the above circulating biomarkers in assessing prognosis are different from those used in the screening/diagnostic setting. For example, for prostate cancer, recent joint guidelines published by the American Urological Association, American Society for Radiation Oncology, and Society of Urologic Oncology [26] stated that men with PSA concentrations of <10 μg/L should be classified in a group with a very low or low risk of recurrence, those with PSA concentration $10-<20$ μg/L should be regarded as being at intermediate risk of recurrence, while those with values ≥ 20 μg/L should be placed at high risk of recurrence. In patients with nonseminomatous testicular cancer, subjects that have AFP $<$ 1000 μg/L, hCG $<$ 5000 IU/L, and LDH $<$ 1.5 times upper limit of normal are classified as having a good prognosis, while those with any of the following: AFP $>$ 10,000 μg/L or hCG $>$ 50,000 IU/L, or LDH $>$ 10 times upper limit tend are classified as having a poor outcome. As mentioned earlier, for prognostic classification, the above biomarkers are not used alone but are combined with the relevant clinical and pathological features [25,26].

Although concentrations of several other serum biomarkers at time of initial diagnosis [e.g., CEA (carcinoembryonic antigen) in CRC, CA125 in ovarian cancer, and CA15-3 in breast cancer] are associated with patient outcome, they are not widely used for everyday prognostic purposes [1,2].

13.2.3 Surveillance following curative surgery for primary cancers

Following curative-intent surgery (with or without adjuvant systemic treatment), a common practice is to follow-up patients at regular intervals, at least for the first 3–5 years after diagnosis. This practice is based on the widely held assumption that the early detection of recurrent/metastatic disease followed by the initiation of therapy enhances the chance of cure or results in improved outcome. This is a reasonable assumption as it might be expected that treating patients with a low volume of recurrent disease would be preferable than waiting for radiological or clinical evidence of relapse [27,28]. With low volume recurrence, the clonal complexity, mutation prevalence, and intratumor heterogeneity are likely to be less than when bulk disease is present which theoretically should make tumors more amenable to treatment. Clinical evidence for the latter is multiple studies across different cancer types which showed that adjuvant treatment (which is believed to eradicate low tumor load and/or micrometastases) of early cancer improves outcome [27]. On the other hand, increasing tumor size and mutation burden have been associated with resistance to therapy [29].

Although it might be expected that the early detecting of emerging recurrences would improve outcome, there is a lack of high-level evidence such as data from prospective randomized trials that postoperative follow-up using serial biomarker measurements and/or regular imaging is beneficial. Despite this, most expert panels recommend the regular measurement of biomarkers in the postoperative surveillance of patients with a range of different cancer. Thus, following a literature review of the United States and European guidelines, Merkow et al. [30] found that biomarkers were universally recommended for surveillance in patients with CRC (CEA), prostate cancer (PSA), thyroid cancer (thyroglobulin), and testicular germ cell cancers (AFP and hCG). In contrast, most guidelines recommended against the use of biomarkers during postoperative surveillance in patients with breast or lung cancer. The recommendations against the use of biomarkers in postoperative surveillance of breast or lung cancer patients are not based on evidence showing a lack of clinical utility in this setting. Rather, for these two common cancers, the potential clinical utility of using biomarkers in surveillance has never been addressed in an adequately powered randomized prospective trial.

Although biomarkers are widely measured during follow-up surveillance in certain cancer types (Table 13.2), there are several caveats for their use in this setting. These include:

- Uncertainty with regard to frequency of measurement, magnitude of change in biomarker concentration that is clinically relevant or when to stop testing.
- The available biomarkers are rarely elevated in patients with small localized recurrence (e.g., in local lymph nodes) and

Table 13.2 Biomarkers that may be used in postoperative surveillance or monitoring therapy in different cancer types.

Cancer	Biomarker(s)
Breast	CA15-3, CEA
Colorectal	CEA
Lung (SC)	proGRP, NSE
Lung (NSC)	CYFRA 21-1, CEA
Hepatocellular	AFP
Melanoma	S-100, LDH
Ovarian	CA125
Pancreatic	CA19.9
Testicular (NS)	AFP, hCG
Testicular (seminoma)	hCG, LDH
Thyroid (differentiated)	Thyroglobulin

NS, *nonseminomatous;* NSC, *nonsmall cell;* SC, *small cell.*

usually only become elevated in the presence of distant metastases.

- With the possible exception of hCG in patients with gestational trophoblastic disease/choriocarcinoma, none of the existing biomarkers are increased in all patients with a specific cancer type.
- Certain benign lesions can also give rise to elevated biomarkers levels, although these elevations are usually less than that found in patients with distant metastatic disease.
- Lack of high-level evidence such as from randomized clinical trials that the measurement of biomarkers in postoperative surveillance enhances patient outcome or increases quality of life.

13.2.4 Monitoring the effectiveness of systemic therapy in advanced cancer

The goal of therapy in patients with advanced cancer is to extend survival and increase quality of life. Currently, several different types of systemic therapy are available for patients with advanced cancer, including cytotoxic chemotherapy, endocrine therapy (e.g., for breast or prostate cancer), targeted therapy directed against cancer driver genes, immunotherapy, and antiangiogenic therapy. Of these therapies, biomarkers have been best validated for evaluating response to cytotoxic drugs. In contrast, the use of biomarkers for monitoring response to targeted therapy, antiangiogenic therapy or immunotherapy is less well studied.

As the effectiveness of a specific therapy is highly variable from patient to patient, it is important to be able to monitor response and identify progression or regression as early as possible. Although radiological imaging is the gold standard method for monitoring response to therapy, the use of blood-based biomarkers is more convenient for patients and is less expensive. In most (but not all) situations, blood levels of biomarkers decrease with tumor regression and increase with progression. Confirmation of tumor progression or regression, however, usually requires radiological imaging.

A caveat in the use of biomarkers in monitoring treatment with cytotoxic therapy, especially if bulk metastases are present, is that transient increases or spikes in levels may occur within the first few weeks of treatment [31–34]. These short-term increases are believed to result from tumor cell death (necrosis or apoptosis) in response to the initial treatment and are not usually associated with disease progression. These transient increases have been found with cytotoxic therapies in different solid tumors [31,32], with endocrine therapy in breast cancer [33] and following radiotherapy or brachytherapy in patients with prostate cancer [34]. It is unclear if the newer therapies such as immunotherapies or antiangiogenic agents can cause such spikes. Frequently used biomarkers for monitoring response to therapy, especially chemotherapy, are listed in Table 13.2.

13.3 Future trends in cancer biomarker applications and ongoing research

The vast majority of cancer biomarkers currently in clinical use are proteins or glycoproteins and entered clinical practice in the 1970s, 1980s, and 1990s. As few new clinically useful protein-based biomarkers have been discovered in recent years, research into cancer biomarkers has shifted from proteins to a group of tests that are frequently referred to as liquid biopsies. Liquid biopsy tests include the detection of circulating tumor DNA (ctDNA), circulating RNAs such as micro-RNAs (miRs), circulating tumor cells (CTCs), and extracellular vesicles (EVs). These tests are referred to as liquid biopsies, as they were assumed to provide similar information to that available from tissue biopsies. In reality, however, none are as informative as tissue pathology, at least in the diagnostic setting. Nevertheless, some of these liquid biopsy tests, especially the measurement of ctDNA, have the potential to be a transformative cancer biomarker.

13.3.1 Circulating tumor DNA

Similar to protein biomarkers, DNA is released or shed from tumors into the circulation where it is present in a cell-free state. As DNA is also shed from normal healthy cells, the relative amount of ctDNA (variant allele frequency) in cancer patients is usually low ($<1\%$ of total), although it has been reported to vary widely from a low of $<0.01\%$ to a high of $>90\%$ of the total circulating-free DNA (cfDNA) present [35–37]. The amount of ctDNA present is determined by the rate at which DNA is released from tumor tissue relative to the rate that it is cleared by the kidneys and liver. The extent of shedding from tumor depends on factors such as tumor load/volume, location of tumor, proximity of tumor to the vasculature, extent of necrosis, the presence or absence of lymphovascular invasion, and rate of malignant cell turnover [35–37]. For clinical purposes, ctDNA can be differentiated from normal cell DNA by the presence of specific genetic alteration such as mutations, altered gene copy numbers, altered methylation patterns, or the presence of viral sequences. A further difference is that ctDNA tends to exist in smaller size fragments than DNA shed from normal cells [38,39].

13.3.1.1 Advantages of circulating tumor DNA over proteins as cancer biomarkers

As biomarkers for cancer, ctDNA has several advantages over the proteins presently available [40]. These advantages include:

- Provides more information regarding tumor biology.
- Have a wider range of clinical applications such as the ability to predict response to specific therapies and identify mechanism of therapeutic resistance.
- Potentially more sensitive and specific for cancer.
- Have shorter half-life, allowing close to real-time monitoring of response to surgery or systemic therapy.

Disadvantages of measuring ctDNA compared to protein biomarkers include higher costs, slower turn-around-time, less availability, difficulty in automation, and a requirement for highly skilled staff and specialized equipment.

13.3.1.2 Clinical uses of circulating tumor DNA

13.3.1.2.1 Upfront selecting therapy

Irrespective of the type of systemic therapy administered to patients, only a proportion benefit. Indeed, for some therapies, especially certain forms of targeted therapy and immunotherapy, only a minority of unselected patients exhibit a clinically meaningful response. In addition to this low response rate, several different anticancer drugs cause severe toxicity and occasionally death. Finally, anticancer drugs, especially targeted therapies and immunotherapies are expensive, costing in some situations in excess of $100,000 per year. Clearly, therefore, for the rational and most cost-effective administration of anticancer therapies, biomarkers that upfront identify those patients that are likely to benefit or be resistant to a specific treatment is a necessity. Such biomarkers are termed predictive biomarkers or companion diagnostics. For this chapter, both types of these biomarkers are referred to as predictive biomarkers.

Predictive biomarkers have traditionally been measured on tumor tissue obtained by biopsy or surgery. Increasing, however, such biomarkers are being measured using ctDNA. One of the first approved clinical applications of ctDNA involved measurement of a predictive biomarker, that is, *EGFR* mutations for predicting response to EGFR tyrosine kinase inhibitors (TKIs) (e.g., erlotinib, gefitinib, afatinib, and osimertinib) in patients with advanced nonsmall cell lung cancer (NSCLC) [41]. Activating mutations such as in-frame deletions in exon 19 (exon19del) or the point mutations (L858R) in exon 21 of *EGFR* are associated with response to these drugs. Overall, approximately 70% of patients with advanced NSCLC containing these activating *EGFR* mutations respond to EGFR TKIs [41], whereas patients lacking such mutations rarely benefit from these inhibitors [42]. Furthermore, in patients with localized NSCLC (stages IB, II, and IIIA) positive for *EGFR* mutations, adjuvant treatment with osimertinib was found to significantly prolong disease-free survival compared with placebo [43].

The availability of EGFR TKIs has transformed the outcome for patients with *EGFR* mutation-positive advanced NSCLC, in recent years. Thus administration of these inhibitors has resulted in the median progression-free survival for patients with advanced or metastatic lung adenocarcinoma increasing from 5.6 months with chemotherapy to 11 months with the first-generation TKI inhibitors (erlotinib and gefitinib) and up to 38 months with the third-generation TKI inhibitor, osimertinib [44,45].

The gold standard method for evaluating the mutational status of *EGFR* in patients with NSCLC requires tumor tissue which necessitates an invasive procedure. However, because of the risk of harm associated with performing an invasive procedure, especially in old or frail patients, obtaining tumor tissue is not always possible [46]. Even if a biopsy can be performed, the amount of tissue obtained may be inadequate or unsuitable for mutation testing (i.e., contain too few malignant cells), especially when multigene testing is being performed.

Compared with tissue mutation testing, the use of ctDNA has several advantages. These advantages include minimally invasive sampling, cheaper to obtain, and more rapid turn-around-time for results. Furthermore, the use of ctDNA allows serial testing which would be impractical with tissue biopsies. Finally, the use of ctDNA would be expected to provide a more comprehensive picture of the overall mutation profile in spatially heterogeneous tumors compared with a single-site needle biopsy [41,44].

Overall, good agreement has been found between the *EGFR* mutational status of plasma and corresponding tumor tissue in patients with advanced NSCLC [44]. In particular, ctDNA has high specificity and high PPV (approaching 100%) for detecting clinically relevant *EGFR* mutations and importantly provides similar predictive information to that of tissue in selecting patients for EGFR TKI treatment [47,48].

The high specificity and equivalent predictive value suggest that a positive ctDNA result may be used to identify patients with advanced NSCLC likely to be responsive to EGFR TKIs. However, the sensitivity of ctDNA for detecting tissue mutations is relatively low (70%–80%). Due to this relatively low sensitivity, the absence of a ctDNA mutation should not be interpreted that the tumor tissue is also mutation-negative. In such a situation, tissue should be evaluated for *EGFR* mutations, when feasible.

Since strong concordance is found, especially the high specificity between tissue and plasma mutational status, measurement of *EGFR* mutation using ctDNA is now widely endorsed by multiple expert panels for predicting response to anti-EGFR inhibitors in patients with NSCLC when tissue is unavailable or unsuitable for mutational analysis [49–51]. The ability to determine *EGFR* mutations in plasma should result in an increased number of patients with this malignancy receiving anti-EGFR TKIs which should in turn lead to an improved quality of life and better outcome vis-à-vis treatment with chemotherapy.

Several commercially available ctDNA tests (companion diagnostics) for identifying patients with advanced NSCLC who are likely to benefit from anti-EGFR therapy have been approved by the U.S. Food and Drug Administration (FDA) and European Medicines Agency (EMA). One of these tests, known as the cobas EGFR Mutation Test v2 (Roche Diagnostics, Indianapolis, IN, USA) detects 42 different mutations in exons 18–21 of the *EGFR* gene, including L858R, exon 19 deletions, L861Q, and T790M. The cobas test may be used when a tissue biopsy cannot be performed. In addition to this PCR test, two next-generation sequencing (NGS) tests have also been approved for ctDNA *EGFR* mutation testing, that is, the Guardant360 CDx (Guardant Health, Redwood City, CA, USA) and the FoundationOne Liquid CDx (Foundation Medicine, Cambridge, MA, USA) assays. It should be stated that these NGS tests can detect several other mutated genes in addition to *EGFR*. In Europe, a different commercially available test, that is, therascreen EGFR Plasma RGQ PCR Kit version 2 (Qiagen,

Manchester, United Kingdom), has received the Conformité Européenne (CE) mark from the European Union for detecting *EGFR* mutation in ctDNA when tissue is unavailable. This test detects exon 19 deletions, mutations at L858R, L861Q, G719X, S768I, exon 20 insertions, and T790M in the *EGFR* gene using real-time PCR.

Other ctDNA-based therapy predictive biomarker approved by the FDA includes: mutational analysis of *PIK3CA* for identifying postmenopausal breast cancer patients with hormone-receptor positive-HER2-negative disease that are likely to benefit from combined treatment with alpelisib (a PIK3CA inhibitor) and fulvestrant (therascreen PIK3CA RGQ PCR Kit, Qiagen, and the FoundationOne Liquid CDx test), mutations in *BRCA1* and *BRCA2* for predicting response to treatment with the PARP inhibitor rucaparib in patients with prostate or ovarian cancer (FoundationOne Liquid CDx), and *ALK* rearrangements to identify patients with NSCLC eligible for treatment with the ALK inhibitor, alectinib (FoundationOne Liquid CDx) [52,53].

Promising but not yet approved ctDNA predictive biomarkers include *RAS* mutation analysis for identifying patients with advanced CRC who are resistant to anti-EGFR antibodies (cetuximab, panitumumab) [54], tumor mutation burden (TMB) for predicting response to immunotherapy in patients with lung cancer [55], and *BRCA1/2* mutations for predicting response to PARP inhibitor, olaparib in patients with breast, ovarian or pancreatic cancer [56].

13.3.1.2.2 Detecting mutations associated with the development of acquired therapy resistance

Irrespective of the type of systemic therapy administered, effectively all patients with advanced cancer eventually develop acquired resistance against it. Acquired resistance is frequently due to secondary mutations in the target gene of the drug or in genes downstream of the target gene. Being able to upfront identity resistance-conferring mutations may enable decision-making with respect to rationally planning second-line or further lines of therapy. For many of the targeted therapies in everyday clinical use, the identity of the resistance-conferring mutations is now known (Table 13.3).

As with predictive biomarkers, the use of ctDNA to identify resistance mutations is also best validated in patients with NSCLC

Table 13.3 Mutations in ctDNA that have been associated with response or resistance to different therapies.

ctDNA mutation	Therapy	Response/ resistance	Cancer
EGFR (exon 19 deletion, L858R)	Gefitinib, erlotinib, osimertinib	Response	Lung (NSC)
EGFR T790M	Gefitinib, erlotinib	Resistance	Lung (NSC)
EGFR T790M	Osimertinib	Response	Lung (NSC)
EGFR C797S	Osimertinib	Resistance	Lung (NSC)
BRCA1/2 (reverse)	PARP inhibitors	Resistance	Breast, ovarian
PIK3CA	Alpelisib + fulvestrant	Response	Breast
ESR1	Aromatase inhibitors	Resistance	Breast
KRAS/NRAS	Cetuximab, panitumumab	Resistance	CRC
AR	Antiandrogen	Resistance	Prostate
MSI/dMMR	Immunotherapy (anti-PD-1)	Response	Multiple
TMB (high)	Immunotherapy (anti-PD-1/PD-L1)	Response	Lung (NSC)

AR, *androgen receptor;* CRC, *colorectal cancer;* dMMR, *defective mismatch repair;* ESR1, *estrogen receptor gene;* MSI, *microsatellite instability;* NSC, *nonsmall cell;* TMB, *tumor mutational burden.*
Data reviewed from refs. [35–37].

[57]. While *EGFR* exon 19 deletions and exon 21 point mutations are associated with response to first-generation EGFR TKIs such as gefitinib or erlotinib, mutations in exon 20 of *EGFR* (T790M) frequently result in acquired resistance to these inhibitors [57]. Approximately 60%–70% of patients with the T790M mutation, however, respond to the third-generation EGFR TKI, osimertinib [44]. Similar to the situation with the commonly occurring *EGFR* activating mutations (e.g., exon 19 deletions and L858R in exon 21), T790M mutations can also be detected at high specificity using ctDNA [41,58]. For example, across 19 studies, the ability of ctDNA to detect the *EGFR* T790M mutation had a positive and negative predictive value of 89% and 61%, respectively, a

sensitivity of 68%, and specificity of 86% [58]. Importantly, response to osimertinib and duration of therapy were shown to be similar irrespective of whether testing was performed using tissue or plasma (overall response rate, 62.5% for tissue, 66.7% for plasma) [48]. Thus, if the ctDNA T790M mutation is positive, osimertinib may be administered. However, testing of tumor tissue is recommended if the ctDNA result is negative. The cobas EGFR Mutation Test v2 mentioned above has also been approved by the FDA for detecting the T790M mutation in plasma when tissue is unavailable.

13.3.1.2.3 Postoperative surveillance and monitoring response to therapy

For postoperative surveillance following curative-intent surgery, emerging evidence suggests that ctDNA provides more potent prognostic information and has enhanced sensitivity for the early detection of recurrent disease than standard biomarkers [35–37]. Thus, in a preliminary prospective study involving the follow-up of stage II CRC patients who developed recurrence, ctDNA was found to be more frequently positive than CEA at the time of recurrence (85% vs 41%, $P = .002$) [59]. Furthermore, the median time interval between ctDNA becoming positive and radiological evidence of recurrence was significantly longer than that found with increasing CEA concentrations (167 days vs 61 days; $P = .04$).

Similarly, in the postoperative follow-up of women who had undergone curative surgery for breast cancer, ctDNA was reported to be more sensitive in detecting recurrent disease than the standard biomarker, CA15-3 [60]. In a preliminary prospective cohort study, Coombes et al. [60] found that ctDNA was elevated prior to clinical or radiological evidence of recurrence in 16 of the 18 relapsed patients (sensitivity, 89%). Recurrent disease was predicted with a median lead time of 8.9 months (range, 0.5–24.0 months). None of the 31 nonrelapsing patients were ctDNA-positive at any time point during the study (specificity, 100%). In contrast to ctDNA, CA15-3 was elevated prior to recurrence in only 6 of 13 patients that had the biomarker measured. On average, ctDNA detected recurrences >6 months ahead of CA15-3.

ctDNA assay is also under investigation for monitoring response to various types of therapy, including chemotherapy, endocrine therapy, targeted therapy, and immunotherapy, especially in patients with metastatic cancers [35–37]. As with standard biomarkers, serial alterations in ctDNA generally mirror response to therapy but is apparently more sensitive in detecting tumor progression than the standard biomarkers [35–37,61]. Importantly, because of its short half-life (<2.5 hours), ctDNA should allow an earlier indication of response to therapy vis-à-vis standard protein biomarkers. A further advantage of ctDNA over standard biomarkers in postoperative surveillance and therapy monitoring is its ability to identify new actionable mutation and resistance-conferring mutations, thus potentially enabling the rational planning of further lines of treatments, see above. A limitation of ctDNA in monitoring response to therapy is that similar to protein biomarkers, ctDNA may exhibit transient increases following the initiation of therapy [62]. However, because of its short half-life, these transient increases should not persist as long as standard biomarkers.

13.3.1.2.4 Early detection of cancer

One of the potentially most transformative uses of ctDNA is its ability to detect early cancer [63–69]. Preliminary results suggest that ctDNA can detect early cancer with high specificity and moderate sensitivity [63–69] (Table 13.4). It is important to point out, however, that most of the so-far published studies analyzed patients with a previous diagnosis of cancer, while the control subjects were apparently asymptomatic. Despite these reservations, tests such as CancerSEEK and GRAIL appear to have high specificity for cancer, that is, >99% [65,66]. High specificity is important for any screening test in order to minimize false-positive findings that could lead to unnecessary anxiety and unnecessary treatment.

Sensitivity of ctDNA for cancer detection, however, is variable, depending on tumor type and tumor stage. For example, the median overall sensitivity of CancersEEK was 70%, ranging from 98% for ovarian cancer to 33% for breast cancer [65]. As with standard protein biomarkers, sensitivity for CancersEEK increased with increasing stage of the cancer, that is, the median sensitivity was 43%,

Table 13.4 ctDNA tests under investigation for screening/early detection of cancer.

Test	Parameters detected	Target cancer
Epi proColon	Methylated Sept9 gene	CRC
EBV	EBV DNA	Nasopharyngeal
CancerSEEK	Mutated DNA + 8 protein biomarkers	Pan cancer
GRAIL	Altered methylated DNA	Pan cancer
DELFI	Altered DNA fragment patterns	Pan cancer
Lung-CLiP	Mutated DNA[a]	Lung

CRC, *colorectal cancer;* DELFI, *DNA evaluation of fragments for early interception;* EBV, *Epstein–Barr virus;* Lung Clip, *lung cancer likelihood in plasma.*

[a]*Used with a machine learning model that estimates the probability that a cfDNA mutation is tumor rather than derived from clonal hematopoiesis of indeterminate origin.*

73%, and 78% for patients with stages I, II, and III disease, respectively. Similarly, using the GRAIL test, sensitivity increased with increasing stage, that is, 39% in stage I, 69% in stage II, 83% in stage III, and 92% in stage IV [66].

One of the first ctDNA screening tests to have undergone evaluation in a prospective study is Epi proColon (Epigenomics, Berlin) [63]. This test uses real-time PCR to detect aberrant methylation at the promoter region of the *SEPT9* gene. Epi proColon was prospectively evaluated for CRC screening in 7941 asymptomatic individuals ≥50 years old [69]. Overall sensitivity for CRC was 48.2%, while specificity was 91.5%. According to the FDA, Epi proColon is indicated for screening subjects 50 years or older at average risk for CRC, who have been offered and have a history of not completing CRC screening [63]. Patients with a positive Epi proColon test result, however, should be referred for diagnostic colonoscopy.

Although promising, there are several caveats in using ctDNA as a pan-cancer screening test. One of these is the inability to consistency localize the tissue of origin of the cancer. Thus further work-up, including expensive imaging, may be required to localize the tumor site. Application of a supervised machine learning algorithm to the CancerSEEK findings, however, enabled the tissue of origin to be correctly identified in a median of 63% of the patients evaluated [65]. The accuracy for predictions tumor site varied by tumor type and was highest for colorectal, ovarian, and pancreatic cancers (approximately 80%), but lower for lung, liver, and upper gastrointestinal tract cancers (approximately 40%). The GRAIL test was reported to correctly identify the tissue of origin in 93% of cases [66].

Although CancersEEK and GRAIL have reported specificities >99%, false-positive findings with ctDNA tests may occur in patients with certain benign lesions, especially clonal hematopoiesis (CH) (a benign expansion of a clonal population of blood cells) [70]. CH can release mutant-like DNA into plasma in healthy individuals as well as in cancer patients [71]. Potential strategies to reduce or eliminate the problem of CH-derived mutations include simultaneous analysis of white blood cell DNA, fractionating DNA based on fragment size [68,72] or selecting mutations that have not been associated with CH.

Finally, as mentioned above, it is important to bear in mind that almost all the ctDNA tests investigated to-date used patients with a previous diagnosis of cancer while the controls were healthy subjects rather than symptomatic populations. The sensitivity for detecting cancer in asymptomatic subjects is likely to be lower than that in patients with a previous clinical diagnosis of cancer. Furthermore, because these studies did not investigate subjects with relevant comorbidities (benign or inflammatory diseases), specificity may also have been overestimated.

13.3.2 Circulating tumor cells

CTCs are malignant cells shed from tumors into the bloodstream, (see also Chapter 10 for a discussion of CTC). As CTCs are derived from tumors and contain a diverse range of tumor-associated molecules, they have been investigated as potential cancer biomarkers. Similar to ctDNA, CTC can be obtained relatively noninvasively and they enable serial monitoring of disease. However, the isolation and analysis of CTC is more difficult than that of ctDNA [73]. In particular, the measurement of CTC requires specialized equipment which is not widely available, especially in clinical diagnostic laboratories.

A further problem is that CTCs are a highly heterogeneous population of cells, consisting of epithelial tumor cells, epithelial-to-mesenchymal (EMT) cells, hybrid epithelial/EMT cells, and cancer stem cells. Furthermore, only a minority of CTC gives rise to metastasis, as the vast majority die in the circulation due to anoikis, attack by immune cells or shear forces [74,75]. As CTCs constitute a very small fraction of cells in the vasculature (<10 cells per one million white cells and one billion red cells), assays for their detection require initial enrichment steps [76].

Three main types of enrichment have been described, immunoaffinity-based methods, size-density-based methods, and the use of microfluidic or chip devices [74–76]. Of these, the most widely used is the immunoaffinity-based platform, known as the CELLSEARCH Circulating Tumor Cell Kit (Menarini Silicon Biosystems, Huntingdon Valley, Philadelphia, PA, USA). For this method, CTCs are initially enriched by binding to epithelial cell adhesion molecule (EpCAM) that is assumed to be present only on epithelial-derived cells. Further selection is based on cells being cytokeratin-positive, CD45-negative, and possessing a cell nucleus. The CellSearch method has undergone analytical validation and is approved by the FDA for determining prognosis in patients with advanced breast cancer, CRC, and prostate cancer [74,75]. One of the problems with CellSearch, however, is that it is unable to detect CTC that have undergone epithelial mesenchymal transition (EMT) as such cells loose expression of EpCAM.

Compared with ctDNA, the use of CTC as a cancer biomarker has been less investigated [77]. Furthermore, in contrast to ctDNA, the clinical uses of CTC are currently more restricted and essentially confined to determining prognosis/monitoring response to therapy in advanced cancers. An advantage of CTC over ctDNA is that it allows the measurement of tumor-specific alterations not only at the DNA level but also at the mRNA and protein levels. Although not yet exploited for clinical use, CTC could potentially be used for functional assays such as predicting response to treatment and establishing metastatic potential.

To broaden the clinical application of CTC, research is currently focusing on their molecular characterization and biomarker profile. One of the most extensively studied CTC-based biomarker is the androgen receptor splice variant 7 (AR-V7) [78,79]. The presence of AR-V7 mRNA or protein in CTC has been shown to correlate with resistance to antiandrogen therapy and poor prognosis in patients with advanced prostate cancer [78,79]. Although AR-V7-positive patients are resistant to specific antiandrogens, they respond to taxane-based chemotherapy [78]. A caveat in the measurement of AR-V7 is that its levels should be related to the CTC count [80].

Other genetic alteration possessing biomarker potential detected in CTC includes *HER2* amplification in breast cancer [81], the *EML4-ALK* translocation in NSCLC [82], and the *TMPRSS-ERG* translocation in prostate cancer [83].

13.3.3 Micro-RNA

miRs are small nonprotein coding RNAs molecules (19–25 nucleotides) that are involved in the regulation of gene expression at a posttranscriptional level [84]. As with ctDNA, miRs are shed by tumors into the circulation [84]. In the circulation, miRs are present in exosomes (see Section 13.3.4) or in association with the argonaute protein, AGO2. As the expression of multiple miRs is deregulated in cancer, they have been widely investigated as potential cancer biomarkers.

Although studied in multiple cancer types, one of the most relevant clinical findings on miRs has been in testicular cancer. Multiple studies, including prospective multicenter trials, reported that miR-371a-3p and miR-302/367 levels had superior sensitivity and specificity than the standard biomarkers for this malignancy, that is, AFP, hCG, and LDH, for review, see ref. [85]. These miRs may thus be of use in aiding diagnosis, surveillance, and early detection of recurrences in testicular cancer patients [85].

Measurement of circulating miRs may also be useful in aiding the early diagnosis of lung cancer. In a multicenter cohort study involving 606 patients with NSCLC, 593 patients with nontumor lung diseases, 883 patients with diseases not affecting the lung, and 964 unaffected controls, a 14 miR signature was found to differentiate patients with lung cancer from those with nontumor lung diseases with an accuracy of 92.5% (95% CI, 92.1%–92.9%), sensitivity of 96.4% (95% CI, 95.9%–96.9%), and specificity of

88.6% (95% CI, 88.1%–89.2%) [86]. Furthermore, the signature distinguished patients with early-stage lung cancer from subjects without lung cancer with an accuracy of 95.9% (95% CI, 95.7%–96.2%), sensitivity of 76.3% (95% CI, 74.5%–78.0%), and specificity of 97.5% (95% CI, 97.2%–97.7%).

13.3.4 Extracellular vesicles and exosomes

EVs are membrane-bound particles produced in the endosomal compartment of normal and malignant cells and subsequently secreted into surrounding fluids such as the bloodstream [87]. EV can exist in different sizes, ranging from a few nanometers to a few micrometers. A subtype of EV known as exosomes exists in two different sizes, small exosomes 60–80 nm (*Exo*-S) and large exosomes 90–120 nM (*Exo*-L) [87]. In patients with cancer, EV and exosomes are potentially valuable as a source of biomarkers, as they are rich in tumor-derived DNA, RNA (including miRs), and protein.

Commercially available plasma exosome-based biomarker assays such as ExoDx Lung ALK, (Exosomes Diagnostics, Waltham, MA, USA) has been used for the detection of specific *ALK* fusions for predicting response to ALK inhibitors in patients with advanced NSCLC [88] and the ExoDx Prostate Test (Exosome Diagnostics) which is a urine-based assay is indicated for risk management in men 50 years of age and older with a PSA level of 2–10 μg/L [89]. The latter recently received the CE mark for use in the European Union.

Finally, a recent preliminary finding suggests that protein profiling of circulating EV may be value in the early detection cancer [90]. Thus, in a study involving 43 healthy controls and 77 cancer patients with 16 different cancer types, EV protein profiling was able to detect malignancy with 95% sensitivity and 90% specificity [89]. Clearly, these promising findings require confirmation using larger numbers of patients and controls. A practical problem in using exosomes is that they are difficult to isolate, requiring sequential steps of centrifugation and ultracentrifugation. The requirement for ultracentrifugation may make it difficult to introduce these assays into routine practice.

13.4 Conclusion and future directions

Going forward, it is likely that biomarkers will play an increasing important role in the detection and management of patients with cancer. While traditionally the main applications of protein biomarkers was in postoperative surveillance and monitoring therapy, newer tests such as those based on ctDNA are likely to be most useful in therapy decision-making and identifying mechanisms of therapy resistance. One of the most exciting possibilities for the future is the use of ctDNA for the early detection of cancer. Tests such as CancerSEEK and GRAIL are promising for this possibility but clearly require further validation, especially in large prospective randomized trials focusing on asymptomatic subjects. The long sought holy grail of a cancer biomarker, that is, a simple blood test to detect early cancer, may now be finally on the horizon.

13.5 Summary points

- The combination of low sensitivity, low specificity, and the low prevalence of cancers in asymptomatic populations limit the use of most existing protein biomarkers in cancer screening.
- PSA may be used in screening men 55–70 years of age for prostate cancer after being informed of the advantages and disadvantages of undergoing the test.
- Although high levels of evidence of clinical utility are largely lacking, serum biomarkers are widely used in surveillance following curative-intent surgery and monitoring therapy in advanced disease.
- The emergence of ctDNA as a biomarker is likely to be the most important practice-changing development in the near future use of cancer biomarkers.
- ctDNA assays are currently in clinical use for planning therapy in patients with advanced NSCLC and advanced breast cancer.

Acknowledgment

This work was supported by the Cancer Clinical Research Trust.

Conflict of interest

MJD has no conflicts of interest.

References

[1] Duffy MJ. Tumor markers in clinical practice: a review focusing on common solid cancers. Med Princ Pract 2013;22:4–11.

[2] Duffy MJ. Clinical uses of tumor markers: a critical review. Crit Rev Clin Lab Sci 2001;38:225–62.

[3] Brawley OW, Kramer BS. Cancer screening in theory and in practice. J Clin Oncol 2005;23:293–300.

[4] Glasziou PP, Bell KJ, Barratt AL. Estimating the magnitude of cancer overdiagnosis in Australia. Med J Aust 2020;213:189 189.e1.

[5] Smith RA, Andrews KS, Brooks D, Fedewa SA, et al. Cancer screening in the United States, 2019: a review of current American Cancer Society guidelines and current issues in cancer screening. CA Cancer J Clin 2019;69:184–210.

[6] Duffy MJ. Use of biomarkers in screening for cancer. Adv Exp Med Biol 2015;867:27–39.

[7] Andriole GL, Crawford ED, Grubb 3rd RL, Buys SS, et al. Prostate cancer screening in the randomized Prostate, Lung, Colorectal, and Ovarian Cancer Screening Trial: mortality results after 13 years of follow-up. J Natl Cancer Inst 2012;104:125–32.

[8] Schröder FH, Hugosson J, Roobol MJ, Tammela TL, et al. Screening and prostate cancer mortality: results of the European Randomised Study of Screening for Prostate Cancer (ERSPC) at 13 years of follow-up. Lancet. 2014;384:2027–35.

[9] Martin RM, Donovan JL, Turner EL, Metcalfe C, et al. Effect of a low-intensity PSA-based screening intervention on prostate cancer mortality: the CAP randomized clinical trial. J Am Med Assoc 2018;319:883–95.

[10] Welch HG, Albertsen PC. Reconsidering prostate cancer mortality – the future of PSA screening. N Engl J Med 2020;382:1557–63.

[11] Shoag JE, Schlegel PN, Hu JC. Prostate-specific antigen screening: time to change the dominant forces on the pendulum. J Clin Oncol 2016;34:3499–501.

[12] Duffy MJ. Biomarkers for prostate cancer: prostate-specific antigen and beyond. Clin Chem Lab Med 2020;58:326–39.

[13] Fenton JJ, Weyrich MS, Durbin S, Liu Y, et al. Prostate-specific antigen-based screening for prostate cancer: evidence report and systematic review for the US Preventive Services Task Force. J Am Med Assoc 2018;319:1914–31.

[14] Trinidad CV, Tetlow AL, Bantis LE, Godwin AK. Reducing ovarian cancer mortality through early detection: approaches using circulating biomarkers. Cancer Prev Res (Phila) 2020;13:241–52.

[15] Bast Jr RC, Urban N, Shridhar V. Early detection of ovarian cancer: promise and reality. Cancer Treat Res 2002;107:61–97.

[16] Nash Z, Menon U. Ovarian cancer screening: current status and future directions. Best Pract Res Clin Obstet Gynaecol 2020;65:32–45.

[17] Buys SS, Partridge E, Black A, Johnson CC, et al. Effect of screening on ovarian cancer mortality: the Prostate, Lung, Colorectal and Ovarian (PLCO) Cancer Screening Randomized Controlled Trial. JAMA 2011;305:2295–303.

[18] Menon U, Gentry-Maharaj A, Burnell M, Singh N, et al. Ovarian cancer population screening and mortality after long-term follow-up in the UK Collaborative Trial of Ovarian Cancer Screening (UKCTOCS): a randomised controlled trial. Lancet. 2021; S0140–6736(21):00731–5.

[19] Global Burden of Disease Liver Cancer Collaboration, Akinyemiju T, Abera S, Ahmed M, Alam N, et al. The burden of primary liver cancer and underlying etiologies from 1990 to 2015 at the global, regional, and national level: results from the Global Burden of Disease Study 2015. JAMA Oncol 2017;3:1683–91.

[20] Kanwal F, Singal AG. Surveillance for hepatocellular carcinoma: current best practice and future direction. Gastroenterology 2019;157:54–64.

[21] Zhang B-H, Yang B-H, Tang ZY. Randomized controlled trial of screening for hepatocellular carcinoma. J Cancer Clin Oncol 2004;130:417–22.

[22] Chen J-G, Parkin DM, Chen Q-G, Lu J-H, et al. Screening for liver cancer: results of a randomised controlled trial in Qidong, China. J Med Screen 2003;10:204–9.

[23] Singal AG, Pillai A, Tiro J. Early detection, curative treatment, and survival rates for hepatocellular carcinoma surveillance in patients with cirrhosis: a *meta*-analysis. PLoS Med 2014;11: e1001624.

[24] Tzartzeva K, Obi J, Rich NE, Parikh ND, et al. Surveillance imaging and alpha fetoprotein for early detection of hepatocellular carcinoma in patients with cirrhosis: a *meta*-analysis. Gastroenterology. 2018;154:1706–18 e1.

[25] Paner GP, Stadler WM, Hansel DE, Montironi R, et al. Updates in the eighth edition of the tumor-node-metastasis staging classification for urologic cancers. Eur Urol 2018;73:560–9.

[26] Sanda MG, Cadeddu JA, Kirkby E, Chen RC, et al. Clinically localized prostate cancer: AUA/ASTRO/SUO guideline. Part I: risk stratification, shared decision making, and care options. J Urol 2018;199:683–90.

[27] Luskin MR, Murakami MA, Manalis SR, Weinstock DM. Targeting minimal residual disease: a path to cure? Nat Rev Cancer 2018;18:255–63.

[28] Duffy MJ, McDermott EW, Crown J. Blood-based biomarkers in breast cancer: from proteins to circulating tumor cells to circulating tumor DNA. Tumour Biol 2018;40(5) 1010428318776169.

[29] Vasan N, Baselga J, Hyman DM. A view on drug resistance in cancer. Nature. 2019;575:299–309.

[30] Merkow RP, Korenstein D, Yeahia R, Bach PB, et al. Quality of cancer surveillance clinical practice guidelines: specificity and consistency of recommendations. JAMA Intern Med 2017;177:701–9.

[31] Sørbye H, Dahl O. Transient CEA increase at start of oxaliplatin combination therapy for metastatic colorectal cancer. Acta Oncol 2004;43:495–8.

[32] Søgaard CH, Lindegaard JC, Havsteen H, Nielsen OS, et al. Chemotherapy-induced changes of CA 125 in patients with epithelial ovarian cancer. Gynecol Oncol 2005;97:410–12.

[33] Kreienberg R. Droloxifene-induced spikes of tumor markers predict benefit of therapy. Am J Clin Oncol 1991;14(Suppl 2):S56–61.

[34] Hanlon AL, Pinover WH, Horwitz EM, Hanks GE. Patterns and fate of PSA bouncing following 3D-CRT. Int J Radiat Oncol Biol Phys 2001;50:845–9.

[35] Siravegna G, Mussolin B, Venesio T, Marsoni S, et al. How liquid biopsies can change clinical practice in oncology. Ann Oncol 2019;30:1580–90.

[36] Heitzer E, Haque IS, Roberts CES, Speicher MR. Current and future perspectives of liquid biopsies in genomics-driven oncology. Nat Rev Genet 2019;20:71–88.

[37] Kilgour E, Rothwell DG, Brady G, Dive C. Liquid biopsy-based biomarkers of treatment response and resistance. Cancer Cell 2020;37:485–95.

[38] Mouliere F, Chandrananda D, Piskorz AM, Moore EK, et al. Enhanced detection of circulating tumor DNA by fragment size analysis. Sci Transl Med 2018;10(466) eaat4921.

[39] Marass F, Stephens D, Ptashkin R, Zehir A, et al. Fragment size analysis may distinguish clonal hematopoiesis from tumor-derived mutations in cell-free. DNA Clin Chem 2020;66:616–18.

[40] Duffy MJ. Circulating tumour DNA as a cancer biomarker. Ann Clin Biochem 2019;56:42–8.

[41] Yang CY, Yang JC, Yang PC. Precision management of advanced non-small cell lung cancer. Annu Rev Med 2020;71:117–36.

[42] Mok TS, Wu YL, Thongprasert S, Yang CH, et al. Gefitinib or carboplatin-paclitaxel in pulmonary adenocarcinoma. N Engl J Med 2009;361:947–57.

[43] Wu YL, Tsuboi M, He J, John T, et al. Osimertinib in resected EGFR-mutated non-small-cell lung cancer. N Engl J Med 2020;383:1711–23.

[44] Pennell NA, Arcila ME, Gandara DR, West H. Biomarker testing for patients with advanced non-small cell lung cancer: real-world issues and tough choices. Am Soc Clin Oncol Educ Book 2019;39:531–42.

[45] Ramalingam SS, Vansteenkiste J, Planchard D, Cho BC, et al. Overall survival with osimertinib in untreated, EGFR-mutated advanced NSCLC. N Engl J Med 2020;382:41–50.

[46] Sholl LM, Aisner DL, Varella-Garcia M, Berry LD, et al.On behalf of the LCMC Investigators Multi-institutional oncogenic driver mutation analysis in lung adenocarcinoma: the lung cancer mutation consortium experience. J Thorac Oncol 2015;10:768–77.

[47] Sacher AG, Komatsubara KM, Oxnard GR. Application of plasma genotyping technologies in non-small cell lung cancer: a practical review. J Thorac Oncol 2017;12:1344–56.

[48] Mack PC, Banks KC, Espenschied CR, Burich RA, et al. Spectrum of driver mutations and clinical impact of circulating tumor DNA analysis in non-small cell lung cancer: analysis of over 8000 cases. Cancer 2020;126:3219–28.

[49] Kalemkerian GP, Narula N, Kennedy EB, Biermann WA, et al. Molecular testing guideline for the selection of patients with lung cancer for treatment with targeted tyrosine kinase inhibitors: American Society of Clinical Oncology, Endorsement of the College of American Pathologists/International, Association for the Study of Lung Cancer/Association for Molecular Pathology Clinical Practice Guideline Update. J Clin Oncol 2018;36:911–19.

[50] NCCN Guidelines Version 5.21. Non-small cell lung cancer. <https://www.nccn.org/professionals/physician_gls/pdf/nscl.pdf>. Accessed 25 July 2021.

[51] Wu YL, Planchard D, Lu S, Sun H, et al. Pan-Asian adapted clinical practice guidelines for the management of patients with metastatic non-small-cell lung cancer: a CSCO-ESMO initiative endorsed by JSMO, KSMO, MOS, SSO and TOS. Ann Oncol 2019;30:171–210.

[52] André F, Ciruelos E, Rubovszky G, Campone M, et al. Alpelisib for *PIK3CA*-mutated, hormone receptor-positive advanced breast cancer. N Engl J Med 2019;380:1929–40.

[53] El-Deiry WS, Goldberg RM, Lenz HJ, Shields AF, et al. The current state of molecular testing in the treatment of patients with solid tumors, 2019. CA Cancer J Clin 2019;69:305–43.

[54] Salem ME, Puccini A, Tie J. Redefining colorectal cancer by tumor biology. Am Soc Clin Oncol Educ Book 2020;40:1–13.

[55] Aggarwal C, Thompson JC, Chien AL, Quinn KJ, et al. Baseline plasma tumor mutation burden predicts response to pembrolizumab-based therapy in patients with metastatic non-small cell lung cancer. Clin Cancer Res 2020;26:2354–61.

[56] Vidula N, Rich TA, Sartor O, Yen J, et al. Routine plasma-based genotyping to comprehensively detect germline, somatic, and reversion BRCA mutations among patients with advanced solid tumors. Clin Cancer Res 2020;26:2546–55.

[57] Diao Z, Han Y, Zhang R, Li J. Circulating tumour DNA: A new biomarker to monitor resistance in NSCLC patients treated with EGFR-TKIs. Biochim Biophys Acta Rev Cancer 2020;1873:188363.

[58] Ontario Health (Quality). Cell-free circulating tumour DNA blood testing to detect EGFR T790M mutation in people with advanced non-small cell lung cancer: a health technology assessment. Ont Health Technol Assess Ser 2020;20:1–176.

[59] Tie J, Wang Y, Tomasetti C, Li L, et al. Circulating tumor DNA analysis detects minimal residual disease and predicts recurrence in patients with stage II colon cancer. Sci Transl Med 2016;8:346–92.

[60] Coombes RC, Page K, Salari R, Hastings RK, et al. Personalized detection of circulating tumor DNA antedates breast cancer metastatic recurrence. Clin Cancer Res 2019;25:4255–63.

[61] Davis AA, Iams WT, Chan D, Oh MS, et al. Early assessment of molecular progression and response by whole-genome circulating tumor DNA in advanced solid tumors. Mol Cancer Ther 2020;19:1486–96.

[62] Kato K, Uchida J, Kukita Y, Kumagai T, et al. Transient appearance of circulating tumor DNA associated with de novo treatment. Sci Rep 2016;6:38639.

[63] Shirley M. Epi proColon® for colorectal cancer screening: a profile of its use in the USA. Mol Diagn Ther 2020;24:497–503.

[64] Chan KCA, Woo JKS, King A, Zee BCY, et al. Analysis of plasma Epstein-Barr virus DNA to screen for nasopharyngeal cancer [published correction appears in N Engl J Med. 2018;378:973]. N Engl J Med 2017;377:513–22.

[65] Cohen JD, Li L, Wang Y, Thoburn C, et al. Detection and localization of surgically resectable cancers with a multi-analyte blood test. Science 2018;359:926–30.

[66] Liu MC, Oxnard GR, Klein EA, Swanton C, et al. Sensitive and specific multi-cancer detection and localization using methylation signatures in cell-free DNA. Ann Oncol 2020;31:745–59.

[67] Cristiano S, Leal A, Phallen J, Fiksel J, et al. Genome-wide cell-free DNA fragmentation in patients with cancer. Nature. 2019;570:385–9.

[68] Chabon JJ, Hamilton EG, Kurtz DM, Esfahani MS, et al. Integrating genomic features for non-invasive early lung cancer detection. Nature. 2020;580:245–51.

[69] Church TR, Wandell M, Lofton-Day C, Mongin SJ, et al. Prospective evaluation of methylated SEPT9 in plasma for detection of asymptomatic colorectal cancer. Gut. 2014;63:317–25.

[70] Gibson CJ, Steensma DP. New insights from studies of clonal hematopoiesis. Clin Cancer Res 2018;24:4633–42.

[71] Razavi P, Li BT, Brown DN, Jung B, et al. High-intensity sequencing reveals the sources of plasma circulating cell-free DNA variants. Nat Med 2019;25:1928–37.

[72] Lo YMD, Han DSC, Jiang P, Chiu RWK. Epigenetics, fragmentomics, and topology of cell-free DNA in liquid biopsies. Science. 2021;372(6538):eaaw3616.

[73] Alix-Panabières C, Pantel K. Liquid biopsy: from discovery to clinical implementation. Mol Oncol 2021;15:1617–21.

[74] Aoki M, Shoji H, Kashiro A, Takeuchi K, et al. Prospects for comprehensive analyses of circulating tumor cells in tumor biology. Cancers (Basel) 2020;12:E1135.

[75] Reimers N, Pantel K. Liquid biopsy: novel technologies and clinical applications. Clin Chem Lab Med 2019;57:312–16.

[76] Haber DA, Velculescu VE. Blood-based analyses of cancer: circulating tumor cells and circulating tumor DNA. Cancer Discov 2014;4:650–61.

[77] Tian C, Xu X, Wang Y, Li D, et al. Development and clinical prospects of techniques to separate circulating tumor cells from peripheral blood. Cancer Manag Res 2020;12:7263–75.

[78] Armstrong AJ, Halabi S, Luo J, Nanus DM, et al. Prospective multicenter validation of androgen receptor splice variant 7 and hormone therapy resistance in high-risk castration-resistant prostate cancer: the PROPHECY study. J Clin Oncol 2019;37:1120–9.

[79] Scher HI, Graf RP, Schreiber NA, Garcia JA, et al. Assessment of the validity of nuclear-localized androgen receptor splice variant 7 in circulating tumor cells as a predictive biomarker for castration-resistant prostate cancer. JAMA Oncol 2018;4:1179–86.

[80] Sharp A, Welti JC, Lambros MBK, Dolling D, et al. Clinical utility of circulating tumour cell androgen receptor splice variant-7 status in metastatic castration-resistant prostate cancer. Eur Urol 2019;76:676–85.

[81] Jacot W, Cottu P, Berger F, Dubot C, et al. Actionability of HER2-amplified circulating tumor cells in HER2-negative metastatic breast cancer: the CirCe T-DM1 trial. Breast Cancer Res 2019;21:121.

[82] Manicone M, Scaini MC, Rodriquenz MG, Facchinetti A, et al. Liquid biopsy for monitoring anaplastic lymphoma kinase inhibitors in non-small cell lung cancer: two cases compared. J Thorac Dis 2017;9(Suppl 13):S1391–6.

[83] Stott SL, Lee RJ, Nagrath S, Yu M, et al. Isolation and characterization of circulating tumor cells from patients with localized and metastatic prostate cancer. Sci Transl Med 2010;2:25 ra23.

[84] Terrinoni A, Calabrese C, Basso D, Aita A, et al. The circulating miRNAs as diagnostic and prognostic markers. Clin Chem Lab Med 2019;57:932–53.

[85] Regouc M, Belge G, Lorch A, Dieckmann KP, et al. Non-coding micro RNAs as novel potential tumor markers in testicular cancer. Cancers (Basel) 2020;12:749.

[86] Fehlmann T, Kahraman M, Ludwig N, Backes C, et al. Evaluating the use of circulating microRNA profiles for lung cancer detection in symptomatic patients. JAMA Oncol 2020;6:1–10.

[87] Zijlstra A, Di Vizio D. Size matters in nanoscale communication. Nat Cell Biol 2018;20:228–30.

[88] Aguado C, Giménez-Capitán A, Karachaliou N, Pérez-Rosado A, et al. Fusion gene and splice variant analyses in liquid biopsies of lung cancer patients. Transl Lung Cancer Res 2016;5:525–31.

[89] McKiernan J, Donovan MJ, O'Neill V, Bentink S, et al. Thompson IM Jr, Carroll P. A novel urine exosome gene expression assay to predict high-grade prostate cancer at initial biopsy. JAMA Oncol 2016;2:882–9.

[90] Hoshino A, Kim HS, Bojmar L, Gyan KE, et al. Extracellular vesicle and particle biomarkers define multiple human cancers. Cell. 2020;182:1044–61 e18.

Index

Note: Page numbers followed by "*f*" and "*t*" refer to figures and tables, respectively.

M

R

S

T

U

V

W

9780128243022